An Introduction
to Stochastic
Processes

D. Kannan is currently Associate Professor of Mathematics at the University of Georgia. Professor Kannan received his MSc from Annamalai University, and his MA and PhD from Wayne State University. He also taught at Annamalai University and Calicut University (both in India), University of Guelph (Canada), and New York University.

NORTH HOLLAND SERIES IN

Probability and Applied Mathematics

A. T. Bharucha-Reid, *Editor*

Wayne State University

An Introduction to Stochastic Processes

D. KANNAN
Department of Mathematics
University of Georgia

NORTH HOLLAND • NEW YORK
New York • Oxford

Elsevier North Holland, Inc.
52 Vanderbilt Avenue, New York, New York 10017

Distributors outside the United States and Canada:

Thomond Books
(A Division of Elsevier/North-Holland Scientific Publishers, Ltd.)
P.O. Box 85
Limerick, Ireland

Library of Congress Cataloging in Publication Data

Kannan, D
        An introduction to stochastic processes.
        (North Holland series in probability and applied mathematics)

        Bibliography: p.
        Includes index.
        1. Stochastic processes. I. Title. II. Series.
QA274.K35   519.2   79-11
ISBN 0-444-00301-0

Manufactured in the United States of America

Dedicated to:
Chidambaram Ramalingam,
Dandapani, Amaravathy,
Kamala, and Viji

Contents

Contents   ix

Preface

Stochastic processes play a basic role in several problems of many branches of physical, biological, and social sciences, business and economics, and engineering. Having recognized the applicatory value of the theory of stochastic processes, almost all colleges and universities offer a course on this subject at least for a semester or quarter to suit the needs of students majoring in various areas. This textbook is based on my lectures given at the University of Georgia and University of Guelph. The composition of my classes varied from year to year with students from the junior level to the first year graduate level and majors in mathematics, statistics, physics, chemistry, zoology, ecology, and business. With the exception of a couple of sections, all the material in the book has been covered at one time or another.

The book treats most of the major areas of stochastic processes. The first objective of the book is to present various techniques used in the study of stochastic processes. The notion of stopping time is introduced at a very early stage (Chapter 2) and is used throughout the book. At several places we have explained the intuition behind these techniques. The material covered in this book will prepare a student for further advanced study of stochastic processes. Our next objective is to point out the applicatory value of this area of mathematics. Almost all the topics have been motivated and illustrated by drawing examples and applications from various branches of sciences.

Chapter 1 gives the prerequisite material from probability theory at a level higher than what is needed in the following chapters. Chapter 2 treats random walk and introduces and uses stopping time. An instructor who may not have enough time to cover all the material in the book can start his or her course with

Chapter 3 on Markov chains. There, too, random walk is discussed and stopping time introduced. Strong Markov property is pointed out, recurrence, transience, and stationary distribution are studied in some detail, and elements of branching chain are presented. Chapter 4 is an introduction to Poisson process pointing out its potential value in application. Chapter 5 studies essentially the birth–death processes. In addition to several examples, recurrence and transience are discussed. Chapter 6 is written in the spirit of calculus. It presents sample path properties such as continuity, differentiability, and integrability. Stationary processes are the subject matter of Chapter 7. Spectral representation and ergodic theorems are treated there. Chapter 8 gives a detailed account of martingales. Optional sampling theorem and convergence theorems are presented. Chapter 9 is on Brownian motion and diffusion stochastic processes. Path properties of Brownian motion are given in detail, certain diffusion equations are solved, and hitting time distribution is studied. Numerous examples are provided in each chapter to illustrate the theory and to point out possible applications.

This book is meant as a textbook for the senior undergraduate and first year graduate level students. A potential reader should have had calculus, probability theory, and differential equations. The entire book can be covered in a one-year course. For a two-quarter or one-semester course, several plans are possible using this book. Interdependency between various chapters is kept at a minimal level. Selecting basic material from each chapter one can offer a one-quarter course. I feel that every course should cover Chapter 3, 5, and 7–9.

Acknowledgments

In writing this book, I received constant encouragement from Professor A. T. Bharucha-Reid, my colleagues, and students, and I am deeply indebted to all of them.

A work of this sort cannot be completed without a source of inspiration. For providing me with the necessary inspiration, I am grateful to my wife and especially my daughter, whose main interest is to see her name on the dedication page. With great pleasure I humbly dedicate this book to my chosen guru Chidambaram Ramalingam. I received excellent typing service from various secretaries. I thank them all and in particular Ms. Susan Kay, who typed most of the final version.

An Introduction
to Stochastic
Processes

1

Introduction

1.1 Notion of Stochastic Processes

The origin of the theory of stochastic processes can be traced into the field of statistical physics. The first examples and several basic concepts of stochastic processes were initially considered in statistical physics at the turn of this century. These examples arose and further examples still arise as a consequence of fluctuations and noises of different types in the physical system. A *physical process* is a physical phenomenon, the evolution of which is studied as a function of time. Consider the position $X(t)$ and velocity $V(t)$, in the X-direction and at time t, of a tagged particle that is undergoing an observable motion in a fluid due to almost continuous bombardment by the molecules of the media. Then $(X(t), V(t))$ is a physical process. Because of the randomness involved in the collisions, $(X(t, \omega)V(t, \omega))$ is a random vector for each t. Here ω is a sample point. Thus the randomly evolving physical process $(X(t, \omega), V(t, \omega))$ is a family of random vectors indexed by the time parameter t, where $t \geqslant 0$ or $-\infty < t < \infty$, say. A family of random variables indexed by a parameter set is roughly known as a *stochastic process*.

The theory of stochastic processes can be considered as a mathematical foundation of statistical physics. The modern quantum field theory depends heavily on (noncommutative) stochastic processes. In an axiomatic approach to the quantum field theory (based on the concepts of observables), one takes the random variables as axiomatic elements, and the so-called weak stochastic processes form the founding elements. (However, this text introduces the reader only to the mathematical theory of stochastic processes and its applications.)

Stochastic processes arise not only in physical systems, but also in engineering, biological, medical, and social systems. It is increasingly becoming a dictum that models must be stochastic and not deterministic. This stresses the importance and applicatory value of the theory of stochastic processes.

We described the motion of a tagged particle as a stochastic process with "time" t as the indexing parameter. But the index t need not necessarily denote the time. Consider now a geological study of rock formation. Here $X(t, \omega)$, $t = 0, 1, 2, \ldots$, may denote the rock component (e.g., lignite, shale, sandstone, siltstone) of the tth layer of the rock. Here t is a space variable and is discrete. As another example, consider the fluctuation problem of electron–photon cascade. Let $X(e; t, \omega)$ denote the number of particles (photon, electron, etc.) with the energy value less than e at an arbitrary thickness t of the absorber (matter). Here t is a continuous variable.

Next we present some examples of a stochastic process in its rough sense (as a collection of random variables). Here the index set could be any one of $0 \leqslant t < \infty$, $-\infty < t < \infty$, $t = n = 0, \pm 1, \pm 2, \ldots, n = 0, 1, \ldots$, and so on.

1. The family $\{X(t), t \geqslant 0\}$ denoting the number of impulses registered by a Geiger–Muller counter during the period $[0,t)$ is a stochastic process.
2. The process T_n, $n = 0, 1, \ldots$, denotes the time elapsed between the $(n - 1)$st and nth registration of impulses in the example in the preceding paragraph.
3. Let $X(t)$ be the fluctuating voltage across the end of a resistor in an electric circuit at time $t \geqslant 0$. The fluctuation arises due to the random motion of conduction electrons, and the process $X(t)$ is called the *thermal noise*.
4. In a unimolecular reaction, we can take $X(t)$ to denote the concentration at time $t \geqslant 0$ of a reactant ρ that is irreversibly converted into a product π.
5. (Enzyme amplifier system). In the gellation process of blood clotting, the conversion of proenzymes to the final stage of enzyme fibrin occurs in several stages. Let $X(t)$ denote the concentration of fibrin at time $t \geqslant 0$.
6. In an insect, animal, or human population, the population size fluctuates randomly due to environmental stochasticity in birth, death, immigration, and emigration rates. Take $X(t)$ as the population size at time $t \geqslant 0$ or at generation $t = 1, 2, \ldots$.

7. Consider a predator learning to hunt advantageously in a model–mimic system. Let $X(t)$ denote the last encountered prey (a model or a mimic) before or at time $t \geqslant 0$. Then $X(t)$ is known as the *predator encounter process*.

8. (Epidemiology). In the stochastic theory of epidemics one considers a vector process $(S(t), I(t), R(t))$, $t \geqslant 0$, where $S(t)$, $I(t)$, and $R(t)$ denote the numbers of susceptibles, infectives, and removed, respectively, at time t.

9. (Genetics). Consider a randomly mating population of N diploid parents. Let the alleles A and a occur with frequencies x and $1 - x$, respectively, where x varies in [0,1]. Define a process $X(t) = $ the frequency of allele A. A problem of interest is the probability of fixation of allele A.

10. (Congestion). In a queuing system the following processes arise, among others: $X(t)$ is the number of customers who arrived during $[0,t)$ at a service counter, and $Y(t)$ is the number of customers served during the period $[0,t)$. (Such counting processes arise in various fields. In nuclear physics, for example, one considers the occurrences of collisions between beam particles and target particles in an ideal accelerator bombardment of an amorphous target.)

We conclude this list of general examples here and give further examples later under proper setup. In the remainder of this chapter we review very briefly the necessary background material from probability theory.

1.2. Probability Space

Probability theory is concerned with the study of experiments whose outcomes are random; that is, the outcomes cannot be predicted with certainty. The collection Ω of all possible outcomes of a random experiment is called a *sample space*. An element ω of Ω is called an *elementary event* (or *a sample point*).

Definition 1.2.1. A collection $\mathcal{C}$ of subsets of Ω is called a σ-*algebra* if (a) $\Omega \in \mathcal{C}$, (b) $A \in \mathcal{C}$ implies that its complement $A^C \in \mathcal{C}$, and (c) for any countable collection $\{A_n, n \geqslant 1\} \subset \mathcal{C}$, we have $\cup_{n \geqslant 1} A_n \in \mathcal{C}$. The elements of $\mathcal{C}$ are called *events*.

If $\mathcal{C}$ is a σ-algebra of events, then (1) $\phi \in \mathcal{C}$ and (2) $\cap_{n \geqslant 1} A_n \in \mathcal{C}$ whenever $A_n \in \mathcal{C}$ for all $n \geqslant 1$.

Given an arbitrary collection $\mathcal{C}$ of subsets of Ω, there is at least one σ-algebra, namely, the power set of Ω, which contains $\mathcal{C}$. The smallest σ-algebra $\sigma(\mathcal{C})$ that contains $\mathcal{C}$ is called the σ-*algebra generated by* $\mathcal{C}$.

Definition 1.2.2. Let $\Omega = R$. The σ-algebra generated by the collection of all half-infinite intervals $(-\infty, x)$ is called the *Borel σ-algebra* on R. The sets belonging to this Borel σ-algebra, denoted by $\mathcal{R}$ or $\mathcal{B}(R)$, are called *Borel sets*.

Now let $\Omega = R^n$. Arbitrarily fix a point $\mathbf{x}^0 = (x_1^0, \ldots, x_n^0) \in R^n$. The set of all $\mathbf{x} = (x_1, \ldots, x_n) \in R^n$ such that $x_k < x_k^0$, $1 \leqslant k \leqslant n$, is called the *open negative orthant* at $\mathbf{x}^0$. Let $\mathcal{O}$ denote the collection of all open negative orthants as $\mathbf{x}^0$ varies in R^n. Then $\sigma(\mathcal{O})$ is the Borel σ-algebra on R^n.

If Ω is an at most countable set, we shall always take the power set of Ω as the associated σ-algebra.

Definition 1.2.3. Let Ω be a sample space and $\mathcal{C}$ a σ-algebra of subsets of Ω. A function $P(\cdot)$ defined on $\mathcal{C}$ and taking values in the unit interval $[0,1]$ is called a *probability measure* if (1) $P(\Omega) = 1$, (2) $P(A) \geqslant 0$ for all $A \in \mathcal{C}$, and (3) for an at most countable family A_n, $n \geqslant 1$, of mutually disjoint events, we have

$$P\left\{ \bigcup_{n \geqslant 1} A_n \right\} = \sum_{n \geqslant 1} P(A_n).$$

The triple $(\Omega, \mathcal{C}, P)$ is called a *probability space*. The pair $(\Omega, \mathcal{C})$ is called a *probabilizable space*.

The first basic properties of a probability measure are collected in the following theorem.

Theorem 1.2.4. (i) *If A and B belong to $\mathcal{C}$ then $P(A \cup B) + P(A \cap B) = P(A) + P(B)$.*

   (ii) *If $A, B \in \mathcal{C}$ and $A \subset B$, then $P(A) \leqslant P(B)$ (the monotonicity property).*

   (iii) *If $A, B \in \mathcal{C}$ and $A \subset B$, then $P(B \backslash A) = P(B) - P(A)$.*

   (iv) *If $A_n \in \mathcal{C}$, $n \geqslant 1$, then $P\{\bigcup_{n \geqslant 1} A_n\} \leqslant \sum_{n \geqslant 1} P(A_n)$ (subadditivity).*

   (v) *If $A_n \in \mathcal{C}$ and $A_n \uparrow A$, that is, $A_n \subset A_{n+1}$, $n \geqslant 1$, and $\bigcup_n A_n = A$, then $P(A) = \lim_{n \to \infty} P(A_n)$ (continuity from below).*

   (vi) *If $A_n \in \mathcal{C}$ and $A_n \downarrow A$, that is, $A_n \supset A_{n+1}$, $n \geqslant 1$, and $\bigcap_n A_n = A$, then $P(A) = \lim_{n \to \infty} P(A_n)$ (continuity from above).*

Definition 1.2.5. Let $A_n \in \mathcal{C}$, $n \geqslant 1$. The set of all ω that belong to infinitely many of the sets A_n is called the *limit superior* of the sets A_n and is denoted by $\limsup A_n$. More precisely,

$$\limsup_{n \to \infty} A_n = \bigcap_{k=1}^{\infty} \bigcup_{n=k}^{\infty} A_n.$$

The *limit inferior* of the sets A_n, denoted by lim inf A_n, is the set of all ω that belong eventually to A_n for all large n. That is,

$$\liminf_{n \to \infty} A_n = \bigcup_{k=1}^{\infty} \bigcap_{n=k}^{\infty} A_n.$$

Theorem 1.2.6. (First Borel–Cantelli lemma). *Let $\{A_n\}$, $n \geqslant 1$, be a sequence of events and $A = \limsup_{n \to \infty} A_n$. If $\Sigma_n P(A_n) < \infty$, then*

$$P\{A_n \, \mathrm{IO}\} = P(A) = 0,$$

where IO *represents infinitely often. (Sometimes the notation $\{A_n \, \mathrm{IO}\}$ is used to denote the set* lim sup A_n.)

To avoid some undesirable situations, one completes a probability measure as follows. Let $\mathfrak{N}$ be the family of all subsets of Ω that are contained in the P-null events, that is, in the events of probability zero. The σ-algebra generated by the collection $\{\mathcal{C}, \mathfrak{N}\}$ is called the *completion of* $\mathcal{C}$ and is denoted by $\bar{\mathcal{C}}$. Every set B in $\bar{\mathcal{C}}$ can be written as $B = A \cup N$ with $A \in \mathcal{C}$, $N \in \mathfrak{N}$, and $A \cap N = \phi$. Now extend P to $\bar{\mathcal{C}}$ by defining

$$\bar{P}(B) = \bar{P}(A \cup N) = P(A).$$

It is easy to verify that $\bar{P}$ is a probability measure on $\bar{\mathcal{C}}$. The set function $\bar{P}$ is called the *completion of* P. A probability space $(\Omega, \mathcal{C}, P)$ is called a *complete probability space* if every subset of a P-null event is also an event. Throughout this textbook we assume that P is a complete measure.

1.3. Random Variables and Distribution Functions

Let $(\Omega, \mathcal{C}, P)$ be a (complete) probability space and $(S, \mathcal{S})$ an arbitrary probabilizable space.

Definition 1.3.1. A function $X: \Omega \to S$ is called a *random element* in S (or a *random variable* in case $S = R$ and $\mathcal{S} = \mathcal{R}$) if

$$\{\omega: X(\omega) \in B\} \in \mathcal{C} \qquad \text{for every } B \in \mathcal{S}.$$

Recall that if S is discrete (i.e., at most countable), then $\mathcal{S} = 2^S$, the power set of S. Now a mapping $X: \Omega \to S$ is called a *(discrete) random variable* if

$$\{\omega \in \Omega: X(\omega) = s\} \in \mathcal{C} \qquad \text{for every } s \in S.$$

We abbreviate the term *random variable* by RV.

Definition 1.3.2. Two RVs X and Y are said to be *equivalent* if $P\{\omega \in \Omega: X(\omega) \neq Y(\omega)\} = 0$.

We do not distinguish between two equivalent RVs.

Let A be an arbitrary subset of Ω. The function defined by $I_A(\omega) = 1$ if $\omega \in A$, and 0 otherwise, is called the *indicator function* of A. Let now $A \in \mathcal{C}$. Then $I_A(\omega)$ is clearly an RV. If $A \notin \mathcal{C}$, then I_A is not an RV. Thus there are real-valued functions on Ω that are not RVs.

Theorem 1.3.3. *The following statements are equivalent*:

1. X *is a real-valued* RV.
2. $\{\omega: X(\omega) \geqslant a\} \in \mathcal{C}$ *for all* $a \in R$.
3. $\{\omega: X(\omega) > a\} \in \mathcal{C}$ *for all* $a \in R$.
4. $\{\omega: X(\omega) \leqslant a\} \in \mathcal{C}$ *for all* $a \in R$.
5. $\{\omega: X(\omega) < a\} \in \mathcal{C}$ *for all* $a \in R$.

It is customary to shorten the notation $\{\omega \in \Omega: X(\omega) \leqslant a\}$, say, by $\{X \leqslant a\}$. We follow this practice here.

Theorem 1.3.4. (i) *If X and Y are two RVs, then the sets $\{X < Y\}$, $\{X \leqslant Y\}$, $\{X = Y\}$, and $\{X \neq Y\}$ all lie in $\mathcal{C}$.* (ii) *If X and Y are two RVs, then so are $X \pm Y$ and XY.* (iii) *If $\{X_n\}$ is a sequence of RVs, then $\sup_n X_n$, $\inf_n X_n$, $\limsup_{n \to \infty} X_n$, $\liminf_{n \to \infty} X_n$ are all RVs.*

A mapping $\mathbf{X}: \Omega \to R^d$, represented by $\mathbf{X} = (X_1, \ldots, X_d)$, is called a *random vector* if for every k, $1 \leqslant k \leqslant d$, and every $a \in R$ the set $\{X_k(\omega) \leqslant a\} \in \mathcal{C}$. A complex-valued RV Z is defined as the linear combination $X + iY$ of two real RVs X and Y.

Given an RV X, it induces a σ-algebra on Ω as follows. The σ-algebra induced or generated by X is the smallest σ-algebra that contains all the sets of the form $\{X \leqslant a\}$, $a \in R$. This σ-algebra is denoted by $\sigma(X)$. A similar definition is obtained for the σ-algebra $\sigma(X_1, \ldots, X_n)$ generated by the RVs $X_1, \ldots, X_n$.

Let $X: \Omega \to S$ be a discrete RV and $\{s_1, s_2, \ldots\}$ an enumeration of the points of S. Define a sequence $\{p_k\}$ of reals by $p_k = P\{X = s_k\}$. Then $0 \leqslant p_k \leqslant 1$ and $\Sigma_k p_k = 1$. The sequence $\{p_k\}$ of probabilities is known as the *probability distribution of X*. Now, let $\mathbf{X}: \Omega \to R^d$ and $\mathbf{X} = (X_1, \ldots, X_d)$. The ($d$-variate) distribution function of $\mathbf{X}$ or the joint distribution function of $X_1, \ldots, X_d$ is defined by

$$F(x_1, \ldots, x_d) = P\{X_1 \leqslant x_1, \ldots, X_d \leqslant x_d\}$$

for $d \geqslant 1$, $x_k \in R$, $1 \leqslant k \leqslant d$. (The term *distribution function* is abbreviated DF.)

Theorem 1.3.5. *If $F(x_1, \ldots, x_d)$ is a joint* DF, *then*

(i) $F(x_1, \ldots, x_d)$ *is monotomically increasing in each argument.*
(ii) $F(x_1, \ldots, x_d)$ *is right continuous in each argument.*
(iii) *for any $h_k > 0$, $1 \leqslant k \leqslant d$,*

$$0 \leqslant F(x_1 + h_1, \ldots, x_d + h_d)$$

$$- \sum_{k=1}^{n} F(x_1 + h_1, \ldots, x_{k-1} + h_{k-1}, x_k, x_{k+1} + h_{k+1}, \ldots, x_d$$

$$+ h_d)$$

$$+ \sum_{j<k} \sum F(x_1 + h_1, \ldots, x_j, \ldots, x_k, \ldots, x_d + h_d)$$

$$- \sum_{i<j<k} \sum \sum F(x_1 + h_1, \ldots, x_i, \ldots, x_j, \ldots, x_k, \ldots, x_d + h_d)$$

$$+ \cdots + (-1)^d F(x_1, \ldots, x_d).$$

If $f(x_1, \ldots, x_d) = (\partial^d F / \partial x_1 \cdots \partial x_d)$ exists for all $(x_1, \ldots, x_d) \in R^d$, then the function $f(x_1, \ldots, x_d)$ is called the *joint density function* of $F(x_1, \ldots, x_n)$ or of $(X_1, \ldots, X_d)$, and

$$F(x_1, \ldots, x_d) = \int_{-\infty}^{x_1} \cdots \int_{-\infty}^{x_n} f(t_1, \ldots, t_n) \, dt_n \cdots dt_1 .$$

Let $F(x_1, \ldots, x_d)$ be the joint DF of $X_1, \ldots, X_d$, and $1 \leqslant k_1 < k_2 < \cdots < k_n \leqslant d$. Then the *marginal distribution* $F_{k_1, \ldots, k_n}(x_{k_1}, \ldots, x_{k_n})$ of $(X_{k_1}, \ldots, X_{k_n})$ is defined by

$$F_{k_1, \ldots, k_n}(x_{k_1}, \ldots, x_{k_n})$$
$$= F(\infty, \ldots, \infty, x_{k_1}, \infty, \ldots, \infty, x_{k_2}, \infty, \ldots, \infty, x_{k_n}, \infty, \ldots, \infty).$$

1.3.6. Some Standard Distributions

1. A probability distribution $\{p_1, \ldots, p_n\}$ is called (*discrete-*) *uniform* if

$$p_k = 1/n, \qquad k = 1, \ldots, n.$$

2. Fix an n and $0 \leqslant p \leqslant 1$. A distribution $\{p_k\}$, $0 \leqslant k \leqslant n$, is called *binomial with parameters n and p*, if

$$p_k = \binom{n}{k} p^k q^{n-k}, \qquad q = 1 - p, \quad 0 \leqslant k \leqslant n.$$

3. A *geometric distribution* $\{p_k\}$, $k \geqslant 1$, is given by

$$p_k = pq^{k-1}, \qquad k \geqslant 1.$$

4. A *Poisson distribution* $\{p_k\}$, $k \geqslant 0$, with mean parameter $\lambda > 0$, is $p_k = e^{-\lambda}\lambda^k/k!$, $k = 0, 1, \ldots$.

5. A (*continuous*) *uniform distribution* on $[a, b]$ is defined by its density function $f(x) = [b - a]^{-1}$, on $a \leqslant x \leqslant b$, and 0 elsewhere.

6. A *normal density* (with parameters μ and σ^2) is defined by $(\sigma\sqrt{2\pi})^{-1}$ $\cdot \exp[-(x - \mu)^2/2\sigma^2]$, $x \in R$. To say that an RV X is normally or Gaussian distributed, we use the notation $X \in N(\mu, \sigma^2)$.

7. A *gamma density* with parameters $\alpha > -1$ and $\lambda > 0$ is given by

$$f(x) = \begin{cases} [\lambda/\Gamma(\alpha + 1)](\lambda x)^\alpha e^{-\lambda x}, & x > 0 \\ 0, & x \leqslant 0 \end{cases},$$

where Γ is the gamma function

$$\Gamma(x) = \int_0^\infty e^{-y} y^{x-1} \, dy, \qquad x > 0.$$

8. Let $\alpha = 0$ and $\lambda > 0$ in (7). The resulting density

$$f(x) = \begin{cases} \lambda e^{-\lambda x}, & x > 0 \\ 0, & x \leqslant 0 \end{cases},$$

is known as the *exponential density*.

9. Take $\alpha = (n - 2)/2$, where n is a positive integer, and $\lambda > \frac{1}{2}$ in (7). Then the density

$$f(x) = [2^{n/2}\Gamma(n/2)]^{-1} x^{(n-2)/2} e^{-x/2}, \qquad x > 0$$

is known as the χ^2-*density with n degrees of freedom*.

10. Let $\boldsymbol{\mu} = (\mu_1, \ldots, \mu_d)$, $\boldsymbol{\Sigma}$ as positive definite symmetric matrix of order d,

and $|\Sigma|$ the determinant of Σ. A density function $f(x_1, \ldots, x_d)$ on R^d is called a d-dimensional normal density if f is given by

$$f(x_1, \ldots, x_d) = (2\pi)^{-d/2} |\Sigma|^{-1/2} \exp\left[-\tfrac{1}{2}(\mathbf{x} - \boldsymbol{\mu})' \Sigma^{-1} (\mathbf{x} - \boldsymbol{\mu})\right].$$

1.4. Expectation and Moments

Definition 1.4.1.

1. The *expectation* $E[X]$ of a discrete RV X taking the values $\{s_k\}$ is defined by

$$E[X] = \sum_k s_k p_k = \sum_k s_k P\{X = s_k\},$$

   provided that $\Sigma |s_k| p_k < \infty$.

2. The *expectation* $E[X]$ of an RV $X: \Omega \to R$ with DF $F(x)$ is defined by

$$E[X] = \int_{-\infty}^{\infty} x \, dF(x) = \int_{-\infty}^{\infty} x f(x) \, dx,$$

   provided that $\int_{-\infty}^{\infty} |x| \, dF(x) < \infty$.

3. Let $g: R^d \to R^d$ be a Borel measurable function and $F(x_1, \ldots, x_d)$ the DF of $(X_1, \ldots, X_d)$. Then

$$E[g(X_1, \ldots, X_d)] = \int_R \cdots \int_R g(x_1, \ldots, x_d) \, dF(x_1, \ldots, x_d).$$

4. In (3) choose $g(X_1, \ldots, X_d) = X_1^{k_1} \cdots X_d^{k_d}$, $k_i \geq 0$, $1 \leq i \leq d$. Then $E[X_1^{k_1} \cdots X_d^{k_d}]$ is called the $(k_1, \ldots, k_d)$-*moment* of $(X_1, \ldots, X_d)$.

5. The kth *central moment of an* RV X is defined by $m_k = \int_R (x - \mu)^k \, dF(x)$, where $\mu = E[X]$. The second central moment is called the *variance* of X and is denoted by σ^2 instead of m_2.

6. For any two RVs X and Y with finite variances σ_X^2 and σ_Y^2, the *correlation* of X and Y is defined by $E[XY]$, and the *covariance* $\operatorname{cov}(X, Y)$ is defined by $\operatorname{cov}(X, Y) = E[(X - \mu_X)(Y - \mu_Y)]$.

7. By $L^p(\Omega)$, $p \geq 1$, we denote the collection of all (equivalence classes of) RVs such that $E[|X|^p] < \infty$.

The moments m_k of a RV X can also be defined in terms of the probability measure P. Let us now recall the theory of integration with respect to P.

A *simple function* h on Ω is a finitely valued nonnegative function. If h is a simple function, then there are real values $a_k \geqslant 0$, $1 \leqslant k \leqslant n$, and a partition $A_k \in \mathcal{C}$, $1 \leqslant k \leqslant n$, of Ω such that $h = \Sigma_1^n a_k I_{A_k}$. (By a "partition" we mean sets A_k such that $A_i \cap A_j = \phi$, if $i \neq j$, $1 \leqslant i, j \leqslant n$, and $\cup_1^n A_k = \Omega$.) Let $\mathbf{S}$ denote all simple functions $h \colon \Omega \to R_+$. If $h \in \mathbf{S}$, define its integral with respect to (abbreviated WRT) P by

$$\int h \, dP = \sum_1^n a_k P(A_k) = E[h].$$

The integral $\int h \, dP$ is independent of the representation $\Sigma_{k=1}^n a_k I_{A_k}$ of h.

Let $\{h_n\}$ be an increasing sequence in $\mathbf{S}$, and h an element of $\mathbf{S}$ such that $h \leqslant \sup_n h_n$. It can then be shown that

$$E[h] = \int_\Omega h \, dP \leqslant \sup_n \int_\Omega h_n \, dP = \sup_n E[h_n].$$

Consequently, if g_n and h_n, $n \geqslant 1$, are increasing sequences in $\mathbf{S}$, then $\sup_n g_n = \sup_n h_n$ implies that $\sup_n E[g_n] = \sup_n E[h_n]$.

Let $\mathbf{S}^*$ be the collection of all numerical functions $X \geqslant 0$ on Ω for which there is an increasing sequence $h_n \in \mathbf{S}$ such that $X = \sup_n h_n$. Then, by the preceding remark, $0 \leqslant \sup E[h_n] \leqslant \infty$. Now define

$$E[X] = \sup_n E[h_n].$$

This expectation of X is independent of the choice of the increasing sequence $\{h_n\}$. The following result holds: $\mathbf{S}^*$ is the collection of all nonnegative RVs on Ω.

Let X be an arbitrary RV on Ω. Define

$$X^+ = \max(X, 0) \quad \text{and} \quad X^- = -\min(X, 0).$$

Note that $X^+ \geqslant 0$, $X^- \geqslant 0$, $X = X^+ - X^-$, and $|X| = X^+ + X^-$. Now the $E[X]$ of X is defined by $E[X] = E[X^+] - E[X^-]$.

A property Π of points of Ω is said to hold *almost surely* (abbreviated AS) or *with probability one* if the set of all ω for which the property Π does not hold is of probability zero.

Theorem 1.4.2.

(i) *Let $X \in \mathbf{S}^*$. Then $E[X] = 0$ if and only if $X = 0$ AS.*

(ii) *If $E[X]$ exists, that is, $\int |X| \, dP < \infty$, and $Y = X$ AS, then $E[Y]$ exists and $E[X] = E[Y]$.*

(iii) *If X and Y are two RVs such that $E[Y]$ exists and $|X| \leqslant Y$, then $E[X]$ exists.*

(iv) *If $E[X]$ exists, then X is finite AS.*

(v) *If X is nonnegative integer valued, $E[X] = \Sigma_n P\{X > n\}$.*

Theorem 1.4.3. (B. Levi's monotone convergence theorem). *Let $\{X_n\}$ be an increasing sequence in $\mathbf{S}^*$. Then*

$$\sup_n X_n \in \mathbf{S}^* \quad and \quad E[\sup_n X_n] = \sup_n E[X_n].$$

Consequently, $\Sigma_1^\infty X_n \in \mathbf{S}^*$ and $E[\Sigma_1^\infty X_n] = \Sigma_1^\infty E[X_n]$, for every sequence $\{X_n\} \subset \mathbf{S}^*$

Theorem 1.4.4. (Fatou's lemma). *For any sequence $\{X_n\} \subset \mathbf{S}^*$ we have*

$$E[\liminf_{n\to\infty} X_n] \leqslant \liminf_{n\to\infty} E[X_n] \leqslant \limsup_{n\to\infty} E[X_n] \leqslant E[\limsup_{n\to\infty} X_n].$$

Theorem 1.4.5. (Lebesgue's dominated convergence theorem.) *Let $\{X_n\}$ be a sequence from $L^p(\Omega)$ that converges AS on Ω, and Y a nonnegative function from L^p such that $|X_n| \leqslant Y$ for all $n \geqslant 1$. Then there is an RV X on Ω such that $X_n(\omega) \to X(\omega)$ for almost all ω, $X \in L^p$, and $E[|X_n - X|^p] \to 0$, as $n \to \infty$.*

1.5. Conditioning and Independence

Let B be an event, that is, $B \in \mathcal{C}$, such that $P(B) > 0$. Then the mapping $P_B(\cdot): \mathcal{C} \to [0, 1]$ defined by

$$P_B(A) = P(A \cap B)/P(B)$$

is a measure on $\mathcal{C}$ such that $P_B(B) = 1$, even though $P(B)$ need not be equal to one. The value $P_B(A)$ is called the *conditional probability* of A given B and is customarily written as $P(A|B)$.

Formula of Total Probability 1.5.1. Let $\{B_n\}$ be a partition of Ω such that $P(B_n) > 0$ for all n. If $A \in \mathcal{C}$, then

$$P(A) = \sum_n P(B_n)P(A|B_n).$$

Bayes' Formula 1.5.2. If $P(A) > 0$, then under the total probability formula, we have

$$P(B_n|A) = [P(B_n)P(A|B_n)] \Big/ \Big[\sum_k P(B_k)P(A|B_k)\Big], \quad n \geqslant 1.$$

1. Introduction

The *conditional expectation* of X given an event B with $P(B) > 0$ is defined by

$$E[X|B] = \int X\, dP_B = [P(B)]^{-1} \int_B X\, dP = [P(B)]^{-1} E[XI_B].$$

Theorem 1.5.3. *Let X be an* RV *with $E[|X|] < \infty$. Then, for every σ-subalgebra $\mathscr{B} \subset \mathscr{C}$, there exists* AS *a unique* RV *X^* with $E[|X^*|] < \infty$ and such that X^* is a $\mathscr{B}$-RV, that is, $\{X^* \leqslant a\} \in \mathscr{B}$ for all $a \in R$ and $E[X^* I_B] = E[XI_B]$ for all $B \in \mathscr{B}$. This* RV *X^* is called the conditional expectation of X given $\mathscr{B}$ and is denoted by $X^* = E[X|\mathscr{B}] = E_{\mathscr{B}}[X]$. Now*

$$\int_B E[X|\mathscr{B}]\, dP = \int_B X\, dP \qquad \text{for all } B \in \mathscr{B}.$$

Theorem 1.5.4.

(i) $E[E[X|\mathscr{B}]] = E[X]$, *if X is integrable (i.e., $X \in L^1$).*

(ii) *If X is a $\mathscr{B}$-RV, then $E[X|\mathscr{B}] = X$,* AS, *$(X \in L^1)$.*

(iii) *If $X = Y$* AS, *and is integrable, then $E_{\mathscr{B}}[X] = E_{\mathscr{B}}[Y]$,* AS.

(iv) *If $a, b \in R$ and $X, Y \in L^1$, then*

$$E_{\mathscr{B}}[aX + bY] = aE_{\mathscr{B}} X + bE_{\mathscr{B}} Y.$$

(v) *If $X, Y \in L^1$ and $X \leqslant Y$* AS, *then $E_{\mathscr{B}} X \leqslant E_{\mathscr{B}} Y$,* AS.

(vi) *If X_n, $n \geqslant 1$, is an increasing sequence of nonnegative of* RVs, *then $E_{\mathscr{B}}[\sup_n X_n] = \sup_n E_{\mathscr{B}}[X_n]$,* AS.

(vii) *If $\{X_n\}$ is a sequence of* RVs *such that $X_n(\omega) \to X(\omega)$ for all ω except for a set of probability zero and there is a $Y \in L^1(\Omega)$ with $|X_n| \leqslant Y$ for all n, then $\lim_{n \to \infty} E_{\mathscr{B}}(X_n) = E_{\mathscr{B}}(X)$,* AS.

(viii) *If $\mathscr{B}_1$ and $\mathscr{B}_2$ are two σ-algebras of subsets of Ω such that $\mathscr{B}_1 \subset \mathscr{B}_2 \subset \mathscr{C}$, then*

$$E_{\mathscr{B}_1}(E_{\mathscr{B}_2}(X)) = E_{\mathscr{B}_2}(E_{\mathscr{B}_1}(X)) = E_{\mathscr{B}_1}(X),\ \text{AS.}$$

Definition 1.5.5. Let $f(x_1, \ldots, x_d)$ be the joint density function of the RVs $X_1, \ldots, X_d$. The *conditional density* $f_{1, \ldots, k}(u_1, \ldots, u_k | x_{k+1}, \ldots, x_d)$ of $X_1, \ldots, X_k$ given $X_{k+1}, \ldots, X_d$ is defined by

$$f_{1, \ldots, k}(u_1, \ldots, u_k, x_{k+1}, \ldots, x_d)$$

$$= \frac{f(u_1, \ldots, u_k, x_{k+1}, \ldots, x_d)}{\int_R \cdots \int_R f(y_1, \ldots, y_k, x_{k+}, \ldots, x_d)\, dy_1 \cdots dy_k}.$$

Definition 1.5.6.

1. A family $\{A_i, i \in I\}$ of events from $\mathcal{C}$ is called an *independent family* (WRT P) if for every finite subset $\{i_1, \ldots, i_k\} \neq \phi$ of I, we have

$$P\left\{\bigcap_{j=1}^{k} A_{i_j}\right\} = \prod_{j=1}^{k} P(A_{i_j}). \tag{1.5.1}$$

2. A family $\{\mathcal{C}_i, i \in I\}$ of sub-σ-algebras of $\mathcal{C}$ is said to be *independent* if for every finite subset $\{i_1, \ldots, i_k\}$ of I and $A_{i_j} \in \mathcal{C}_{i_j}$ we have relation (1.5.1).

3. A family $\{X_i, i \in I\}$ of RVs on Ω is said to be *independent* if the family $\{\sigma(X_i)\}$, $i \in I$, is independent.

It is easy to show that the RVs $X_1, \ldots, X_n$ are independent if and only if their joint distribution function factors as

$$F(x_1, \ldots, x_n) = F_{X_1}(x_1) \cdots F_{X_n}(x_n).$$

Theorem 1.5.7.

(i) Let $X_1, \ldots, X_n$ be independent and belong to L^1. Then $E[\prod_{k=1}^{n} X_k] = \prod_{k=1}^{n} E[X_k]$.

(ii) Let $X_1, \ldots, X_n \in L^2$ and be independent. Then $\mathrm{var}(\Sigma_1^n X_k) = \Sigma_1^n \mathrm{var}(X_k)$, where $\mathrm{var}(\cdot)$ denotes the variance.

Second Borel–Cantelli Lemma 1.5.8. *If $\{A_n\}$ is a sequence of independent events with $\Sigma_1^\infty P(A_n) = \infty$, then $P\{A_n \, IO\} = 1$.*

Definition 1.5.9. Let X_n, $n \geq 1$, be a sequence of RVs. Let $\mathcal{B}_k = \sigma[X_k, X_{k+1}, \ldots]$ be the σ-algebra generated by $X_k, X_{k+1}, \ldots$. The sequence $\mathcal{B}_n$ is nonincreasing. The intersection $\mathcal{T} = \bigcap_{n \geq 1} \mathcal{B}_n$ is called the *tail σ-algebra* of the sequence $\{X_n\}$.

Theorem 1.5.10. (Kolmogorov's Zero–One Law). *The probability of any event belonging to the tail σ-algebra of a sequence of independent RVs is either 0 or 1.*

1.6. Convergence Concepts

Definition 1.6.1.

1. A sequence $\{X_n\}$ of RVs is said to *converge almost surely* to an RV X if

$$P\{\omega \in \Omega: X(\omega) = \lim_{n \to \infty} X_n(\omega)\} = 1.$$

2. A sequence X_n of RVs is said to *converge in probability* to a RV X if, for every $\epsilon > 0$,

$$\lim_{n \to \infty} P\{|X_n - X| > \epsilon\} = 0.$$

3. A sequence $\{X_n\} \subset L^p$, $(p \geqslant 1)$, is said to *converge in pth mean* to $X \in L^p$ if

$$\lim_{n \to \infty} E|X_n - X|^p = 0.$$

4. Let X_n, $n \geqslant 1$, be a sequence of RVs with DFs F_n, respectively. The sequence X_n is said to *converge in distribution* if there is a distribution function $F(x)$ such that

$$\lim_{n \to \infty} F_n(x) = F(x)$$

at all continuity points x of $F(\cdot)$.

Theorem 1.6.2.

(i) *If $X_n \to X$ AS, then $X_n \to X$ in probability. The converse need not hold.*

(ii) *A sequence $X_n \to X$ AS if and only if, for every real $\epsilon > 0$, $\lim_{n \to \infty} P\{\sup_{m \geqslant n}|X_m - X| \geqslant \epsilon\} = 0$.*

(iii) *A sequence $X_n \to X$ in probability if and only if every subsequence of $\{X_n\}$ contains a further subsequence that converges AS to X.*

(iv) *Convergence in the pth mean implies the convergence in probability. The converse is not true.*

1.7. Transforms

In this section we collect some basic properties of the Fourier transform (characteristic function), the Laplace–Stieltjes transform, and the probability-generating functions of a distribution or an RV.

Definition 1.7.1. The *characteristic function* (CF) of an RV X or its DF $F(x)$

is defined as the Fourier transform

$$\hat{f}(\theta) = E[e^{i\theta X}] = \int_R e^{i\theta x} dF(x)$$

$$= u(\theta) + iv(\theta) = \int_R \cos \theta x \, dF(x) + i \int_R \sin \theta x \, dF(x).$$

Theorem 1.7.2. *Let* $\hat{f} = u + iv$ *be the* CF *of an* RV X. *Then* (i) $\hat{f}$ *is continuous,* (ii) $\hat{f}(0) = 1$, (iii) $|\hat{f}(\theta)| \leqslant 1$ *for all* θ, (iv) *the* CF *of* $aX + b$ *is* $e^{ib\theta}\hat{f}(a\theta)$, (v) $u(\theta)$ *is even and* $v(\theta)$ *is odd,* (vi) $\hat{f}$ *is real if and only if* F *is a symmetric distribution,* (vii) $|\hat{f}(\theta_2) - \hat{f}(\theta_1)|^2 \leqslant 2[1 - u(\theta_2 - \theta_1)]$, *and* (viii) $u^2(\theta) \leqslant 2^{-1}[1 + u(2\theta)]$.

Theorem 1.7.3.

(i)   *If* $\hat{f}_1$ *and* $\hat{f}_2$ *are the* CFs *of the independent* RVs X_1 *and* X_2, *then the* CF *of* $X_1 + X_2$ *is* $\hat{f}_1\hat{f}_2$.
(ii) *Distinct* DFs *have distinct* CFs.
(iii) *For any two continuity points* a *and* b *of* F *we have*

$$F(b) - F(a) = \lim_{T \to \infty} \frac{1}{2\pi} \int_{-T}^{T} (i\theta)^{-1}[e^{-i\theta a} - e^{-i\theta b}]\hat{f}(\theta) \, d\theta.$$

(iv) *A continuous function that is the pointwise limit of a sequence of* CFs *is a* CF.
(v)   *Continuity theorem. A sequence* $\{F_n\}$ *of* DFs *converges to a* DF $F(x)$ *if and only if the corresponding* CFs $\hat{f}_n$ *converges at all* θ *to a function* $\hat{f}$ *that is continuous at* 0. *Moreover,* $\hat{f}$ *is the* CF *of* F.

Theorem 1.7.4. (Bochner.) *A function* g *on* R *is a characteristic function if and only if it is nonnegative definite and continuous.*

Definition 1.7.5. The *Laplace transform* of a DF $F(\cdot)$ is defined by

$$\phi(\theta) = \int_R e^{-\theta x} dF(x) = E[e^{-\theta X}], \qquad \theta > 0,$$

wherever the integral exists.

In discussing Laplace transforms, let us assume that the DF is concentrated on $[0, \infty)$.

Theorem 1.7.6.

15

1. Introduction

    (i)   *Distinct DFs have distinct Laplace transforms.*

    (ii) *The nth derivative $\phi^{(n)}(\theta)$, $n \geqslant 1$, is given by $\phi^{(n)}(\theta)$*
       $= (-1)^n \int_0^\infty e^{-\theta x} x^n \, dF(x).$

    (iii) $\int_0^\infty e^{-\theta x}[1 - F(x)] \, dx = \theta^{-1}[1 - \phi(\theta)].$

    (iv) *A function ϕ on $(0, \infty)$ is the Laplace transform of a DF F if and only if $\phi(0) = 1$, the derivatives $\phi^{(n)}$ of all orders exist, and $(-1)^n \phi^{(n)}(\theta) \geqslant 0$.*

    (v) *Inversion formula. If ϕ is the Laplace transform of a DF F, then at all points of continuity of F, $F(x) = \lim_{a \to \infty} \sum_{n \leqslant ax} (n!)^{-1} (-a)^n \phi^{(n)}(a).$*

Definition 1.7.7. The *(probability) generating function* (PGF) $g(s)$ of a probability distribution $\{p_k\}$ is defined by

$$f(s) = \sum_k p_k s^k.$$

Theorem 1.7.8.

    (i) *Let $\{p_k, k \geqslant 0\}$ be the probability distribution of a nonnegative RV. Then the corresponding PGF $g(s)$ converges for all s with $|s| \leqslant 1$.*

    (ii) *If $g_1, \ldots, g_n$ are the PGFs for n independent RVs $X_1, \ldots, X_n$, then the PGF of $X_1 + \cdots + X_n$ is $g_1 \cdots g_n$.*

    (iii) *Let $\{X_n\}$, $n \geqslant 1$, be a sequence of nonnegative integral-valued RVs with the probability distributions $\{p_k^{(n)}\}$ and PGFs $\{g_n(s)\}$. For a random index $Z(\omega)$, that is, for an RV Z with values in $\{1, 2, \ldots\}$, define $Y(\omega) = X_{Z(\omega)}(\omega)$. Let $\{q_k\}$ be the probability distribution of Z. Then the PGF of Y is given by $g_Y(s) = \sum_{k=1}^\infty q_k g_k(s)$, $0 \leqslant s \leqslant 1$.*

    (iv) *The limit $\lim_{s \uparrow 1} d^n g / ds^n$ and the sum $\sum_{k \geqslant n}(k! \, p_k/(k - n)!)$ are either both finite or both infinite. If they are both finite, they are equal.*

1.8. Limit Theorems

Theorem 1.8.1. (Generalized Chebyshev Inequality.) *Let X be an RV and f an even nonnegative function on R such that $f(\cdot)$ is nondecreasing for $x \geqslant 0$ and $E[f(X)]$ exists. Then for any $a \geqslant 0$, the following inequality holds:*

$$P\{|X| \geqslant a\} \leqslant [f(a)]^{-1} E[f(X)].$$

By choosing $f(x) = |x|^\alpha$, $\alpha \geqslant 0$, one obtains the Markov inequality

$$P\{|X| \geqslant a\} \leqslant a^{-\alpha} E[|X|^\alpha], \qquad a > 0.$$

16

Theorem 1.8.2. (Weak Law of Large Numbers.) *If $\{X_n\}$ is a sequence of independent and identically distributed* (IID) *RVs with mean μ and variance σ^2, then*

$$\lim_{n \to \infty} P\left\{\left| n^{-1} \sum_{k=1}^{n} X_k - \mu \right| \geq \epsilon \right\} = 0$$

for every $\epsilon > 0$.

Theorem 1.8.3. (Kolmogorov Inequality.) *Let $X_1, \ldots, X_n$ be independent real L^2-RVs. Then, for every $a > 0$,*

$$P\left\{ \sup_{1 \leq k \leq n} \left| \sum_{i=}^{k} (X_i - \mu_i) \right| \geq a \right\} \leq a^{-2} \sum_{k=1}^{n} \mathrm{var}(X_k),$$

where $\mu_i = E(X_i)$.

Theorem 1.8.4. (Kolmogorov's Strong Law of Large Numbers.) *Let $\{X_n\}$ be an independent sequence of L^2-RVs. If $\Sigma_{n \geq 1} n^{-2} \mathrm{var}(X_n) < \infty$, then*

$$\lim_{n \to \infty} n^{-1} \sum_{k=1}^{n} (X_k - EX_k) = 0, \qquad \text{AS.}$$

Definition 1.8.5. We say that the *central limit theorem* holds for an independent sequence of L^2-RVs with $\mathrm{var}(X_n) > 0$ if the sequence F_{S_n} of DFs of the sums

$$S_n = [\mathrm{var}(X_1 + \cdots + X_n)]^{-1/2} \sum_{k=1}^{n} (X_k - EX_k)$$

converges to the DF of $N(0, 1)$.

Theorem 1.8.6. *Let $\{X_n\}$ be a sequence of IID RVs with a finite, positive variance σ^2. Then*

$$\lim_{n \to \infty} P\left\{ [\sigma\sqrt{n}]^{-1} \left[\left(\sum_{k=1}^{n} X_k \right) - n\mu \right] \leq x \right\} = (2\pi)^{-1/2} \int_{-\infty}^{x} e^{-u^2/2} \, du,$$

where $\mu = E[X_n]$.

Theorem 1.8.7. (Lyapunov.) *Let $\{X_n\}$ be an independent sequence of RVs with mean zero and such that for some $a > 0$, the moments $E|X_k|^{2+a}$ exist and*

$$\lim_{n \to \infty} \left[\sum_{1}^{n} EX_k^2 \right]^{-(2+a)/2} \sum_{1}^{n} E|X_k|^{2+a} = 0.$$

Then the central limit theorem holds for the sequence $\{X_n, n \geq 1\}$.

2

Random Walk

2.1. Introduction

The first mathematical foundation of the theory of random walk (RW) seems to be due to Pearson, who in 1905 posed an RW model of a motion similar to that of the bacterium *Escherichia coli*. The general theory was initially developed by Markov, Smoluchowski, and Pólya, among others. Currently the field of RW has a rich theory (Spitzer 1964). One can find applications of RW theory in such diverse areas as polymer physics, solid-state physics, kinetic theory of chemical reactions, astronomy, population growth, neural networks, carcinogenesis, stockmarket trends, business risk, and storage theory. This chapter presents an introduction to such an important area of study. For most part we study only the simple lattice model.

Let us motivate the definition by means of two examples. Consider a gambler starting out with an initial fortune X_0. At the end of each game he either wins or loses a dollar with probabilities p (a measure of his gambling skill) and q, respectively, where $0 \leqslant p, q \leqslant 1$, $p + q = 1$. Each game is played independently of other games. Let J_n, $n \geqslant 1$, denote the gambler's winning in the nth game. Then $J_1, J_2, \ldots$ are independent and identically distributed (IID) random variables (RVs) with the common distribution

$$P\{J_n = 1\} = p, \qquad P\{J_n = -1\} = q, \qquad p + q = 1.$$

Let $X_n = (X_0 + J_1 + \cdots + J_n)$. Then $\{X_n, n \geqslant 0\}$ is a discrete-time, discrete-space stochastic process denoting the gambler's cumulative fortune at the end of game n, $n \geqslant 0$.

Next consider the example of the escape of comets from the solar system (Kendall 1961). Consider the motion of a comet around the earth. During each revolution, the energy of the comet undergoes a change due to the penetration and passage through the planetary zone. In successive revolutions the changes J_n, $n \geqslant 1$, in the energy of the comet are assumed to be IID RVs. If X_0 denotes the initial energy of the comet, then $X_n = X_0 + J_1 + \cdots + J_n$ denotes the energy of the comet at the end of nth revolution. The problem studied by Kendall is that of the escape of the comet from the solar system. For positive energy the comet is bound and follows an elliptical orbit. Once the energy hits the level 0, the comet escapes from the solar system (which automatically kills, for our purpose, the process $\{X_n\}$ at this random time of hitting 0). Here $X = \{X_n, n \geqslant 0\}$ is a discrete time and continuous-state space stochastic process.

All our RVs can be defined on a supporting probability space $(\Omega, \mathcal{C}, P)$. We often suppress the sample variable ω in $X(\omega)$, $\omega \in \Omega$, and simply write X. Abstracting the ideas common to the preceding examples, we have the following definition.

Definition 2.1.1. Let $\{J_n, n \geqslant 1\}$ be a sequence of IID RVs taking values in the d-dimensional Euclidean space R^d, and X_0 a fixed vector in R^d. The stochastic process $X = \{X_n, n \geqslant 0\}$ defined by

$$X_n = X_0 + J_1 + \cdots + J_n, \qquad n \geqslant 1,$$

is called a *d-dimensional random walk*. If the vector X_0 and the RVs J_n take values in I^d, where I is the set of all integers, then $\{X_n\}$ is called a *d-dimensional lattice random walk*. In the lattice walk case, if we allow only the jumps J_n from $\mathbf{x} = (x_1, \ldots, x_d)$ to $\mathbf{y} = (x_1 + \epsilon_1, \ldots, x_d + \epsilon_d)$, where $\mathbf{x} \in I^d$ and $\epsilon_k = -1$ or 1, $1 \leqslant k \leqslant d$, then the corresponding walk is called a *simple random walk*. If each of the $2d$ moves at any given jump in a simple RW occurs with equal probability $p = (1/2d)$, then X is called a *symmetric random walk*. In all these cases, if the jumps J_n are only independent but not necessarily identically distributed, then X is called a *nonhomogeneous random walk*.

We study only the simple one-dimensional lattice walk in this chapter.

Examples 2.1.2

EXAMPLE 1. *Photosynthesis.* Consider the photosynthesis in plants. This is the use of light energy for the production of sugar and carbohydrates from CO_2 and H_2O. This photosynthetic reaction takes place in the so-called photosynthetic unit, which contains a lattice of chlorophyll molecules. In his study of exciton trapping on photosynthetic units, Montroll (1969) treated the transfer of exciton as an RW from one chlorophyll molecule to a neighboring one on the lattice. Consider the lattice of chlorophyll molecules with one trap for every T molecules. In the photosynthetic reaction, the photon is absorbed (with equal probability) by any chlorophyll molecule and then walks on the lattice by exciton transfer until it hits a trap, at which time it triggers the chemical reaction. For further details, see Montroll (1969). A similar situation is described in Example 2.

EXAMPLE 2. *Spike Activity of a Neuron.* The *excitability* of a neuron is the ability of the neuron to respond to stimuli. An excitation with sufficiently strong stimuli is propagated along the nerve fiber (a motion of ions along the axon membrane). This propagation is called *impulse,* and the corresponding electrical manifestation is known as the *action potential.* Here we describe an RW model of the spike activity of a single neuron; the electrical state of polarization of the (somatic − dendritic) membrane of the neuron is specified by a single number. As the electrical state of the membrane varies (relative to time), the state point walks (back and forth) along a line. Fix a point on this line as a *resting potential.* Fix another point, called *threshold* level, a constant unit away from the resting potential. Once the electrical state reaches the threshold, the neuron fires, producing an action potential. Let each incoming excitatory postsynaptic potential (resp. inhibitory postsynaptic potential) move the state point one unit toward (resp. one unit away) from the threshold. Let the steps toward and away from the threshold occur with equal probability ($\frac{1}{2}$). Once the state point hits the threshold, it returns to the resting potential and begins its walk afresh. For further details, see Gerstein and Mandelbrot (1964).

EXAMPLE 3. *Radiation Damage.* Bharucha-Reid and Landau (1951) have suggested an RW model for the transmission of radiation damage through a biological system. The proposed physical mechanism is as follows. A control molecule is present in an organism, and chain macromolecules are connected to the control molecule. A *hit* in this control molecule causes the initial damage that is transmitted through the system by the chain depolymerization of the macromolecules. The complete depolymerization of the macromolecules is assumed to be responsible for the observed damage to the system. This depolymerization process is treated as an RW on the state space $S = \{0, 1, \ldots, D\}$. Initially the macromolecules are assumed to be intact (i.e., state 0). Following a hit, the system is in state 1 and the transmission begins. A unit step forward describes further depolymerization, and a unit step backward describes recovery. State D denotes the completion of the depolymerization and thereby causing an observable damage. The return to 0 represents complete recovery.

EXAMPLE 4. *Storage.* In this example we treat the content of a dam as a random walk. Let X_k denote the amount of water in a dam at the end of kth day. During day $k + 1$ a random amount I_{k+1} units of water flow into the dam due to rainfall and so on. If $X_k + I_{k+1}$ is larger than a fixed amount a, then a units of water is released during the day. If $(X_k + I_{k+1}) \leqslant a$, the dam is drained. To accommodate the overflow, let c denote the full capacity of the dam. If $(X_k + I_{k+1} - a) > c$, an overflow occurs. Let $J_k = (I_k - a)$. Then $X_n = (X_{n-1} + J_n)$ for $0 < X_{n-1} + J_n < b$, $X_n = 0$ for $(X_{n-1} + J_n \leqslant 0)$, or $X_n = c$ in the case of overflow. If $\{J_n\}$ is a sequence of IID RVs, then $\{X_n, n \geqslant 0\}$ is an RW.

2.2. Gambler's Ruin

Let J_0 be a fixed positive integer and J_n, $n \geqslant 1$, be the independent and identically distributed jump variables in a random walk $\{X_n, n \geqslant 0\}$ such that

$$X_n = J_0 + J_1 + \cdots + J_n.$$

The random walk $\{X_n, n \geqslant 0\}$ is called a *simple* random walk provided that

$$J_n = \begin{cases} 1 & \text{with probability } p \\ -1 & \text{with probability } q, \\ 0 & \text{with probability } r \end{cases}$$

where $(p + q + r) = 1$ and $0 < p, q < 1$; $0 \leqslant r < 1$. When $p = \frac{1}{2} = q$ (so that $r = 0$) we call it a *symmetric random walk*. Throughout this section we consider only the simple random walk on the integer lattice.

Proposition 2.2.1. Let $X = \{X_n; n \geqslant 0\}$ be a (simple) random walk and $\nu > 0$ a fixed integer. Define

$$Y = \{Y_n = X_{n+\nu} - X_\nu, \quad n \geqslant 0\}.$$

Then, Y is also a (simple) random walk. This proposition says that an RW starts from scratch at any given time ν.

PROOF. By definition, $Y_0 = 0$ and

$$Y_n = X_{n+\nu} - X_\nu = 0 + J_1^* + J_2^* + \cdots + J_n^*,$$

where $J_k^* = J_{k+\nu}$. Since the RVs J_n are independent and identically distributed (IID), the J_k^* are also independent and identically distributed. Therefore, Y_n is the sum of n IID RVs for every n. Hence $\{Y_n, n \geqslant 0\}$ is a (simple) random walk. $\square$

Consider a gambler, call him Tom, with initial fortune of $\$a > 0$ playing against a gambling house with initial fortune $\$b > 0$. We view the game in terms of Tom so that X_n will denote the cumulative fortune of Tom at the end of game n. Tom's winnings in each game will be either 1, -1, or 0, so that $\{X_n, n \geq 0\}$ is a simple RW. The game ends whenever either Tom or the house goes broke. In physical terms, the particle is absorbed either at 0 or at $a + b$. To analyze the situation, define the *time of absorption* or the *time of ruin* as follows:

$$T = \min\{n \geq 0 : X_n \leq 0 \quad \text{or} \quad X_n \geq a + b\}.$$

We assume throughout that $0 \leq r < 1$, $(P(J_1 = 0) < 1)$.

A *random* or *stopping time* T with respect to $\{X_n\}$ is an extended positive integer-valued RV such that, for each n, the event $\{T = n\}$ depends only on $\{X_1, \ldots, X_n\}$. Clearly, the absorption time T is a stopping time. (See also Section 3.4).

Proposition 2.2.2.

(i) $P\{T < \infty\} = 1$.

(ii) $E[T] < \infty$.

In other words, with probability 1 eventually Tom either goes broke or breaks the house. The expected duration for absorption is finite.

PROOF. Let A be the event that $T = \infty$; that is, A is the event that $0 < X_n < (a + b)$ for every $n \geq 0$. We show that $P(A) = 0$. Define

$$A_n = \{0 < X_k < a + b, 0 < k \leq n\}.$$

Then, $A = \cap\, A_n$ and $P(A) \leq P(A_n)$ for every $n \geq 0$. Put $c = (a + b)$ and

$$\xi_k = X_{kc} - X_{(k-1)c} = J_{kc-c+1} + \cdots + J_{kc}.$$

Since $\xi_{k+1} = [J_{kc+1} + \cdots + J_{(k+1)c}]$, it follows from the fact that J_n are IID RVs that ξ_k, $k \geq 1$, are IID RVs. Now, noting that $X_0 \equiv J_0 \equiv a$,

$$P\{X_{a+b} \geq 2a + b \quad \text{or} \quad X_{a+b} \leq -b\}$$

$$= P\{J_k = 1 \quad \text{for all } 0 < k \leq a + b$$

$$\text{or } J_k = -1 \quad \text{for all } 0 < k \leq a + b\}$$

$$= p^{a+b} + q^{a+b} = p^c + q^c (= d, \text{ say}),$$

and consequently $P\{\xi_k \geq 2a + b \text{ or } \xi_k \leq -b\} = d(< 1)$. Then,

$$P(A_{nc}) = P\left\{ \bigcap_{k=1}^{nc} \{0 < X_k < a + b\} \right\}$$

$$\leqslant P\left\{ \bigcap_{k=1}^{n} \{0 < X_{kc} < a + b\} \right\}$$

$$\leqslant P\left\{ \bigcap_{k=1}^{n} \{-b < \xi_k < 2a + b\} \right\}$$

$$= (1 - d)^n = s^n,$$

where $s = 1 - d < 1$. Hence

$$0 \leqslant P(A) \leqslant P(A_{nc}) \leqslant s^n, \qquad \text{for every } n,$$

and consequently (as $n \to \infty$), $P(A) = 0$. This proves (i).

For every $n \geqslant 0$, there is a $k \geqslant 1$ such that $(k - 1)c \leqslant n < kc$. Then,

$$P\{T > n\} = P(A_n) \leqslant P(A_{kc-c}) \leqslant s^{k-1}.$$

Since T is nonnegative integer valued,

$$E[T] = \sum_n P\{T > n\} \leqslant c \sum_{k=1}^{\infty} s^k = \frac{cs}{1 - s} < \infty. \qquad \square$$

This proposition tells us an important fact: that Tom eventually either goes broke or breaks the house AS. So we have Theorem 2.2.3.

Theorem 2.2.3. (Gambler's Ruin). *Let Tom and Dick be two gamblers starting a game with initial fortunes a and b, respectively. The game is such that the sequence $\{X_n, n \geqslant 0\}$ of cumulative fortunes of Tom, say, is a simple random walk. Let the probability that any given game ends in a tie be strictly less than 1, $(0 \leqslant r < 1)$. Then the probability of either one of them ever breaking the other is 1. The expected duration of the game is finite.*

Proposition 2.2.4. (Wald's First Identity). Let the jump variates J_n have a (common) mean μ, $(E[J_k] = \mu)$, and $r < 1$. Then

$$E[X_T] = a + \mu E[T],$$

where $a = X_0 \equiv J_0$.

$$E[X_T] = E\left[a + \sum_{k=1}^{T} J_k \right]$$

$$= a + E\left[\sum_{k=1}^{\infty} J_k I_{\{T \geqslant k\}}\right],$$

where $I_{\{T \geqslant n\}}$ is the indicator RV of the event $\{T \geqslant n\}$,

$$= a + \sum_{k=1}^{\infty} E[J_k I_{\{T \geqslant k\}}],$$

where the interchange of E and $\sum$ can be justified,

$$= a + \sum_{k=1}^{\infty} E[J_k (1 - I_{\{T < k\}})]$$

$$= a + \sum_{k=1}^{\infty} E[J_k] E[1 - I_{\{T < k\}}], \quad \text{since} \quad \{T<k\} \text{ is determined}$$

by $J_1, \ldots, J_{k-1}$, which are independent of J_k,

$$= a + \sum_{k=1}^{\infty} \mu(1 - P\{T < k\})$$

$$= a + \mu \sum_{k=1}^{\infty} P\{T \geqslant k\}$$

$$= a + \mu E[T], \quad \text{since T is } (>0) \text{ integer valued.}$$

If Tom and Dick gamble, it follows from Theorem 2.2.3 that one of them will eventually be ruined. Let $\{X_n, n > 0\}$ be Tom's random walk (gambling). Next, we find the probabilities of Tom breaking Dick and Tom going broke.

Theorem 2.2.5. (Probability of a Gambler's Ruin). *Let a and b be the initial fortunes of Tom and Dick, respectively, and $\{X_n, n \geqslant 0\}$ be the random walk corresponding to Tom's cumulative fortune. Let $r < 1$. If $p = q$, then:*

(i) $P\{X_T = a + b\} = P\{\text{Dick is ruined}\} = a/(a + b)$.
(ii) $P\{X_T = 0\} = P\{\text{Tom is ruined}\} = b/(a + b)$.

If $p \neq q$, then:

(i*) $P\{X_T = a + b\} = (1 - s^a)/(1 - s^{a+b})$.
(ii*) $P\{X_T = 0\} = (s^a - s^{a+b})/(1 - s^{a+b})$, *where $s = (q/p)$.*

PROOF.   Because of Theorem 2.2.3, it suffices to prove (i) and (i*). Noting that $X_T = 0$ or $X_T = (a + b)$, we have

$$E[X_T] = 0 \cdot P\{X_T = 0\} + (a + b)P\{X_T = a + b\}.$$

2. Random Walk

Let $p = q$. Then, $\mu = E[J_1] = 1P\{J_1 = 1\} + (-1)P\{J_1 = -1) + 0P\{J_1 = 0\} = p - q = 0$. Consequently, from Wald's first identity,

$$a = E[X_T] = (a + b)P\{X_T = a + b\},$$

from which (i) follows.

Now let $p \neq q$. Then, $\mu = p - q \neq 0$ and Wald's identity cannot be used because we do not have an expression for $E[T]$, the expected duration of the game. The method utilized below is illustrative of the use of difference equations in random walk problems.

For $J_0 = a$, an integer in $[0, a + b]$, define $\pi(a)$ as the probability that $X_T = a + b$. Then

$$\pi(a) = P\{X_T = a + b\}$$
$$= pP\{X_T = a + b | J_1 = 1\} + qP\{X_T = a + b | J_1 = -1\}$$
$$+ rP\{X_T = a + b | J_1 = 0\},$$

by the so-called *first step decomposition method*. Because the game starts from scratch every time (Proposition 2.2.1), Tom effectively starts from $a + 1$ provided $J_1 = 1$. Therefore, we have $P\{X_T = a + b | J_1 = 1\} = \pi(a + 1)$ and similar expressions for the cases $J_1 = -1$ and $J_1 = 0$. Hence, $\pi(a)$ satisfies the following difference equation:

$$\pi(a) = p\pi(a + 1) + q\pi(a - 1) + r\pi(a), \qquad 0 < a < a + b. \quad (2.2.1)$$

We need to solve this equation with the obvious boundary conditions

$$\pi(0) = 0; \quad \pi(a + b) = 1. \quad (2.2.2)$$

From (2.2.1) and $p + q = (1 - r)$, we get, setting $(q/p) = s$,

$$\pi(a + 1) - \pi(a) = s[\pi(a) - \pi(a - 1)], \qquad 0 < a < a + b. \quad (2.2.3)$$

Let $\alpha = \pi(1) = \pi(1) - \pi(0)$ [see boundary conditions (2.2.2)]. Then, from (2.2.3),

$$\pi(a + 1) - \pi(a) = s[\pi(a) - \pi(a - 1)]$$
$$= s^2[\pi(a - 1) - \pi(a - 2)]$$
$$\vdots \qquad\qquad\qquad (2.2.4)$$
$$= s^a[\pi(1) - \pi(0)] = \alpha s^a.$$

Next, from (2.2.4) we get

26

$$\pi(a) = \pi(a) - \pi(0) = \sum_{x=0}^{a-1} [\pi(x+1) - \pi(x)]$$

$$= \sum_{x=0}^{a-1} \alpha s^x = \frac{\alpha(1-s^a)}{1-s}, \qquad 0 \leqslant a \leqslant a+b.$$

(2.2.5)

Because $\pi(a+b) = 1$, from (2.2.5) we get

$$\alpha = \frac{1-s}{1-s^{a+b}}$$

and hence,

$$P\{X_T = a+b\} = \pi(a) = \frac{1-s^a}{1-s^{a+b}}. \; \square$$

This proves (i*) and hence the theorem. $\square$

Corollary 2.2.6. (i) *If* $p = q$, *then* $E[X_T] = a$. (ii) *If* $p \neq q$, *then*

$$E[X_T] = \frac{(a+b)(1-s^a)}{1-s^{a+b}}, \qquad s = \frac{q}{p}.$$

PROOF.

(i) $E[X_T] = a + \mu E[T]$. If $p = q$, then $\mu = 0$ and $E[X_T] = a$.
(ii) $E[X_T] = 0 \cdot P\{X_T = 0\} + (a+b)P\{X_T = a+b\}$
$= (a+b)(1-s^a)/(1-s^{a+b})$. $\square$

Next let us consider Tom's luck while he plays against an infinitely rich house. So, we let $b \to \infty$. Then it is not hard to see that $P\{X_T = a+b\}$ converges to $P\{X_n > 0$, for all $n \geqslant 0\}$. It follows from Theorem 2.2.5 (i*) that

$$P\{X_n > 0 \text{ for all } n \geqslant 0\} = \begin{cases} 1 - (q/p)^a & \text{if } q < p \\ 0 & \text{if } q > p \end{cases}.$$

If $p = q$, then from Theorem 2.2.5 (i),

$$P\{X_n > 0 \quad \text{for all } n \geqslant 0\} = 0.$$

Similarly, one can consider $P\{X_n < a+b$ for all $n \geqslant 0\}$. We summarize these observations in the following proposition.

27

Proposition 2.2.7. For $0 \leqslant a = X_0$:

$$\text{(i)} \quad P\{X_n > 0, n \geqslant 0\} = \begin{cases} 1 - (q/p)^a & \text{if } q < p \\ 0 & \text{if } q \geqslant p \end{cases}.$$

$$\text{(ii)} \quad P\{X_n < a + b, n \geqslant 0\} = \begin{cases} 1 - (p/q)^b & \text{if } p < q \\ 0 & \text{if } p \geqslant q \end{cases}.$$

2.3. Expected Duration of the Game

In Proposition 2.2.2 we have shown that $E[T] < \infty$. In the present section we want to find an expression for the expected duration of the game. Let $p \neq q$. From the Corollary 2.2.6 and Wald's first identity (Proposition 2.2.4) we get, noting $\mu = p - q$,

$$E[T] = \frac{1}{\mu}(E[X_T] - a)$$

$$= \frac{1}{p - q}\left[(a + b)\frac{1 - s^a}{1 - s^{a+b}} - a\right], \quad s = \frac{q}{p}.$$

If $p = q$, then $\mu = 0$, and we cannot use Wald's first identity. To remedy the situation, we proceed to Wald's second identity.

Proposition 2.3.1. (Wald's Second Identity). Let σ^2 be the common variance of the jump variates J_n, $n \geqslant 1$. Then,

$$\text{var}(X_T) = \sigma^2 E[T] = (1 - r)E[T].$$

PROOF. When $p = q$, $\mu = E[J_n] = 0$, and $E[X_T] = a$. Now we proceed as in the proof of Wald's first identity.

$$E[(X_T - a)^2] = E\left[\left(\sum_{k=1}^{\infty} J_k(1 - I_{\{T<k\}})\right)^2\right] \tag{2.3.1}$$

$$= \sum_{i=1}^{\infty}\sum_{j=1}^{\infty} E[J_i(1 - I_{\{T<i\}})J_j(1 - I_{\{T<j\}})].$$

First let $i = j$. Then, for each $i \geqslant 1$,

$$E[J_i^2(1 - I_{\{T<i\}})^2] = E[J_i^2(1 - I_{\{T<i\}})]$$

$$= E[J_i^2]E[(1 - I_{\{T<i\}})] \tag{2.3.2}$$

$$= \text{var}(J_i)(1 - P\{T < i\})$$

$$= \sigma^2 P\{T \geqslant i\}.$$

Now let $i \neq j$ and $i < j$ (the case $j < i$ follows similarly). Because $J_i(1 - I_{\{T \le i\}})(1 - I_{\{T < j\}})$ is determined by $J_1, \ldots, J_{j-1}$, which are independent of J_j and $E[J_j] = 0$,

$$E[X_i(1 - I_{\{T < i\}})X_j(1 - I_{\{T < j\}})] = 0. \tag{2.3.3}$$

From (2.3.1)–(2.3.3)

$$\mathrm{var}(X_T) = \sum_{i=1}^{\infty} \sigma^2 P\{T \ge i\} = \sigma^2 E[T].$$

□

Theorem 2.3.2. *Under the conditions of a simple random walk and that $r < 1$, we have*

(i) $E[T] = \dfrac{a + b}{p - q} \dfrac{1 - (q/p)^a}{1 - (q/p)^{a+b}} - \dfrac{a}{p - q}$, *provided that $p \neq q$.*

(ii) $E[T] = ab/(1 - r)$, *provided that $p = q$.*

PROOF. We have proved (i) already. Now, if $p = q$,

$$\mathrm{var}(X_T) = E[X_T^2] - a^2, \quad \text{from Corollary 2.2.6 (i),}$$

$$= 0^2 \cdot P\{X_T = 0\} + (a + b)^2 P\{X_T = a + b\} - a^2$$

$$= (a + b)^2 \frac{a}{a + b} - a^2, \quad \text{from Theorem 2.2.5 (i),}$$

$$= ab.$$

Hence, from Wald's second indentity,

$$E[T] = \frac{1}{(1 - r)} \mathrm{var}(X_T) = \frac{ab}{(1 - r)}.$$

□

2.4. Recurrence and First Passage

First we answer the recurrence problem for the one-dimensional RW. We defer a complete solution until we talk about the recurrence of Markov chains. Next we give some sample first passage properties for the RW. The first passage problems are very basic in the theory of stochastic processes, and we see more of them in the following chapters. In developing this

2. Random Walk

section we use the intuition from gambling and physical motion of a particle. For simplicity, we take $J_0 = X_0 = 0$. (Then X_n denotes the cumulative winnings, rather than the cumulative fortunes.) Let $r = 0$.

Theorem 2.4.1. (Recurrence). *Let $\{X_n, n \geqslant 0\}$ be a simple random walk. Then*

$$P\{X_n = a \text{ IO}\} = 1, \tag{2.4.1}$$

for any integer a, if and only if the walk is symmetric ($p = \frac{1}{2} = q$).

PROOF (*Necessity*). Assuming the contrary, let $p \neq q, p + q = 1$, and $a = 0$. Define $A_n = \{X_{2n} = 0\}$, $n \geqslant 0$. The event A_n occurs iff there were n jumps each to the right and left. Hence

$$P\{X_{2n} = 0\} = \binom{2n}{n} p^n q^n. \tag{2.4.2}$$

(We need this expression later.) Using Stirling's formula (i.e., $n! \sim \sqrt{2\pi n}\, n^n e^{-n}$, for large n), $\binom{2n}{n} p^n q^n \sim (\pi n)^{-1/2}(4pq)^n$, as $n \to \infty$. Consequently, since $p \neq \frac{1}{2}$ and hence $pq < \frac{1}{4}$, we have

$$\sum_{n=1}^{\infty} P(A_n) \sim \sum_{n=1}^{\infty} (\pi n)^{-1/2}(4pq)^n < \infty. \tag{2.4.3}$$

Therefore, from the first Borel–Cantelli lemma,

$$P\{X_{2n} = 0 \text{ IO}\} = 0,$$

contradicting (2.4.1). □

PROOF (*Sufficiency*). For integers $m, n > 0$, define

$$A_{mn} = \{X_n = a \quad \text{and exactly } (m-1) \text{ of } X_k, k = 1, \ldots, n-1,$$

$$\text{are equal to } a\}$$

$$A_m = \bigcup_{n \geqslant 1} A_{mn}, \quad \text{and} \quad A = \bigcap_{m \geqslant 1} A_m.$$

Under the assumption $p = q = \frac{1}{2}$ we can show that $P(A) = 1$. (By Proposition 2.2.1 we can take $a = 0$.) From Section 2.2 we see, for a simple symmetric random walk, that $P(A_1) = 1$. To proceed inductively, let $P(A_m) = 1$. Then

$$P\{A_{m+1}\} = \sum_{n \geqslant 1} P\{A_{m+1} \cap A_{mn}\}$$

$$= \sum_{n \geqslant 1} P\{A_{mn} \text{ and } Y_k \equiv X_{n+k} - X_n = 0 \text{ for some } k \geqslant 1\}$$

$$= \sum_{n \geqslant 1} [P\{A_{mn}\} - P\{A_{mn} \text{ and } Y_k \neq 0 \text{ for all } k \geqslant 1\}]$$

$$= \sum_{n \geqslant 1} P\{A_{mn}\}, \text{ since } P\{Y_k \neq 0 \text{ for all } k \geqslant 1\} = 0,$$

from Section 2.2,

$$= P\{A_m\} = 1.$$

Hence, noting $A_m \downarrow A$,

$$P(A) = \lim_{n \to \infty} P(A_m) = 1,$$

and this completes the proof. $\qquad\qquad\qquad\qquad\qquad\qquad\qquad\square$

Relation (2.4.2) gives the probability that at time $2n$ the particle is at 0. It is interesting and important to find the first passage probability that the particle returns to 0 for the first time at time $2n$.

Theorem 2.4.2. Let $\{X_n, n \geqslant 0\}$ be a simple random walk ($p + q = 1, r = 0, X_0 \equiv 0$). Then

$$P\{X_{2n} = 0, X_k \neq 0, 1 \leqslant k < 2n\}$$

$$= \begin{cases} 2pq & \text{if } n = 1 \\ \dfrac{1}{2} \cdot \dfrac{3}{2} \cdots \dfrac{2n-3}{2} \dfrac{(4pq)^n}{2(n!)} & \text{if } n \geqslant 2. \end{cases} \qquad (2.4.4)$$

Consequently, the probability that the particle never returns to 0 is $(1 - 4pq)^{1/2}$.

PROOF. Since we are talking about the first return, set $p_0 = P\{\text{the particle returns to 0 at } n = 0\} = 0$. Let p_{2n} denote $P\{X_{2n} = 0, X_k \neq 0, 1 \leqslant k < 2n\}$ and A_m denote the event $\{X_{2m} = 0, X_k \neq 0 \text{ for } 0 < k < 2m, \text{ and } X_{2n} = 0\}$, $1 \leqslant m \leqslant n$. Then A_m are disjoint and

$$P\{A_m\} = \binom{2n - 2m}{n - m} p^{n-m} q^{n-m} p_{2m}. \qquad (2.4.5)$$

Consequently, (2.4.2)–(2.4.5) and $P\{X_{2n} = 0\} = P\{\cup_{m=1}^n A_m\}$ imply

$$\binom{2n}{n} p^n q^n = \sum_{m=1}^n p_{2m} \binom{2n - 2m}{n - m} p^{n-m} q^{n-m}. \qquad (2.4.6)$$

Equation (2.4.6) reduces to the (renewal equation) form

$$a_n = \sum_{m=0}^{n} b_m a_{n-m}, \qquad a_n = \binom{2n}{n} p^n q^n, \qquad b_m = p_{2m}, \qquad n \geqslant 0,$$

so that we can write

$$\sum_{n \geqslant 1} a_n x^n = \left\{ \sum_{n \geqslant 1} b_n x^n \right\} \left\{ \sum_{n \geqslant 0} a_n x^n \right\}, \tag{2.4.7}$$

where each of the series converges for $|x| < 1$. Let $|x| < 1$. Then $|2pqx| < 1$ and

$$\sum_{n \geqslant 0} a_n x^n = \sum_{n \geqslant 0} \binom{2n}{n} p^n q^n x^n = (1 - 4pqx)^{-1/2}.$$

Using this in (2.4.7), we obtain

$$(1 - 4pqx)^{-1/2} - 1 = \left\{ \sum_{n \geqslant 1} b_n x^n \right\} \{1 - 4pqx\}^{-1/2},$$

which, on multiplying both sides by $(1 - 4pqx)^{1/2}$, gives

$$\sum_{n \geqslant 1} p_{2n} x^n = \sum_{n \geqslant 1} b_n x^n$$

$$= 1 - (1 - 4pqx)^{1/2} \tag{2.4.8}$$

$$= \frac{-4pqx}{2(1!)} + \frac{1}{2} \cdot \frac{1}{2} \frac{(4pqx)^2}{2!} + \frac{1}{2} \cdot \frac{1}{2} \cdot \frac{3}{2} \frac{(4pqx)^3}{3!} + \cdots,$$

by Maclaurin's expansion for $(1 - 4pqx)^{1/2}$. Equating the coefficients of equal powers of x on both sides, we get (2.4.4). Finally,

$$P\{\text{the particle never returns to } 0\}$$

$$= 1 - P\{\text{the particle returns to } 0 \text{ at some time}\}$$

$$= 1 - \sum_{n \geqslant 1} p_{2n} = 1 - \left[\sum_{n \geqslant 1} p_{2n} x^n \right]_{x=1}$$

$$= (1 - 4pq)^{1/2}, \text{ from (2.4.8)}.$$

This completes the proof. $\qquad\qquad\qquad\qquad\qquad\qquad\qquad\qquad\qquad\square$

Next result computes the probability of the maximum winnings in the first n games and this has an important application.

Theorem 2.4.3. *Let* $\{X_n, n \geqslant 0\}$ *be a symmetric simple random walk and* $M_n = \max\{X_0, \ldots, X_n\}, n \geqslant 1$. *Then for all positive integers n and α we have*

$$P\{M_n \geqslant \alpha\} = P\{X_n \geqslant \alpha\} + P\{X_n > \alpha\}. \qquad (2.4.9)$$

PROOF. First we note that

$$
\begin{aligned}
P\{M_n \geqslant \alpha\} &= P\{M_n \geqslant \alpha, X_n \geqslant \alpha\} + P\{M_n \geqslant \alpha, X_n < \alpha\} \\
&= P\{X_n \geqslant \alpha\} + P\{M_n \geqslant \alpha, X_n < \alpha\} \qquad (2.4.10)
\end{aligned}
$$

To show that the second term is equal to $P\{X_n > \alpha\}$, we follow Kolmogorov's idea (used to establish Kolmogorov's inequality). Define $A_k = \{X_k \geqslant \alpha, X_m < \alpha, 1 \leqslant m < k - 1\}, 1 \leqslant k \leqslant n$. Clearly, the events A_k are disjoint and $\{M_n \geqslant \alpha\} = \cup_{k=1}^n A_k$. Now

$$
\begin{aligned}
P\{M_n \geqslant \alpha, X_n < \alpha\} &= \sum_{k=1}^n P\{A_k, X_n < \alpha\} \\
&= \sum_{k=1}^n P\{A_k, X_n - X_k < 0\} \\
&= \sum_{k=1}^n P(A_k) P\{X_n - X_k < 0\} \qquad \text{(why?)} \\
&= \sum_{k=1}^n P(A_k) P\{X_n - X_k > 0\} \qquad \text{(why?)} \\
&= \sum_{k=1}^n P\{A_k, X_n - X_k > 0\} \\
&= P\{X_n > \alpha\}.
\end{aligned}
$$

Using this in (2.4.10), we obtain (2.4.9). □

Definition 2.4.4. Let $\{X_n, n \geqslant 0\}$ be a symmetric simple random walk on the integer lattice and α be any integer. The *hitting time* of α or *the first passage time* through α is defined by

$$
T_\alpha = \begin{cases} \min\{n \geqslant 1 : X_n = \alpha\} & \text{if the minimum exists;} \\ \infty & \text{otherwise.} \end{cases}
$$

Clearly, T_α is a random time. From (2.4.1), $P\{T_\alpha = \infty\} = 0$.

Theorem 2.4.5. *Let* $\{X_n, n \geqslant 0\}$ *be a symmetric simple random walk on the integer lattice and* T_α *be the first passage time through* α. *Then*

(i) *the probability generating function of* T_α *is* $\Sigma_{n \geqslant 0} p_n s^n = (2qs)^{-\alpha}$
$\cdot [1 - (1 - 4pqs^2)^{1/2}]^\alpha, 0 < s < 1$.

(ii) *for* $x > 0$, $\lim_{\alpha \to \infty} P\{T_\alpha < \alpha^2 x\} = 2[1 - \Phi(x^{-1/2})]$, *where* Φ *is the standard normal distribution.*

PROOF. We postpone the proof of (i) until we establish the optional sampling theorems for Martingale processes [see Example 8.3.6 (2)]. To see (ii), first observe that $P\{T_\alpha \leqslant n\} = P\{M_n \geqslant \alpha\}$, (the gambler's cumulative winnings at or prior to the nth game would have reached α iff the maximum cumulative winnings at the end of the nth game were at least α). Let $n = [\![\alpha^2 x]\!]$, where $[\![\cdot]\!]$ is the greatest integer function. Appealing to the central limit theorem, we have from Theorem 2.4.3 that

$$P\{T_\alpha \leqslant n\} = P\{M_n \geqslant \alpha\}$$
$$= P\{X_n \geqslant \alpha\} + P\{X_n > \alpha\}$$
$$= P\{n^{-1/2} X_n \geqslant n^{-1/2}\alpha\} + P\{n^{-1/2} X_n > n^{-1/2}\alpha\}$$
$$\to [1 - \Phi(x^{-1/2})] + [1 - \Phi(x^{-1/2})].$$

This completes the proof. □

2.5. Two Examples

The two examples presented here are chosen to illustrate the methods of generating functions and characteristic functions. The RWs considered are more general than the simple walks we studied in the last three sections.

2.5.1 A Random Walk in Carcinogenesis

A cancer-inducing agent is called a *carcinogen*. In the study of carcinogenesis, a *hit* refers to the interaction between the carcinogen and the normal cell, which results in the mutation of that normal cell to a cancer cell. The transmission of a normal cell to a (malignant) cancer cell need not occur in one hit or one stage. The *number of stages* is the number of mutations required to produce a cancer cell. A mutation is said to occur in a given stage if during that stage the mutated cell is subject to reproduction, death, further mutation to the next stage, and so on. To study a simple multistage RW model, let $S = \{0, 1, \ldots, N\}$ be the number of stages. Here 0 represents the state of complete recovery. In a multistage model one postulates several successive mutations, each producing a clone of mutant cells. The state N denotes the completion of the mutation process resulting in malignant cells. Let $\{X_n\}$, $n \geqslant 0$, denote the RW corresponding to the mutation process. A step forward implies further mutation to the next stage and the backward step implies a move toward recovery. Let these transitions occur with the following probabilities.

$$P\{x \to x + 1\} = P_x = x/N \qquad \text{for } x = 1, 2, \ldots, N - 1$$
$$P\{x \to x - 1\} = q_x = (N - x)/N \qquad \text{for } x = 1, 2, \ldots, N - 1.$$
$$P\{N \to N\} = 1 = P\{0 \to 0\}; \; P\{x \to x\} = 0 \qquad \text{for } x = 1, 2, \ldots, N - 1$$

The states 0 and N are absorbing states.

Let π_0, π_N, and π_x denote respectively the probabilities of complete recovery, absorption into the cancerous state N, and the absorption into the cancer state N given that the initial mutation state is x, $1 \leqslant x \leqslant N - 1$. Using the first step decomposition [see the proof of Theorem 2.2.5 (i*)], we obtain the difference equation

$$\pi_x = \frac{x}{N} \pi_{x+1} + \left(1 - \frac{x}{N}\right) \pi_{x-1}, \qquad 1 \leqslant x < N - 1, \qquad (2.5.1)$$

which we consider with the boundary conditions

$$\pi_0 = 0 \quad \text{and} \quad \pi_N = 1. \qquad (2.5.2)$$

Let $g(s) = \Sigma_x \pi_x s^x$ be the (probability) generating function of $\{\pi_x\}$. Now rewriting (2.5.1) as

$$\pi_x = \frac{x + 1}{N} \pi_{x+1} - \frac{1}{N} \pi_{x+1} + \left(1 - \frac{x - 1}{N}\right) \pi_{x-1} - \frac{1}{N} \pi_{x-1},$$

we obtain, taking $\pi_{N+k} = 1$, $k \geqslant 0$,

$$g(s) = \sum_{x=0}^{\infty} \left\{ \frac{x}{N} \pi_x s^{x-1} - \frac{1}{N} \pi_x s^{x-1} + \left(1 - \frac{1}{N}\right) \pi_x s^{x+1} - \frac{x}{N} \pi_x s^{x+1} \right\},$$
$$(2.5.3)$$

which yields the differential equation

$$[g(s)]^{-1} dg/ds = s^{-1} + (1 - s)^{-1} + (N - 1)(1 + s)^{-1}. \qquad (2.5.4)$$

Solving this differential equation, we see that

$$g(s) = Cs(1 + s)^{N+1}(1 - s)^{-1} = Cs\left[\sum_{x=0}^{N-1} C\binom{N-1}{x} s^x\right] \sum_{y \geqslant 0} s^y, \qquad (2.5.5)$$

35

where C is the constant of integration. Using the boundary conditions (2.5.2), we get $1 = C\sum_{y=0}^{x-1}\binom{N-1}{y} = C2^{N-1}$, and hence

$$\pi_x = \sum_{y=0}^{x-1}\binom{N-1}{y}2^{1-N}, \qquad 0 \leqslant x \leqslant N. \qquad (2.5.6)$$

After the initial hit by the carcinogen, the state of the mutation process is normally assumed to be 1. Therefore

$$\pi_N = \pi_1\binom{N-1}{0}2^{1-N} = 2^{1-N}. \qquad (2.5.7)$$

Since the walk is simple,

$$\pi_0 = 1 - \pi_N = 1 - 2^{1-N} \qquad (2.5.8)$$

(see Proposition 2.2.2 and Theorem 2.2.3).

2.5.2. Random Flights

By *random flights* we mean the motion of a particle that undergoes a sequence of displacements $x_1, \ldots, x_n, \ldots$, such that the magnitude and direction of each displacement are independent of all the preceding displacements, and the density function $f_k(x_k)$ of each displacement is assigned *a priori*. A natural question here is: What is the probability $p_n(x)dx$ that after n transitions the particle is in the interval $(x, x + dx)$?

Polymer physics provides us with an example of random flights. A *polymer* is a molecular chain of monomers. These repeating monomers may be identical units as in the case of polyethylene or nonidentical (but chemically similar) units as in nucleic acids and proteins. In organic carbon-chain polymers, side groups such as hydrogen atoms are attached to carbon atoms. In a carbon–carbon bond there is a considerable rotational freedom, constrained by side groups and multibonds, and this freedom allows the polymer chain to assume numerous configurations in space. Any sample of a polymer consists of a set of heterogeneous chains, each with a varying number of monomers. This phenomenon, known as *polydispersity*, can be thought of as a random flight. Consider a polymer of N monomers. This can be treated as a random flight of N steps each of length l, say. Let the successive monomers be independent of each other, randomly oriented, and represented by a vector l_i, $1 \leqslant i \leqslant N$. The length of the chain is given by

$$\mathbf{L}_N = \sum_{i=1}^{N} \mathbf{l}_i.$$

One of the most important problems in polydispersity is finding the distribution of $\mathbf{L}_N$. Using the characteristic function method, we can solve this problem.

Let $f_k(\mathbf{x}_k)$ be the density function of $\mathbf{l}_k$, $\mathbf{x}_k = (x_k^1, \ldots, x_k^d) \in R^d$, $1 \leqslant k \leqslant N$. Let $\psi_k(\mathbf{t})$ be the characteristic function of f_k. Then the characteristic function Ψ_N of $\mathbf{L}_N$ is

$$\Psi_N(\mathbf{t}) = \prod_{k=1}^{N} \psi_k(\mathbf{t}) = \prod_{k=1}^{N} \int_{-\infty}^{\infty} \cdots \int_{-\infty}^{\infty} e^{i\langle \mathbf{t}, \mathbf{x} \rangle} f_k(\mathbf{x}) \, dx^1 \cdots dx^d,$$

where $\langle \mathbf{t}, \mathbf{x} \rangle$ is the inner product in R^d. Inverting the Fourier transform given above and denoting the density of $\mathbf{L}_N$ by $p_N(\mathbf{x})$ we get

$$p_N(\mathbf{x}) = (2\pi)^{-d} \int_{-\infty}^{\infty} \cdots \int_{-\infty}^{\infty} e^{-i\langle \mathbf{t}, \mathbf{x} \rangle} \Psi_N(\mathbf{t}) \, dt^1 \cdots dt^d.$$

Exercises

1. Our gambler Tom makes a series of bets of $1. He has decided to quit playing soon after a net winning of $20 or a net loss of $10. He has a probability of $\frac{1}{2}$ for each event of winning or losing any given game. Find (a) the probability that he will have won $20 when he quits, (b) the probability that he will have lost $10 when he quits, (c) the expected duration of the game, (d) his expected winning, and (e) his expected loss.

2. Do Exercise 1 now with probability 17/36 of winning and probability 19/36 of losing a game.

3. A particle starting from the origin performs a symmetric unrestricted RW. Consider only the first 10 steps. Compute (a) $P\{X_4 \leqslant 3\}$, (b) $P\{X_5 = 0 \text{ but } X_n < 0 \text{ for } n = 1, \ldots, 4\}$, (c) $P\{X_5 \leqslant 0 \text{ and } X_{10} \leqslant 0\}$, (d) $P\{X_k > 0 \text{ for some } k | X_{10} = -1\}$, (e) P {first return to the origin occurs at time $n = 6$}, (f) $P\{X_k > 0, 1 \leqslant k \leqslant 8\}$.

4. Consider a general RW $X_n = x_0 + J_1 + \cdots + J_n$ where the jumps J_k are IID RVs following $N(\mu, \sigma^2)$. Find the distribution of $X_n, n \geqslant 1$, and $P\{X_n \leqslant x \text{ and } X_m \leqslant y\}$, where $n < m$.

5. Consider the RW of Exercise 4 with $x_0 = 0$ and $\mu \neq 0$. Let $a > 0$ and $-b < 0$ be two absorbing barriers for this particle. Show that:

(a) the probability p_a of absorption at a is

$$p_a \simeq \frac{1 - \exp(2\mu b/\sigma^2)}{\exp(-2\mu a/\sigma^2) - \exp(2\mu b/\sigma^2)},$$

(b) the probability p_{-b} of absorption at $-b$ is

$$p_{-b} \simeq \frac{\exp(-2\mu a/\sigma^2) - 1}{\exp(-2\mu a/\sigma^2) - \exp(2\mu b/\sigma^2)},$$

(c) the expected duration $E[T]$ for absorption is

$$E[T] \simeq \frac{(a + b) - a \exp(2\mu b/\sigma^2) - b \exp(-2\mu a/\sigma^2)}{\exp(-2\mu a/\sigma^2) - \exp(2\mu b/\sigma^2)}.$$

6. Consider a simple RW X_n, $n \geqslant 0$. Let A be the event that $X_n = 0$ for some $n \geqslant 1$. Show that $P(A) = 1 - |p - q|$.

7. Establish the following assertion. By doubling the stakes while keeping the initial capitals unchanged, the probability of gambler's ruin is decreased for the weak player whose probability of success $p < \frac{1}{2}$ and the same increases for the strong player.

8. Consider a general RW $X_n = (x_0 + J_1 + \cdots + J_n)$, where the jumps J_k are IID RVs following a probability distribution $\{p_k\}$, $k = 0, \pm 1, \ldots$; that is, the particle jumps from x to $x + k$ with probability p_k. Let π_{x_0}, $0 < x_0 < a$, be the probability that the particle assumes some position $\leqslant 0$ before taking any position $\geqslant a$. Set $\pi_x = 1$ if $x \leqslant 0$ and $\pi_x = 0$ if $x \geqslant a$. Then show that $\pi_{x_0} = \Sigma \pi_x p_{x - x_0}$. In particular, take $p_{-2} = p_{-1} = p_1 = p_2 = \frac{1}{4}$. Now show that

$$\pi_{x_0} = 1 - \frac{x_0}{a} + \frac{(2x_0 - a)(s_0^a - s_1^a) - a(s_0^{2x_0 - a} - s_1^{2x_0 - a})}{a\{(a + 2)(s_0^a - s_1^a) - a(s_0^{a+2} - s_1^{a+2})\}},$$

where $s_0 = s_1^{-1} = (-3 + \sqrt{5})/2$ and $s_1 = s_0^{-1} = (-3 - \sqrt{5})/2$.

9. Consider a simple unrestricted RW on the one-dimensional lattice. For any set A of lattice points and, $x \notin A$ and $y \notin A$, let $_A p(x, y)$ denote the probability of a particle starting from x will hit y before hitting any state in A, and let $_A m(x, y)$ be the mean number of such visits to y.

If $p = q = \frac{1}{2}$ and $a < b$, show that

$$_{\{a,b\}} p(x, y) = \begin{cases} [2(y - a)(b - y) - (b - a)]/2(y - a)(b - y) & \text{if } x = y \\ (x - a)/(y - a) & \text{if } x < y. \\ (b - x)/(b - y) & \text{if } x > y \end{cases}$$

$$_{\{a,b\}} m(x, y) = \begin{cases} [2(y - a)(b - y) - (b - a)]/(b - a) & \text{if } x = y \\ 2(x - a)(b - y)/(b - a) & \text{if } x < y. \\ 2(y - a)(b - x)/(b - a) & \text{if } x > y \end{cases}$$

10. In how many different ways can a particle perform a symmetric RW such that $X(0) = 0 = X(10)$ and $X(k) \neq 0$ for all $k = 1, 2, \ldots, 9$?

11. Consider a simple RW $X_n = J_1 + \cdots + J_n$ with $P\{J = 1\} = p = 1 - P\{J = -1\} = 1 - q$. Let $p_{2n+1} = P\{X_{2n+1} = 1 \text{ and } X_k < 1 \text{ for } 0 < k < 2n + 1\}$. Show that

$$p_{2n+1} = [4(n + 1)!]^{-1} \frac{1}{2} \cdot \frac{3}{2} \cdots (\tfrac{1}{2} + n - 1) 4^{n+1} p^{n+1} q^n.$$

3

Markov Chains

3.1. Definitions, Simple Consequences, and Examples

Several physical, biological, and social systems behave as follows. The system evolves with respect to time. If the present state of the system is given, then the past and future are (conditionally) independent. Such a behavior is called the *Markov property* of the system. In this chapter we treat such systems evolving in a discrete (atmost countable) state space with respect to discrete time. Let $T = \{0, 1, 2, \ldots\}$ be the set of time points indexing a sequence of random variables and S be the state space, where the elements of S are denoted by $s_0, s_1, \ldots$, or $x, y, \ldots$.

Definition 3.1.1. A discrete-time stochastic process $X = \{X_n, n \geq 0\}$ is called a *Markov chain* (MC) if, for any sequence $\{x_0, x_1, \ldots, x_{n+1}\}$ of states,

$$P\{X_{n+1} = x_{n+1} | X_k = x_k, 0 \leq k \leq n\} = P\{X_{n+1} = X_{n+1} | X_n = x_n\}.$$
(3.1.1)

We denote the probability on the right-hand-side (RHS) of the defining relation (3.1.1) by $p(n, x_n, x_{n+1})$. The probabilities $p(n, x, y)$, $(x, y \in S)$, are called *one-step transition probabilities* at time n. Fix an $n \in T$. Then the transition probabilities $p(n, x, y)$, $(x, y \in S)$, can be treated as the elements of a matrix

$$M(n) = [p(n, x, y)]_{x, y \in S}.$$

$M(n)$ is called the *one-step transition matrix* at time n.

Because of the dependence of $p(n, \cdot, \cdot)$ on n, the MC is temporally inhomogeneous. Many of the isolated dynamical systems are inhomogeneous MCs. An important subclass of dynamical systems is the class of conservative systems. Here, if the particle is in state x at time s and moves to state y at time $s + t$, then the motion of the particle is the same as its motion from its initial ($s = 0$) position x to state y at time t. If the MC $\{X_n\}$ has this stationarity then $p(n, x, y)$ is independent of n.

Definition 3.1.2. A Markov chain $\{X_n, n \geqslant 0\}$ is called a *(temporally) homogeneous Markov chain* or a *Markov chain with stationary transition probabilities* if the one-step transition probabilities $p(n, x, y)$, ($n \in T; x, y \in S$), are independent of n; that is, $M(n) = M = [p(x, y)]$ for all $n \in T$.

For the most part we study only the homogeneous case. Whenever we treat the inhomogeneous case it is explicitly stated or is clear from the notations. So, by a Markov chain we essentially mean a temporally homogeneous MC. The matrix $M \equiv M(n)$ is called the *transition matrix* of $X = \{X_n\}$. Let $\{p_0(x), x \in S\}$ denote the *initial distribution* of X; that is,

$$p_0(x) = P\{X_0 = x\}, \qquad x \in S. \tag{3.1.2}$$

Next, we state some easy consequences of these definitions. For the sake of quick reference, we formulate them in the form of a lemma.

Lemma 3.1.3. *Let $X = \{X_n, n \geqslant 0\}$ be a Markov chain. Then*

(i) $p(x, y) \geqslant 0$, $(x, y \in S)$; $\sum_y p(x, y) = 1, x \in S$.

(ii) $p_0(x) \geqslant 0$, $x \in S$; $\sum_x p_0(x) = 1$.

(iii) $P\{X_0 = x_0, X_1 = x_1, \ldots, X_n = x_n\} = P_0(x_0)p(x_0, x_1) \cdots p(x_{n-1}, x_n)$.

(iv) *Relation (3.1.1) is equivalent to* $P\{X_{\nu_{n+1}} = x_{\nu_{n+1}} | X_{\nu_k} = x_{\nu_k}, 0 \leqslant k \leqslant n\} = P\{X_{\nu_{n+1}} = x_{\nu_{n+1}} | X_{\nu_n} = x_{\nu_n}\}$ *for* $n \geqslant 0$, $\nu_k \in T$, $\nu_k < \nu_{k+1}$, $x_{\nu_k} \in S, 0 \leqslant k \leqslant n$.

(v) $P\{X_{\nu_\lambda} = x_{\nu_\lambda}, \ n + 1 \leqslant \lambda \leqslant n + m | X_{\nu_\lambda} = x_{\nu_\lambda}, \ 0 \leqslant \lambda \leqslant n\}$
$= P\{X_{\nu_\lambda} = x_{\nu_\lambda}, \ n + 1 \leqslant \lambda \leqslant n + m | X_{\nu_n} = x_{\nu_n}\}$ *for all* $n \geqslant 0$, $m \geqslant 1$, $\nu_0 < \nu_1 < \cdots < \nu_{n+m}$, $x_{\nu_\lambda} \in S, 0 \leqslant \lambda \leqslant n + m$.

(vi) $P\{X_k = x_k, X_i = x_i, 0 \leqslant k \leqslant n, n + 2 \leqslant i \leqslant n + m | X_{n+1} = x_{n+1}\} = P\{X_k = x_k, 0 \leqslant k \leqslant n | X_{n+1} = x_{n+1}\} P\{X_i = x_i, n + 2 \leqslant i \leqslant n + m | X_{n+1} = x_{n+1}\}$.

(vii) *Time reversibility. For all* $n \geqslant 0$, $m \geqslant 1$, $x_k \in S$, $n + 1 \leqslant k \leqslant n + m$, $P\{X_n = x_n | X_k = x_k, n + 1 \leqslant k \leqslant n + m\} = P\{X_n = x_n | X_{n+1} = x_{n+1}\}$.

The proof is a simple and useful exercise.

Lemma 3.1.3 (iii) shows that the joint distribution of $X_0, \ldots, X_n$ is determined by the initial distribution $p_0(x)$ and the transition probabilities $p(x, y)$, $x, y \in S$. The equation in Lemma 3.1.3 (iv) seems to be more general than (3.1.1) while it is actually equivalent to (3.1.1). Lemma 3.1.3 (vi) is a consequence of (v) and is equivalent to the Markov property. [Lemma 3.1.3 (v) follows, by induction, from Lemma 3.1.3 (iv).] A matrix $M = [p(x, y)]$, $(x, y \in S)$, satisfying the relations in Lemma 3.1.3. (i) is called a *stochastic matrix*.

An MC defines an initial distribution $p_0(x)$ and the transition matrix $M = [p(x, y)]$. Conversely, given a stochastic matrix $[p(x, y)]$ and an initial distribution $p_0(x)$, one can construct a corresponding Markov chain. This will establish the existence of Markov chains. More precisely, we have the following theorem.

Theorem 3.1.4. *Let* $\{M(n), n \geqslant 0\}$ *be a sequence of stochastic matrices, where* $M(n) = [p(n, x, y)]$, $(x, y \in S)$. *Then there exists a probability space* $(\Omega, \mathcal{C}, P)$ *and a (inhomogeneous) Markov chain* $\{X_n, n \geqslant 0\}$ *defined on* Ω *with state space* S *and transition matrices* $\{M(n), n \geqslant 0\}$.

Any reader who knows Kolmogorov's consistency theorem or the corresponding result of Ionescu Tulcea can easily prove this theorem (Chung 1967).

Examples 3.1.5

EXAMPLE 1. *Sequence of Independent Random Variables.* If $X = \{X_n, n \geqslant 0\}$ is a sequence of independent random variables, trivially X is an MC. Actually the notion of an MC is a generalization of a sequence of independent random variables.

EXAMPLE 2. *Random Walk.* Let $I = \{\cdots, -1, 0, 1, \ldots\}$ and I^d, for a positive integer d, be the d-dimensional integer lattice. An MC $X = \{X_n, n \geqslant 0\}$ with state space $S = I^d$ is called a *homogeneous d-dimensional lattice random walk* if the transition probabilities $p(\mathbf{x}, \mathbf{y})$ depend only on $\mathbf{y} - \mathbf{x} = (y_1 - x_1, \ldots, y_d - x_d)$, that is, $p(\mathbf{x}, \mathbf{y}) = p(\mathbf{y} - \mathbf{x})$. This definition looks different from our earlier Definition 2.1.1 $(d = 1)$. But actually the following theorem holds.

Theorem 3.1.6. *An MC* $X = \{X_n, n \geqslant 0\}$ *is a d-dimensional lattice random walk if and only if it is an* I^d-*valued MC with independent and identically distributed increments* $\{X_n - X_{n-1}\}$, $n \geqslant 1$.

41

PROOF. Define $J_n = X_n - X_{n-1}$, $n \geq 1$, $J_0 = X_0$. If $\mathbf{x}$, $\mathbf{x}_k \in I^d$, $0 \leq k \leq n$, and X constitutes a d-dimensional lattice walk, then

$$P\{J_{n+1} = \mathbf{x} \mid J_k = \mathbf{x}_k, 0 \leq k \leq n\}$$

$$= P\left\{X_{n+1} = \mathbf{x} + \sum_{k=0}^{n} \mathbf{x}_k \mid X_k = \sum_{i=0}^{k} \mathbf{x}_i, 0 \leq k \leq n\right\}$$

$$= P\left\{X_{n+1} = \mathbf{x} + \sum_{k=0}^{n} \mathbf{x}_k \mid X_n = \sum_{k=0}^{n} \mathbf{x}_k\right\}$$

$$= p(\mathbf{x}).$$

Therefore, the increments $\{J_n\}$ form a sequence of independent RVs identically distributed according to $P\{J_n = \mathbf{x}\} = p(\mathbf{x})$, $\mathbf{x} \in I^d$. This proves the necessity.

To prove the sufficiency, let

$$X_n = X_0 + \sum_{k=1}^{n} J_k, \qquad n \geq 0,$$

where X_0 is independent of the increments J_k, which are independent and identically distributed according to $p(\mathbf{x})$, $\mathbf{x} \in I^d$, ($p(\mathbf{x}) \geq 0$, $\Sigma_{\mathbf{x} \in I^d} p(\mathbf{x}) = 1$).
Then:

$$P\{X_{n+1} = \mathbf{y} \mid X_n = \mathbf{x}\} = P\{J_{n+1} = \mathbf{y} - \mathbf{x} \mid X_n = \mathbf{x}\}$$

$$= P\{J_{n+1} = \mathbf{y} - \mathbf{x}\} = p(\mathbf{y} - \mathbf{x}),$$

so that $\{X_n\}$ is an RW. This proves the sufficiency and hence the theorem. □

Examples 3.1.7

EXAMPLE 1. *Simple Random Walk.* Let $\{X_n, n \geq 0\}$ be a simple one-dimensional lattice walk (see Section 2.2). It is clear that $\{X_n\}$ is an MC. The transition probabilities $p(x, y)$ are given by

$$p(x, x + 1) = p, \qquad p(x, x - 1) = q, \qquad p(x, x) = r,$$

where $0 \leq p, q, r \leq 1$, $p + q + r = 1$.

EXAMPLE 2. *Random Walks with Barriers.* We now consider the lattice walk of a particle on I under certain constraints. Let B be a subset of I. The elements of B are called *barriers*. If Tom is playing against an infinitely wealthy opponent, $B = \{0\}$. If the initial fortunes of Tom and Dick are a

and b, respectively, then $B = \{0, a + b\}$, and so on. Let $\{X_n, n \geqslant 0\}$ be an MC with state space I, $B \subset I$, and transition probabilities defined as follows: for $x \notin B$, $p(x, y)$ is the same as in Example 1, and for $x \in B$

$$p(x, x + 1) = p_x, \qquad p(x, x - 1) = q_x, \qquad p(x, x) = r_x,$$

where $0 \leqslant p_x, q_x, r_x \leqslant 1$, $p_x + q_x + r_x = 1$. Now we define the random walk with barriers as follows.

Definition 3.1.8. The MC $\{X_n, n \geqslant 0\}$ with transition probabilities defined as above is called a *random walk with*: (1) *absorbing barrier* at $b \in B$ if, $r_b = 1$, (2) *right-reflecting barrier at* $b \in B$ if $p_b = 1$, (3) *left-reflecting barrier at* $b \in B$ if $q_b = 1$, (4) *right-elastic barrier at* $b \in B$ if $q_b = 0$ and $p_b r_b > 0$, (5) *left-elastic barrier at* $b \in B$ if $p_b = 0$ and $q_b r_b > 0$.

Note that an RW with barriers no longer has independent increments. It is easy to write down the transition matrices in the preceding barrier cases.

Examples **3.1.9**

EXAMPLE 1. *Absorbing Barriers at 0 and b.*

	0	1	2	3	·	·	·	$b-2$	$b-1$	b
0	1	0	0	0	·	·	·	0	0	0
1	q	r	p	0	·	·	·	0	0	0
2	0	q	r	p	·	·	·	0	0	0
·	·	·	·	·	·	·	·	·	·	·
·	·	·	·	·	·	·	·	·	·	·
$b-1$	0	0	0	0	·	·	·	q	r	p
b	0	0	0	0	·	·	·	0	0	1

$M =$, where $S = \{0, 1, \ldots, b\}$.

EXAMPLE 2. *Right-Reflecting Barrier at 0.*

	0	1	2	3	·	·
0	0	1	0	0	·	·
1	q	r	p	0	·	·
2	0	q	r	p	·	·
·	·	·	·	·	·	·
·	·	·	·	·	·	·

$M =$, $S = \{0, 1, 2, \ldots\}$.

EXAMPLE 3. *Right-Elastic Barrier at 0.*

$$M = \begin{array}{c} \\ 0 \\ 1 \\ 2 \\ \\ \\ \\ \end{array} \begin{array}{cccccccc} 0 & 1 & 2 & 3 & 4 & \cdot & \cdot & \cdot \\ \left[\begin{array}{cccccccc} r_0 & p_0 & 0 & 0 & 0 & \cdot & \cdot & \cdot \\ q & r & p & 0 & 0 & \cdot & \cdot & \cdot \\ 0 & q & r & p & 0 & \cdot & \cdot & \cdot \\ \cdot & \cdot & \cdot & \cdot & \cdot & \cdot & \cdot & \cdot \\ \cdot & & \cdot & \cdot & \cdot & \cdot & \cdot & \cdot \\ \cdot & & \cdot & \cdot & \cdot & \cdot & \cdot & \cdot \end{array}\right] \end{array}, \qquad S = \{0, 1, 2, \ldots\}.$$

EXAMPLE 4. *Birth–Death Chain.* This is a discrete time version of birth and death processes. (The continuous time version is more realistic; see Chapter 5.) An MC with state space $S = \{0, 1, 2, \ldots\}$ is called a *birth–death chain* if its transition probabilities are given by

$$p(x, x + 1) = p_x, \qquad p(x, x - 1) = q_x, \qquad p(x, x) = r_x,$$

where $0 \leqslant p_x, q_x, r_x \leqslant 1$, $p_x + q_x + r_x = 1$, $x \in S$. If at any time n the population size is x, then at time $(n + 1)$ there will be birth with probability p_x, a death with probability q_x, or no change in the population size with probability r_x, and these are the only possibilities. These probabilities depend on the state of the system.

EXAMPLE 5. *Branching Chain.* In 1873 Galton posed the following demography problem. If a man of a given family has the probability $p(k)$ of producing k male offsprings, $k \geqslant 0$, what is the probability that the family name will eventually die out? Let $X_0 = 1$; that is, the initial generation consists of the given man. Let X_n, $n \geqslant 1$ denote the size of the nth generation of descendants. The family line dies out at the nth generation if $X_k > 0, 0 \leqslant k \leqslant n - 1$, and $X_n = 0$. We formulate an MC $\{X_n\}$ as follows. If at a certain time n the number of males is i, that is, if $X_n = i$, then each one of them independently produce offspring with the same probabilities $p(k)$, $k \geqslant 0$, so that there will be $J_1 + J_2 + \cdots + J_i$ males in the $(n + 1)$th generation, where $J_1, \ldots, J_i$ are independent RVs with the same distribution $P\{J = k\} = p(k)$. Hence the transition probabilities are given by

$$p(x, y) = P\{J_1 + \cdots + J_x = y\}, \qquad x \geqslant 1, y \geqslant 0, p(0, 0) = 1.$$

An expression for $p(x, y)$ is given by Selivanov (1969). This expression is not easy to work with, and generating function is the tool used in the investigation of branching chains.

EXAMPLE 6. *Ehrenfests Diffusion Chain.* A quandary arose, at the turn of this century, when Boltzmann attempted to explain thermodynamics on the basis of kinetic theory. On the one hand, a conservative system is time reversible and, as the system evolves, it should return infinitely often to the neighborhood of any initial state, irrespective of the distance between the initial and equilibrium states. On the other hand, experimental observations and thermodynamical considerations say that, starting with any initial state, the system will move irreversibly toward equilibrium. So it appears, as remarked by Zermelo, that the "irreversibility" from the thermodynamical considerations and "recurrence" property of conservative dynamical systems are irreconcilable. Ehrenfests (and Smoluchowski) clarified this situation. Consider the following Ehrenfests urn model, where $2N$ balls, numbered $1, \ldots, 2N$, are distributed among urns I and II, say, x in I and $(2N - x)$ in II. At time n randomly choose an integer among $1, \ldots, 2N$. Find the ball with that number on it and transfer it to the other container. Continue this process indefinitely. Let X_n denote the number of balls in urn I after the nth transfer. Clearly, $\{X_n, n \geq 0\}$ is an MC with state space $S = \{0, 1, \ldots, 2N\}$. Let us compute the transition probability $p(x, y)$. Suppose that there are x balls in urn I at an arbitrary time n. The probability of any one of these balls being selected for transfer to urn II is $x/2N$. There are $(2N - x)$ balls in urn II. The probability of one of them being transferred to urn I is $(2N - x)/2N$. Therefore,

$$p(x, x - 1) = \frac{x}{2N}, \qquad p(x, x + 1) = 1 - \frac{x}{2N}.$$

(The states 0 and $2N$ are reflecting barriers.) Let x be the initial number of balls in urn I. If the process of transferring balls as described above is continued indefinitely, it is at least intuitively clear that the associated MC $\{X_n\}$ will visit x, and for that matter every possible state, infinitely often. Note also that N is the equilibrium level. Assume that the initial state x is far removed from N. As noted above, no matter how far removed x is from N the chain will visit x infinitely often. But if one computes [as we do later: see Example 3.6.5 (2)] the expected return time between any two visits to x, one can see that this expected time is too large to obviously imply the recurrence. How about the experimental verifications? With large number of balls and with one transition per second, one can show that the expected return time can run into billions of years, and consequently the process appears irreversible. This explains the seeming contradiction. A continuous time version of this model can be found in Bellman and Harris (1951).

EXAMPLE 7. *Pólya Urn Model.* Pólya proposed his urn model as a discrete scheme for the spread of an epidemic. Consider a population initially with i infected and s susceptibles. An infected person adds, by contamination, a

45

more individuals to the list of infected. As a mathematical expediency we induce a symmetry by assuming that each susceptible increases the chance of a susceptibles. Then, $\{X_n = $ number of infected at time n, $n \geq 0\}$ becomes an MC. This scheme is described by an urn model as follows. Consider an urn containing r red and g green balls. Draw a ball at random from the urn and note its color. In addition to returning this ball to the urn add a balls of the same color. Continue this process indefinitely. Let X_n be the number of red balls at the end of the nth trial. Let $X_n = x$, $x \in S = \{0, 1, 2, \ldots\}$, $x \geq r$. Irrespective of how we reached state x, the system will be in state $x + a$ or x at time $(n + 1)$. Clearly, $\{X_n\}$ is a Markov chain. Actually, it is a temporally inhomogeneous MC, as we can see by computing its transition probabilities:

$$P\{X_{n+1} = y \mid X_0 = x_0, \ldots, X_{n-1} = x_{n-1}, X_n = x\}$$
$$= \begin{cases} x/(r + g + na) & \text{if } y = x + a \\ 1 - \{x/(r + g + na)\} & \text{if } y = x \\ 0 & \text{otherwise} \end{cases}$$

Note that this transition probability depends on the (present) time n and, of course, on the present state x.

EXAMPLE 8. *A Queuing Chain.* Queuing theory is a well-developed branch of stochastic processes. We give here a very simple discrete queuing system. Consider a port where cargo ships arrive for service. Assume that it takes one unit of time (a day or a week) to service each ship. During the service time of a ship more ships may arrive and join the waiting line for their turns. Let J_n be the number of cargo ships arriving during the nth unit of time, where we assume that the J_n, $n \geq 1$, are IID RVs with values in $\{0, 1, 2, \ldots\}$. Define a chain $\{X_n, n \geq 0\}$ as follows: $X_0 = $ the initial number of ships waiting to be served and, for $n \geq 1$, $X_n = $ number of ships waiting in the line at the end of servicing n ships. Then $X_{n+1} = J_{n+1}$ if $X_n = 0$, and $X_{n+1} = X_n + J_{n+1} - 1$ if $X_n \geq 1$. Clearly, $\{X_n, n \geq 0\}$ is an MC with $S = \{0, 1, 2, \ldots\}$. If $p(x)$, $x \in \{0, 1, \ldots\}$, denotes the common distribution of the RVs J_n, then the transition probabilities $p(x, y)$ are given by

$$p(x, y) = \begin{cases} p(y - x + 1) & \text{if } x \geq 1 \\ p(y) & \text{if } x = 0 \end{cases}.$$

EXAMPLE 9. *A Two-State* MC. This is one of the simplest kind of Markov chains. Nevertheless, such chains occur in practical situations. Consider, for example, two competing grocery stores located in a shopping mall. Due to competition and special sales, the customers are influenced and switch between stores (randomly). If a customer shops in store I (resp. II) any given week, he or she will shop in store II (resp. I) the next week with probability p (resp. q). Then the one-step transition matrix is given by

$$M = \begin{array}{c} \\ \text{I} \\ \text{II} \end{array} \begin{bmatrix} \overset{\text{I}}{1-p} & \overset{\text{II}}{p} \\ q & 1-q \end{bmatrix}.$$

A similar situation arises in the study of the learning strategy for a predator preying on a model–mimic system. Huheey, in his study of warning coloration and mimicry, proposed a two-state MC model which corresponds to the following encounters of predators with models (unpalatable prey) and mimics (palatable prey that mimics the models to derive protection). Here, taking $S = \{$I $=$ encountering a mimic, II $=$ encountering a model$\}$, define X_n as the nth encounter, and

$$p = P\{X_{n+1} = \text{II}|X_n = \text{I}\}, \qquad q = P\{X_{n+1} = \text{I}|X_n = \text{II}\}.$$

Then, the one-step transition matrix of this MC is given as above. We call this two-state MC the *predator-encounter chain*.

3.2. *n*-Step Transition Matrices

Let $X = \{X_n, n \geqslant 0\}$ be a (temporally homogeneous) MC with state space S. We have seen in Section 3.1 that this chain is completely determined by its one-step transition matrix $M = [p(x,y)]$, $(x,y \in S)$, and an initial distribution $\{p_0(x)\}$, $x \in S$. The so-called n-step transition matrices, $n \geqslant 0$, play a central role in further analysis of an MC; for example, the probability distribution $p_m(x) = P\{X_m = x\}$ of X_m, $m \geqslant 1$, can be computed from $\{p_0(x)\}$ and the m-step transition matrix $M^{(m)} = [p^m(x,y), x,y \in S]$. Define

$$M^{(0)} = [p^0(x,y)] = \begin{cases} 1 & \text{if } x = y \\ 0 & \text{if } x \neq y \end{cases}$$

$$M^{(1)} = [p^1(x,y)] = [p(x,y)] = M \tag{3.2.1}$$

$$M^{(n)} = [p^n(x,y)] = [P\{X_{m+n} = y|X_m = x\}], \qquad n \geqslant 2.$$

Thus $p^n(x,y)$ is the probability of transition from state x to state y in n-steps, and is called the *n-step transition probability*.

Theorem 3.2.1. *Let $\{X_n, n \geqslant 0\}$ be an MC with an initial distribution $\{p_0(x)\}$, $x \in S$, and n-step transition probabilities $p^n(x,y)$, $(x,y \in S, n \geqslant 0)$. Then*

(i)
$$P\{X_n = y\} = \sum_{x \in S} p_0(x)p^n(x,y). \tag{3.2.2}$$

(ii) *Chapman–Kolmogorov relation:*
$$p^{n+m}(x,y) = \sum_{z \in S} p^m(x,z)p^n(z,y). \tag{3.2.3}$$

PROOF.

$$P\{X_n = y\} = \sum_{x \in S} P\{X_n = y, X_0 = x\}$$
$$= \sum_{x \in S} P\{X_0 = x\}\, P\{X_n = y \mid X_0 = x\}$$
$$= \sum_{x \in S} p_0(x)p^n(x,y).$$

$$p^{n+m}(x,y) = P\{X_{n+m} = y \mid X_0 = x\}$$
$$= \sum_{z \in S} P\{X_m = z, X_{n+m} = y \mid X_0 = x\}$$
$$= \sum_{z \in S} P\{X_m = z \mid X_0 = x\}\, P\{X_{n+m} = y \mid X_0 = x, X_m = z\}$$
$$= \sum_{z \in S} p^m(x,z)\, P\{X_{n+m} = y \mid X_m = z\}$$
$$= \sum_{z \in S} p^m(x,z)p^n(z,y). \qquad \square$$

In terms of matrix multiplication, the Chapman–Kolmogorov relation becomes $M^{(n+m)} = M^{(n)} M^{(m)}$. Taking $n = 1 = m$, we have $M^{(2)} = MM = M^2$, so that, in general, $M^{(n)} = M^n$ and $M^{n+m} = M^n M^m$.

Example **3.2.2** The predator-encounter chain (two-state MC) is one of the very few examples for which one can compute M^n easily and explicitly. To do this, let $p + q > 0$ in Example 3.1.9 (9). Before computing M^n, we compute $p_n(x)$, $x = $ I, II.

$$P\{X_{n+1} = \mathrm{I}\} = P\{X_{n+1} = \mathrm{I}, X_n = \mathrm{I}\} + P\{X_{n+1} = \mathrm{I}, X_n = \mathrm{II}\}$$
$$= P\{X_n = \mathrm{I}\}\, P\{X_{n+1} = \mathrm{I} \mid X_n = \mathrm{I}\} + P\{X_n = \mathrm{II}\}\, P\{X_{n+1} = \mathrm{I} \mid X_n = \mathrm{II}\}$$
$$= P\{X_n = \mathrm{I}\}p(\mathrm{I}, \mathrm{I}) + (1 - P\{X_n = \mathrm{I}\})p(\mathrm{II}, \mathrm{I})$$
$$= (1 - p - q)P\{X_n = \mathrm{I}\} + q.$$

Repeatedly using this recursive relation, we have

$$
\begin{aligned}
P\{X_{n+1} = \mathrm{I}\} &= q + (1 - p - q)P\{X_n = \mathrm{I}\} \\
&= q[1 + (1 - p - q)] + (1 - p - q)^2 P\{X_{n-1} = \mathrm{I}\} \\
&= q \sum_{k=0}^{2} (1 - p - q)^k + (1 - p - q)^3 P\{X_{n-2} = \mathrm{I}\} \\
&= \quad \vdots \\
&= q \sum_{k=0}^{n} (1 - p - q)^k + (1 - p - q)^{n+1} P\{X_0 = \mathrm{I}\}.
\end{aligned}
$$

Hence

$$
P\{X_n = \mathrm{I}\} = \frac{q}{p + q} + (1 - p - q)^n \left[p_0(\mathrm{I}) - \frac{q}{p + q} \right], \qquad (3.2.4)
$$

and similarly

$$
P\{X_n = \mathrm{II}\} = \frac{p}{p + q} + (1 - p - q)^n \left[p_0(\mathrm{II}) - \frac{p}{p + q} \right]. \qquad (3.2.5)
$$

Now, to compute $p^n(\mathrm{I}, \mathrm{I})$ and $p^n(\mathrm{I}, \mathrm{II})$, let $p_0(\mathrm{I}) = 1$. From (3.2.4), we obtain

$$
p^n(\mathrm{I}, \mathrm{I}) = P\{X_n = \mathrm{I} \mid X_0 = \mathrm{I}\} = \frac{q}{p + q} + (1 - p - q)^n \frac{p}{p + q}
$$

and from (3.2.5)

$$
p^n(\mathrm{I}, \mathrm{II}) = \frac{p}{p + q} - (1 - p - q)^n \frac{p}{p + q}.
$$

Similarly,

$$
p^n(\mathrm{II}, \mathrm{I}) = \frac{q}{p + q} - (1 - p - q)^n \frac{q}{p + q},
$$

$$
p^n(\mathrm{II}, \mathrm{II}) = \frac{p}{p + q} + (1 - p - q)^n \frac{q}{p + q},
$$

and hence

$$
M^n = \frac{1}{p + q} \begin{bmatrix} q & p \\ q & p \end{bmatrix} + \frac{(1 - p - q)^n}{p + q} \begin{bmatrix} p & -p \\ -q & q \end{bmatrix}.
$$

3.3. Strong Markov Property

In all our expressions thus far we have suppressed the sample variable $\omega \in \Omega$. We continue to do so except in places where the sample paths have to be emphasized or where it will help intuition. Let $(\Omega, \mathcal{C}, P)$ be (the basic) complete probability space and $\{X_n, n \geq 0\}$ an MC with state space S. Let $\mathcal{C}_n^m = \sigma\{X_k, m \leq k \leq n\}$ be the σ-algebra generated by the events $\{X_k \in B\}$, $B \in \mathcal{S} = 2^S$, the power set of S, and $m \leq k \leq n$. Then $\mathcal{C}_n^m$ contains all the information about $\{X_m, \ldots, X_n\}$ part of the MC. Let $\mathcal{C}^m = \mathcal{C}_\infty^m$, $\mathcal{C}_n = \mathcal{C}_n^0$, and, $\mathcal{C} \supseteq \mathcal{C}^0$.

Definition 3.3.1. A random variable $\mathbf{t} \colon \Omega \to \bar{T}$, where $\bar{T} = \{0, 1, 2, \ldots\}$ $\cup \{\infty\}$, is called a *stopping time* (*random time or Markov time*) relative to $\{X_n\}$ if, for each integer $n \geq 0$,

$$\{\omega \colon \mathbf{t}(\omega) = n\} \in \mathcal{C}_n,$$

that is, the event $\{\mathbf{t} = n\}$ is determined by (as a function of) $X_0, \ldots, X_n$.

Examples 3.3.2

EXAMPLE 1. The deterministic time $\mathbf{t} \equiv k$ is a stopping time, for

$$I_{\{\mathbf{t}=n\}}(X_0, \ldots, X_n) = \begin{cases} 1 & \text{if } n = k \\ 0 & \text{if } n \neq k \end{cases}.$$

EXAMPLE 2. Let $A \in \mathcal{S}$. The *hitting time* $\mathbf{t}_A$ of the set A of states is defined by $\mathbf{t}_A(\omega) = \min\{n > 0 \colon X_n(\omega) \in A\}$, and $\mathbf{t}_A$ is a stopping time because

$$I_{\{\mathbf{t}_A=n\}}(X_0, \ldots, X_n)$$
$$= \begin{cases} 1 & \text{if } X_k \notin A \text{ for } k = 0, \ldots, n-1, \text{ and } X_n \in A \\ 0 & \text{otherwise} \end{cases}.$$

EXAMPLE 3. Let τ_x be the time of last visit to state x. Then τ_x is *not* a stopping time because, to determine whether τ_x is the last visiting time, one has to know the entire future.

Let us arbitrarily fix a $K \geq 0$ and define $Y_n = X_{K+n}$, $n \geq 0$. Then $\{Y_n, n \geq 0\}$ is an MC. For

$$P\{Y_{n+1} = y \mid Y_0 = y_0, \ldots, Y_{n-1} = y_{n-1}, Y_n = x\}$$
$$= P\{X_{K+n+1} = y \mid X_K = y_0, \ldots, X_{K+n-1} = y_{K+n-1}, X_{K+n} = x\}$$
$$= P\{X_{K+n+1} = y \mid X_{K+n} = x\}, \quad \text{by Lemma 3.1.3,}$$
$$= P\{Y_{n+1} = y \mid Y_n = x\}.$$

The Markovian motion $\{Y_n\}$ is the same as observing the MC $\{X_n\}$, from scratch, beginning a deterministic time K. It is natural to ask whether the Markov property is preserved if the fixed time K is replaced by a stopping time $\mathbf{t}$. The answer to such a question is in the affirmative. This is known as the *strong Markov property*.

Theorem 3.3.3. *Let* $\{X_n, n \geqslant 0\}$ *be an* MC *with state space S and m-step transition probabilities* $p^m(\cdot, \cdot)$, *and* $\mathbf{t}$ *be a stopping time relative to* $\{X_n\}$ *such that* $\mathbf{t}(\omega) < \infty$ *for all* $\omega \in \Omega$. *Define* $Y_n(\omega) = X_{\mathbf{t}(\omega)+n}(\omega)$, $n \geqslant 0$. *Then* $Y = \{Y_n, n \geqslant 0\}$ *is an* MC *and, for* $0 = n_0 < n_1 < \cdots < n_m$, $x_0, x_1, \ldots, x_m \in S$, *we have*

$$P\{Y(n_k) = x_k, 0 \leqslant k \leqslant m\} = q_0(x_0) \prod_{k=0}^{m-1} p^{n_{k+1}-n_k}(x_k, x_{k+1}), \quad (3.3.1)$$

where q_0 is the initial distribution of Y_0.

PROOF. Define $\mathcal{Q}_\mathbf{t} = \{A \in \mathcal{Q}: A \cap \{\omega: \mathbf{t}(\omega) \leqslant n\} \in \mathcal{Q}_n\}$. Then $\mathcal{Q}_\mathbf{t}$ is a σ-algebra. Let $\mathcal{B} = \sigma\{Y_n, n \geqslant 0\}$. For $A \in \mathcal{Q}_\mathbf{t}$ there is an $A_n \in \mathcal{Q}_n$ such that $A \cap \{\mathbf{t} = n\} = A_n \cap \{\mathbf{t} = n\}$. If $B = \cap_{k=1}^m \{Y(n_k) = x_k\}$ and $B_n = \cap_{k=1}^m \{X(n + n_k) = x_k\}$, then $B \in \mathcal{B}$, $B_n \in \mathcal{Q}^n$, and $B \cap \{\mathbf{t} = n\} = B_n \cap \{\mathbf{t} = n\}$. Now

$P\{A \cap B\}$

$$= \sum_{n \geqslant 0} \sum_{x_0 \in S} P\{A_n \cap \{\mathbf{t} = n\} \cap \{X_n = x_0\} \cap B_n\}$$

$$= \sum_{n \geqslant 0} \sum_{x_0 \in S} P\{A_n \cap \{\mathbf{t} = n\} \cap \{X_n = x_0\}\} P\{B_n | X_n = x_0\},$$

by the definition of $\mathbf{t}$ and Markov property,

$$= \sum_{n \geqslant 0} \sum_{x_0 \in S} \prod_{k=0}^{m-1} p^{n_{k+1}-n_k}(x_k, x_{k+1}) P\{A \cap \{\mathbf{t} = n\} \cap \{X_n = x_0\}\}$$

$$= \sum_{x_0 \in S} \prod_{k=0}^{m-1} p^{n_{k+1}-n_k}(x_k, x_{k+1}) P\{A \cap \{Y_0 = x_0\}\}.$$

Since the event $\{Y_0 = x_0\} \in \mathcal{Q}_\mathbf{t}$, now choose $A = \{Y_0 = x_0\}$. This yields (3.3.1) and proves the Markov property of $\{Y_n, n \geqslant 0\}$. $\square$

We remark here that this theorem does not extend to, say, a continuous time, continuous-state space Markov process. It is very useful for the students to learn to work with the stopping time as early as possible. So we consider few examples here. To simplify writing, we use $P_x(\cdot)$ to denote the probability of various events $(\cdot)$ determined by an MC starting at x, and we use E_x to denote the expectation (integration) with respect to P_x.

Examples 3.3.4

EXAMPLE 1. Let $\{X_n\}$ be a predator-encounter chain. Compute $P_1\{t_I = n\}$, $n > 0$.

We want to compute the probability of the event that a predator who encountered a mimic initially will encounter again a mimic only at time n. This event is equivalent to the event that the predator encounters a model at time $n = 1$ and after $(n - 1)$ units of time will encounter a mimic for the first time, (this is the method of *first-step decomposition*). Therefore

$$P_1\{t_I = n\} = p(I, II)P_{II}\{t_I = n - 1\}$$
$$= p(I, II)p(II, II)P_{II}\{t_I = n - 2\} = \cdots$$
$$= p(I, II)[p(II, II)]^{n-2}P_i\{t_I = 1\}$$
$$= p(I, II)[p(II, II)]^{n-2}p(II, I) = pq(1 - q)^{n-2}.$$

EXAMPLE 2. Following the first-step-decomposition argument, we see that for a general MC we have

$$P_x\{t_y = n + 1\} = \sum_{\substack{z \neq y \\ z \in S}} p(x, z)P_z\{t_y = n\},$$

where $t_y = t_{\{y\}}$, the hitting time of state y. We leave it as an exercise to show that

$$P_x\{t_y \leqslant n + 1\} = p(x, y) + \sum_{z \neq y} p(x, z)P_z\{t_y \leqslant n\}, \qquad n \geqslant 0.$$

EXAMPLE 3. Noting that the birth–death chain is an extension of simple RW, one can pose the following problem. If $\{X_n\}$ is a birth–death chain on the state space $\{0, 1, \ldots, b\}$, what is the probability of hitting one of the boundaries before hitting the other? In the RW case this corresponds to the gambler's ruin problem. Let t_0 and t_b be the hitting times of the boundary points 0 and b, respectively. Find $P_x\{t_0 < t_b\}, 0 < x < b$.

The computation of this probability is another illustration of the method of difference equations that we used for a RW case in a similar problem. The one-step transition probabilities of the above birth–death chain are given by

$$p(x, x + 1) = p_x, \qquad p(x, x - 1) = q_x, \qquad p(x, x) = r_x,$$
$$q_0 = 0, \qquad p_b = 0, \qquad \text{and} \qquad p_x + q_x + r_x = 1, \qquad x \in S.$$

Define $f(x) = P_x\{t_0 < t_b\}$, $0 < x < b$. Let x be the initial state. From the first-step-decomposition argument,

$$f(x) = p_x f(x + 1) + q_x f(x - 1) + r_x f(x), \qquad 0 < x < b$$

or

$$f(x + 1) - f(x) = \frac{q_x}{p_x}[f(x) - f(x - 1)], \qquad \text{since } r_x = 1 - p_x - q_x$$

$$= \frac{q_x}{p_x} \cdot \frac{q_{x-1}}{p_{x-1}}[f(x - 1) - f(x - 2)]$$

$$= \cdots$$

$$= \frac{q_x q_{x-1} \cdots q_1}{p_x p_{x-1} \cdots p_1}[f(1) - f(0)].$$

Set $a_x = (q_x q_{x-1} \cdots q_1 / p_x p_{x-1} \cdots p_1)$, $a_0 = 1$, $f(0) = 1$, and $f(b) = 0$. Then:

$$f(x) - f(x + 1) = a_x[f(0) - f(1)], \qquad 0 \leqslant x < b. \tag{3.3.2}$$

Now summing over $x = 0, 1, \ldots, b - 1$, we get

$$1 \equiv f(0) - f(b) = (f(0) - f(1)) \sum_{x=0}^{b-1} a_x.$$

Using this in (3.3.2), we get

$$f(x) - f(x + 1) = a_x \bigg/ \sum_{y=0}^{b-1} a_y.$$

$$\therefore f(x) = \sum_{y=x}^{b-1} (f(y) - f(y + 1)) = \sum_{y=x}^{b-1} a_y \bigg/ \sum_{y=0}^{b-1} a_y.$$

Hence

$$P_x\{t_0 < t_b\} = \sum_{y=x}^{b-1} a_y \bigg/ \sum_{y=0}^{b-1} a_y. \tag{3.3.3}$$

EXAMPLE 4. Show that

$$p^n(x,y) = \sum_{m=1}^{n} P_x\{t_y = m\} p^{n-m}(y,y), \qquad n \geqslant 1. \tag{3.3.4}$$

$$p^n(x,y) = P\{X_n = y | X_0 = x\} = P_x\{X_n = y\}$$

$$= P_x\left\{ \bigcup_{m=1}^{n} \{\mathbf{t}_y = m, X_n = y\} \right\} = \sum_{m=1}^{n} P_x\{\mathbf{t}_y = m, X_n = y\},$$

the first-entrance decomposition,

$$= \sum_{m=1}^{n} P\{\mathbf{t}_y = m | X_0 = x\} P\{X_n = y | X_0 = x, \mathbf{t}_y = m\}$$

$$= \sum_{m=1}^{n} P\{\mathbf{t}_y = m | X_0 = x\}$$

$$\times P\{X_n = y | X_0 = x, X_k \neq y, 0 \leqslant k < m - 1, X_m = y\}$$

$$= \sum_{m=1}^{n} P_x\{\mathbf{t}_y = m\} p^{n-m}(y,y).$$

There are several interesting RVs defined in terms of the MC or the associated stopping times.

EXAMPLE 5. Fix an $x \in S$. For each $\omega \in \Omega$, let $\mathbf{t}_x^1(\omega) < \mathbf{t}_x^2(\omega) < \cdots < \mathbf{t}_x^m(\omega) < \cdots$ be an increasing sequence of all values of $n \geqslant 1$ for which $X_n(\omega) = x$. Then each $\mathbf{t}_x^m$ is a stopping time known as the mth *hitting time* of x, $m \geqslant 1$. The RV $\rho_x^m = \mathbf{t}_x^{m+1} - \mathbf{t}_x^m$ is called the mth *return time* to x.

EXAMPLE 6. Define $\nu(x) = \nu(x; \omega) = \sum_{n \geqslant 1} I_x(X_n(\omega))$. Then $\nu(x)$ is a RV and denotes the *number of visits* to the state x. Let $\nu(x,n) = \sum_{k=1}^{n} I_x(X_k)$. Then $\nu(x,n)$ defines the *number of visits* to x in the first n transitions. The term *occupation time* is often used in place of "number of visits."

3.4. Decomposition of State Space

Definition 3.4.1. We say that a state x *leads* to state y and write $x \to y$ if there is a positive probability that in a finite number of transitions the Markov particle moves from x to y, that is, if for some $n \geqslant 0$, $p^n(x,y) > 0$. If $x \to y$ and $y \to x$, we say that x and y *communicate*, and we write $x \leftrightarrow y$.

Proposition 3.4.2. The relation "communication" is an equivalence relation in the state space S and hence decomposes S into equivalence classes of states that communicate with each other.

PROOF. The reflexivity and symmetry of $\leftrightarrow$ is trivial. To see the transitivity, let $x \leftrightarrow y$ and $y \leftrightarrow z$. Then there are integers $m, n \geqslant 0$ such that $p^m(x,y)p^n(y,z) > 0$. From this and Chapman–Kolmogorov relation, we get

$$p^{m+n}(x,z) = \sum_{u \in S} p^m(x,u)p^n(u,z) \geqslant p^m(x,y)p^n(y,z) > 0.$$

Hence $x \to z$ and, similarly, $z \to x$. This completes the proof. $\square$

Definition 3.4.3. A set C of states is said to be (*stochastically*) *closed* if $p(x,y) = 0$, for all $x \in C$ and any $y \notin C$; that is, once the particle enters C, it stays there forever. A closed set C is called *irreducible* if it is an equivalence class. An MC is called *irreducible* if its state space is irreducible.

Proposition 3.4.4. (i) Two states x and y communicate if and only if $P_x\{t_y < \infty\} P_y\{t_x < \infty\} < 0$.

(ii) The states x and y communicate if and only if $P_x\{\nu(y) \geqslant 1\} P_y\{\nu(x) \geqslant 1\} > 0$, where $\nu(x)$ is the number of visits to x.

(iii) A set C is closed if and only if $P_x\{t_y < \infty\} = 0$, for $x \in C$ and $y \notin C$.

(iv) A set C is closed if and only if $p^n(x,y) = 0$, for $x \in C, y \notin C$, and all $n \geqslant 1$.

PROOF. Criteria (i)–(iii) are clear. In (iv), take $n = 1$ to prove the sufficiency. To prove the necessity, note that, from the definition, the condition is true for $n = 1$. To proceed inductively, let $p^m(x,y) = 0$, for $x \in C, y \notin C$. Then, if $x \in C, y \notin C$,

$$p^{m+1}(x,y) = \sum_{u \in S} p^m(x,u)p(u,y) = \sum_{u \in C} p^m(x,u)p(u,y) = 0. \quad \square$$

In the following four examples we decompose the state spaces of the MCs corresponding to the given transition matrices.

Examples 3.4.5

EXAMPLE 1.

	1	2	3	4	5
1	0.2	0.3	0.5	0.0	0.0
2	0.7	0.3	0.0	0.0	0.0
3	0.0	1.0	0.0	0.0	0.0
4	0.0	0.0	0.0	0.4	0.6
5	0.0	0.0	0.0	1.0	0.0

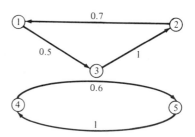

From the directed graph we see that the state space decomposes into two classes $C_1 = \{1, 2, 3\}$ and $C_2 = \{4, 5\}$. Also, for $x \in C_1$ and $y \in C_2$, we see that $x \nrightarrow y$ or $y \nrightarrow x$. Therefore, C_1 and C_2 are closed sets of states.

55

EXAMPLE 2.

$$
\begin{array}{c c c c}
 & 1 & 2 & 3 & 4 \\
\end{array}
$$

$$
\begin{array}{c}
1 \\
2 \\
3 \\
4
\end{array}
\begin{bmatrix}
0.0 & 0.0 & 1.0 & 0.0 \\
1.0 & 0.0 & 0.0 & 0.0 \\
0.3 & 0.7 & 0.0 & 0.0 \\
0.6 & 0.2 & 0.2 & 0.0
\end{bmatrix}
$$

The state space decomposes into $C = \{1, 2, 3\}$ and $A = \{4\}$. Hence, C is closed and A is not closed.

EXAMPLE 3. *Random Walk with Absorbing Barriers at 0 and b.*

$$
\begin{array}{c c c c c c c c c c}
 & 0 & 1 & 2 & 3 & 4 & \cdots & b{-}2 & b{-}1 & b \\
\end{array}
$$

$$
\begin{array}{c}
0 \\
1 \\
2 \\
\\
b-1 \\
b
\end{array}
\begin{bmatrix}
1 & 0 & 0 & 0 & 0 & \cdots & 0 & 0 & 0 \\
q & r & p & 0 & 0 & \cdots & 0 & 0 & 0 \\
0 & q & r & p & 0 & \cdots & 0 & 0 & 0 \\
\cdot & \cdot & \cdot & \cdot & \cdot & \cdot & \cdot & \cdot & \cdot \\
0 & 0 & 0 & 0 & 0 & \cdots & q & r & p \\
0 & 0 & 0 & 0 & 0 & \cdots & 0 & 0 & 1
\end{bmatrix}
$$

The state space decomposes into $\{0\}$, $\{1, 2, \ldots, b-1\}$, and $\{b\}$, where $\{0\}$ and $\{b\}$ are closed.

EXAMPLE 4. Consider N balls numbered 1 through N and two urns, I and II. Perform an experiment as follows. Select a number from $\{1, 2, \ldots, N\}$ at random (all selections equally likely) and then draw the ball corresponding to the selected number. Next select an urn at random, where urn I (resp. urn II) is selected with probability p (resp. $q = 1 - p$). Complete the trial by inserting the drawn ball into the selected urn. Repeat this trial under identical conditions. Let X_n denote the number of balls in urn I at the end of trial n. Then $\{X_n\}$ is an MC with state space $S = \{0, 1, 2, 3, \ldots, N\}$. Once we compute the transition probabilities, one can see that $\{X_n\}$ is an irreducible MC. Let $X_n = x$. At the end of the next trial the state will be $x + 1$, x or $x - 1$. The transition from x to $x + 1$ occurs if we have drawn a ball from urn II and selected urn I for the transfer. Since there are $(N - x)$ balls in urn II, the selection of a ball from this urn occurs with probability $(N - x)/N$. Hence the transition from x to $x + 1$ occurs with probability $p(N - x)/N$. Similarly, the transition from x to $x - 1$ occurs with probability qx/N. There will be no change in the number of balls if a ball is selected from either one of the urns and is inserted back into the same urn. This event occurs with probability $(px/N) + (q(N - x)/N)$. Each state communicates with every other state and the MC is irreducible.

3.5. Recurrence and Transience

Corresponding to an MC $\{X_n, n \geqslant 0\}$ with state space S, we define the following probabilities: (1) $f^n(x,y) = P_x\{t_y = n\}$, (2) $f^*(x,y) = P_x\{t_y < \infty\}$, and (3) $g(x,y) = P_x\{v(y) = \infty\}$. Then $f^n(x,y)$ gives the conditional probability that the first passage from x to y occurs in exactly n steps, $f^*(x,y)$ gives the conditional probability of ever visiting the state y starting the motion from x, and $g(x,y)$ is the conditional probability of infinitely many visits to the state y for the motion starting from x. These conditional probabilities are assumed to exist.

Definition 3.5.1. A state $x \in S$ is called *recurrent* if $f^*(x,x) = 1$; that is, x is recurrent if with probability one the MC, having started from x, will eventually return to x. A state x is called *transient* if $f^*(x,x) < 1$. [Recurrent states (resp. transient states) are also called *persistent* states (resp. *nonrecurrent* states).]

Theorem 3.5.2. *Let* $x, y \in S$. *Then*

(i) $f^*(x,y) = \sum_{n \geqslant 1} f^n(x,y).$

$$(3.5.1)$$

(ii) $g^n(x,y) = P_x\{v(y) = n\} = f^*(x,y)[f^*(y,y)]^{n-1}[1 - f^*(y,y)],$

$\qquad n \geqslant 1.$

$$(3.5.2)$$

(iii) $g(x,y) = f^*(x,y)g(y,y).$

$$(3.5.3)$$

(iv) $g(x,x) = \lim_{n \to \infty} [f^*(x,x)]^n.$

$$(3.5.4)$$

(v) *For any* x, *either* $g(x,x) = 1$ *or* $g(x,x) = 0$.

(vi) $g(x,x) = 1$ *if and only if* $f^*(x,x) = 1$.

(vii) $g(x,x) = 0$ *if and only if* $f^*(x,x) < 1$.

PROOF.

(i) $f^*(x,y) = P_x\{t_y < \infty\} = P_x\{\bigcup_{n \geqslant 1} \{t_y = n\}\} = \sum_{n \geqslant 1} P_x\{t_y = n\}$

$\qquad = \sum_{n \geqslant 1} f^n(x,y).$

(ii) First we claim that

$$P_x\{v(y) \geqslant n\} = f^*(x,y)[f^*(y,y)]^{n-1}.$$

$$(3.5.5)$$

57

For $n = 1$, $P_x\{\nu(y) \geq 1\} = P_x\{t_y < \infty\} = f^*(x,y)$. Next, for $n = 2$,

$$P_x\{\nu(y) \geq 2\} = P_x\{t_y < \infty\} P_y\{\nu(y) \geq 1\} = f^*(x,y)f^*(y,y).$$

Since $P_x\{\nu(y) \geq n\} = P_x\{t_y < \infty\} P_y\{\nu(y) \geq n - 1\}$, inductively we obtain (3.5.5):

$$
\begin{aligned}
g^n(x,y) = P_x\{\nu(y) = n\} &= \sum_{k \geq n} P_x\{\nu(y) = k\} - \sum_{k \geq n+1} P_x\{\nu(y) = k\} \\
&= P_x\{\nu(y) \geq n\} - P_x\{\nu(y) \geq n + 1\} \\
&= f^*(x,y)[f^*(y,y)]^{n-1}[1 - f^*(y,y)].
\end{aligned}
$$

(iii) Using the method of first entrance decomposition (see Example 3.3.4(4),) we have:

$$
\begin{aligned}
g(x,y) &= \sum_{n \geq 1} P_x\{t_y = n\} P_y\{\nu(y) - \nu(y,n) = \infty\} \\
&= \sum_{n \geq 1} f^n(x,y) P_y\{\nu(y) = \infty\} \qquad \text{(by homogeneity of the MC)}, \\
&= f^*(x,y)g(y,y).
\end{aligned}
$$

(iv) Taking $x = y$ in (3.5.5), we see that $g(x,x) = P_x\{\nu(x) = \infty\}$ $= \lim_{n \to \infty} P_x\{\nu(x) \geq n\} = \lim_{n \to \infty} [f^*(x,x)]^n$.

Statements (v)–(vii) follow from (iv). This completes the proof. $\square$

Next we give some criteria in order that a state x be recurrent or transient. In establishing the criteria we need a lemma that we state without proof.

Lemma 3.5.3. *Let $\{a_n, n \geq 0\}$ be a sequence of nonnegative reals such that either $\Sigma a_n < \infty$ or, $\Sigma a_n = \infty$ and a_n is bounded. Also let $\{b_n, n \geq 0\}$ be a convergent sequence of reals with limit b. Then*

$$b = \lim_{n \to \infty} \left\{ \left[\sum_{k=0}^{n} a_k b_{n-k} \right] \Big/ \left[\sum_{k=0}^{n} a_k \right] \right\}. \tag{3.5.6}$$

Theorem 3.5.4. *A state x is recurrent or transient according as $\Sigma_{n \geq 0} p^n(x, x)$ diverges or converges.*

PROOF. First we claim that, for $x, y \in S$, we have

$$f^*(x,y) = \lim_{N \to \infty} \left\{ \left[\sum_{n=1}^{N} p^n(x,y) \right] \Big/ \left[\sum_{n=0}^{N} p^n(y,y) \right] \right\}. \qquad (3.5.7)$$

From (3.3.4) we have

$$\sum_{n=1}^{N} p^n(x,y) = \sum_{n=1}^{N} \sum_{m=0}^{n-1} P_x\{t_y = n - m\} p^m(y,y)$$

$$= \sum_{m=0}^{N-1} p^m(y,y) \sum_{n=m+1}^{N} f^{n-m}(x,y)$$

$$= \sum_{m=0}^{N-1} p^m(y,y) \sum_{n=1}^{N-m} f^n(x,y).$$

To apply Lemma 3.5.3, take $n = N$, $a_m = p^m(x,x)$, $b_0 = 0$, and $b_m = \sum_{k=1}^{m} f^k(x,y) = \sum_{k=1}^{m} P_x\{t_y = k\}$. Now, (3.5.7) follows from Lemma 3.5.3. Next taking $x = y$ in (3.5.7) we obtain the theorem from the definition of recurrence. $\qquad \Box$

The example of Ehrenfests's chain emphasizes the importance of the notion of recurrence in an MC. Several physical, biological, and social processes are modeled to possess Markov property, and the recurrence notion is basic there. Next we present some fundamental properties of recurrent states. By $\mu(x,y)$ we denote the expected number of visits to y for an MC starting at x, that is, $\mu(x,y) = E_x[\nu(y)]$.

Theorem 3.5.5. Let x be a recurrent state. Then $g(x,x) = 1$ and $\mu(x,x) = \infty$; that is, if the particle starts at a recurrent state x, it returns to x infinitely often. If $f^*(u,x) = 0$, then $\mu(u,x) = 0$; that is, if the particle starts at some state u, it is possible that the particle may never hit x. But if $f^*(u,x) > 0$, then $\mu(u,x) = \infty$; that is, if a particle starting from a state u could hit the recurrent state x at some time, it visits x infinitely often.

PROOF. Since x is recurrent, $f^*(x,x) = 1$. From Theorem 3.5.2 (iv), it follows that $g(x,x) = 1$, that is, $P_x\{\nu(x) = \infty\} = 1$. Noting that $\nu(x) = \infty$ with full probability, we have that $\mu(x,x) = E_x[\nu(x)] = \infty$. Next let $f^*(u,x) = 0$. Then $f^n(u,x) = 0$ for all n, and hence from (3.3.4) we see that $p^n(u,x) = 0$ for all n. Now

$$\mu(u,x) = E_u \left[\sum_{n=1}^{\infty} I_x(X_n) \right] = \sum_{n=1}^{\infty} P_u[X_n = x] = \sum_{n=1}^{\infty} p^n(u,x) = 0.$$

(See also the Example 3.5.14.) Finally, let $f^*(u,x) > 0$. From Theorem 3.5.2 (iii) and $g(x,x) = 1$, it follows that

$$P_u\{\nu(x) = \infty\} = g(u,x) = f^*(u,x) > 0.$$

Thus $\nu(x) = \infty$ with positive probability, and hence $\mu(u,x) = \infty$. ☐

Theorem 3.5.6. *Let x be a transient state. Then*

$$P_u\{\nu(x) < \infty\} = 1, \quad \textit{and} \quad \mu(u,x) = \frac{f^*(u,x)}{1 - f^*(x,x)} < \infty, \ u \in S.$$

That is, if x is a transient state, then no matter where the particle starts, it visits x only a finite number of times and the expected number of visits is also finite.

PROOF. Let x be transient. Then $0 \leqslant f^*(x,x) < 1$ and consequently, from Theorem 3.5.2 (iii)–(iv),

$$P_u\{\nu(x) = \infty\} = g(u,x) = f^*(u,x)g(x,x)$$
$$= f^*(u,x) \lim_{n\to\infty} [f^*(x,x)]^n = 0.$$

This gives $P_u\{\nu(x) < \infty\} = 1$. Now using Theorem 3.5.2 (ii),

$$\mu(u,x) = \sum_{n\geqslant 1} nP_u\{\nu(x) = n\}$$
$$= \sum_{n\geqslant 1} nf^*(u,x)[f^*(x,x)]^{n-1}[1 - f^*(x,x)]$$
$$= f^*(u,x)[1 - f^*(x,x)]\frac{1}{[1 - f^*(x,x)]^2} = \frac{f^*(u,x)}{1 - f^*(x,x)}.$$

☐

Theorem 3.5.7. (i) *Let $x, y \in S$ and $x \to y$. If x is recurrent, then y is recurrent and $f^*(x,y) = 1 = f^*(y,x)$.*

(ii) *Let $X = \{X_n\}$ be an irreducible MC with finite state space S. Then X is a recurrent chain; that is, all states of the chain are recurrent.*

PROOF. (i) Let x be recurrent and $x \to y$. There exists an $m > 0$ such that $p^m(x,y) > 0$. Let $M > 0$ be such that $M = \min\{m: p^m(x,y) > 0\}$, and $p^k(x,y) = 0$ for $0 \leqslant k < M$. (Such an M exists.) Now we claim that $f^*(y,x) = 1$. Assume the contrary. Then the particle starting from y has a positive probability $(1 - f^*(y,x))$ of never visiting x. Therefore, if the MC starts from x, then $p(x, x_1) \cdots p(x_{M-1}, y)(1 - f^*(y,x)) > 0$ gives the probability of visiting states $x_1, \ldots, x_{M-1}, y$ successively in the first M transitions and never returning to x after these M steps. This contradicts the recurrence of x and hence $f^*(y,x) = 1$. This gives an $N > 0$ such that $p^N(y,x) > 0$. Therefore

$$p^{N+n+M}(y,y) = \sum_{u,v \in S} p^N(y,u)p^n(u,v)p^M(v,y)$$

$$\geqslant p^N(y,x)p^n(x,x)p^M(x,y)$$

so that

$$\sum_{n=0}^{\infty} p^n(y,y) \geqslant \sum_{n=0}^{\infty} p^{N+n+M}(y,y)$$

$$\geqslant p^N(y,x)p^M(x,y) \sum_{n=0}^{\infty} p^n(x,x)$$

$$= \infty,$$

as follows from $p^N(y,x)p^M(x,y) > 0$ and Theorem 3.5.4. Appealing again to Theorem 3.5.4, we see that y is recurrent.

(ii) Since X is irreducible, each state $x \in S$ communicates with every state $y \in S$. Now part (i) will complete the proof once we show that S contains at least one recurrent state. Suppose on the contrary that all the states are transient. Now from Theorem 3.5.6,

$$\infty > \mu(x,y) = E_x\left[\sum_{n=1}^{\infty} I_y(X_n)\right] = \sum_{n=1}^{\infty} P_x\{X_n = y\} = \sum_{n=1}^{\infty} p^n(x,y)$$

so that $\lim_{n \to \infty} p^n(x,y) = 0$. Therefore

$$0 = \sum_{y \in S} \lim_{n \to \infty} p^n(x,y) = \lim_{n \to \infty} P_x\{X_n \in S\} = \lim_{n \to \infty} 1 = 1.$$

This contradiction implies that not all the states are transient. By irreducibility, X is a recurrent MC. $\qquad \square$

Next we proceed to a finer classification of recurrent states into positive recurrent and null recurrent states. First some notations are in order. Recall from Example 3.3.4 (6) that $\nu(x, n)$ denotes the number of visits by the particle to the state x during the first n transitions. Set

$$\mu^n(x,y) = E_x[\nu(y,n)] = \sum_{m=1}^{n} p^m(x,y).$$

Then $n^{-1}\nu(x,n)$ is the proportion of the time, during the first n transitions, spent by the particle in state x. Let x be a recurrent state and denote by $\mu(x)$ the *mean recurrence time* $E_x[\mathbf{t}_x]$ of state x for a particle starting from x.

Definition 3.5.8. A recurrent state x is called *positive recurrent* (resp. *null recurrent*) if $\mu(x) < \infty$ (resp. $\mu(x) = \infty$).

Theorem 3.5.9. (i) *Let x be a transient state. Then* $\lim_{n\to\infty}\nu(x,n) = \nu(x)$ $< \infty$ *and* $\lim_{n\to\infty} n^{-1}\nu(x,n) = 0$, *AS.* [*Recall AS $=$ almost surely ($=$ with probability one).*] *Also,* $\lim_{n\to\infty}\mu^n(u,x) = \mu(u,x) < \infty$ *and* $\lim_{n\to\infty} n^{-1}\mu^n(u,x) = 0$, $u \in S$.

(ii) *Let x be a recurrent state. Then* $\lim_{n\to\infty} n^{-1}\nu(x,n) = \mu(x)^{-1}I_{\{\mathbf{t}_x<\infty\}}$, *AS, and* $\lim_{n\to\infty} n^{-1}\mu^n(u,x) = f^*(u,x)/\mu(x)$, $u \in S$.

(iii) *Let C be a closed equivalence class of recurrent states. Then* $\lim_{n\to\infty} n^{-1}\mu^n(x,y) = 1/\mu(y)$, $x,y \in C$. *If, furthermore, the particle starts its motion from C AS, then* $\lim_{n\to\infty} n^{-1}\nu(x,n) = 1/\mu(x)$, $x \in C$, *AS.*

(iv) *If x is a positive recurrent state, then* $\lim_{n\to\infty} n^{-1}\mu^n(u,x) = 1/\mu(x)$ > 0.

(v) *If x is a null recurrent state, then* $\lim_{n\to\infty} n^{-1}\mu^n(u,x) = 0$, $u \in S$.

Part (i) follows from Theorem 3.5.6. Parts (iii)–(v) follow from part (ii) of the theorem. As a rigorous proof of (ii) is slightly involved, we omit it. But (ii) is intuitively clear and says the following: once the particle reaches the recurrent state x, it returns to x *on the average every $\mu(x)$ units of time*. If the particle starting at u ever reaches x, then for large n, the proportion of time during the first n transitions that the particle spends in state x is approximately $[\mu(x)]^{-1}$.

Theorem 3.5.10. (i) *Let $x,y \in S$ and $x \to y$. If x is positive recurrent, then y is also positive recurrent.*

(ii) *Every irreducible finite MC is a positive recurrent chain.*

PROOF. (i) As shown in Theorem 3.5.7 (i), there are integers $M, N > 0$ such that $p^M(x,y) > 0$ and $p^N(y,x) > 0$. Also

$$p^{N+m+M}(y,y) \geqslant p^N(y,x)p^m(x,x)p^M(x,y).$$

Summing this relation over $m = 1, \ldots, n$ and dividing by n, we get

$$n^{-1}[\mu^{N+n+M}(y,y) - \mu^{N+M}(y,y)] \geq p^N(y,x)p^M(x,y)n^{-1}\mu^n(x,x).$$

Letting $n \to \infty$, we get

$$[\mu(y)]^{-1} \geq [\mu(x)]^{-1}p^N(y,x)p^M(x,y) > 0.$$

Therefore, $\mu(y) < \infty$, and hence y is positive recurrent.

(ii) First note that

$$1 = \sum_{y \in S} p^m(x,y) = \frac{1}{n}\sum_{m=1}^n \sum_{y \in S} p^m(x,y) = \sum_{y \in S} \frac{1}{n}\mu^n(x,y).$$

From (i) it suffices to show that there is at least one positive recurrent state in S. Assume the contrary. From Theorem 3.5.9 (v) we have

$$0 = \sum_{y \in S} \lim_{n \to \infty} n^{-1}\mu^n(x,y) = \lim_{n \to \infty} \sum_y n^{-1}\mu^n(x,y) = \lim 1 = 1.$$

This contradiction completes the proof. $\qquad\square$

Examples 3.5.11

EXAMPLE 1. Let $X = \{X_n, n \geq 0\}$ be an MC on $S = \{0, 1, 2, \ldots\}$ with transition $p(x,y)$ given by $p(x, x+1) = p$ and $p(x,0) = q$, where $0 < p < 1$ and $p + q = 1$. Show that X is a recurrent chain.

First observe that X is an irreducible chain. Let x and y be two arbitrary states. If $x < y$, then $x \xrightarrow{p} x + 1 \xrightarrow{p} \cdots \xrightarrow{p} y - 1 \xrightarrow{p} y$. If $x > y$, then $x \xrightarrow{q} 0 \xrightarrow{p} 1 \xrightarrow{p} \cdots \xrightarrow{p} y - 1 \xrightarrow{p} y$. Thus $x \to y$ and, similarly, $y \to x$. Hence X is an irreducible chain. Consequently, if one of the states, say, 0, is recurrent, then X is recurrent. Now

$$f^*(0,0) = \sum_{n=1}^\infty P_0\{t_0 = n\} = \sum_{n=1}^\infty qp^{n-1} = \frac{q}{1-p} = 1.$$

Hence X is a recurrent chain.

EXAMPLE 2. A Markov particle moves on the state space $S = \{1, 2, \ldots, \alpha, \alpha + 1, \ldots, \alpha + \beta\}$, where α and β are positive integers. Starting from one of the first α states (resp. last β states), the particle jumps in one transition to a state chosen uniformly from the last β states (resp. first α states). The transition matrix M is given by

$$
M = \begin{array}{c} \\ 1 \\ 2 \\ 3 \\ \cdot \\ \cdot \\ \cdot \\ \alpha \\ \alpha+1 \\ \alpha+2 \\ \cdot \\ \cdot \\ \cdot \\ \alpha+\beta \end{array}
\begin{array}{c}
\begin{array}{ccccccccc} 1 & 2 & 3 & \cdots & \alpha & \alpha+1 & \alpha+2 & \cdots & \alpha+\beta \end{array} \\
\left[\begin{array}{ccccccccc}
0 & 0 & 0 & \cdots & 0 & p & p & \cdots & p \\
0 & 0 & 0 & \cdots & 0 & p & p & \cdots & p \\
0 & 0 & 0 & \cdots & 0 & p & p & \cdots & p \\
\cdot & & & & & & & & \cdot \\
\cdot & & & & & & & & \cdot \\
\cdot & & & & & & & & \cdot \\
0 & 0 & 0 & \cdots & 0 & p & p & \cdots & p \\
q & q & q & \cdots & q & 0 & 0 & \cdots & 0 \\
q & q & q & \cdots & q & 0 & 0 & \cdots & 0 \\
\cdot & & & & & & & & \cdot \\
\cdot & & & & & & & & \cdot \\
\cdot & & & & & & & & \cdot \\
q & q & q & \cdots & q & 0 & 0 & \cdots & 0
\end{array}\right]
\end{array}
$$

where $p = (1/\beta)$ and $q = (1/\alpha)$. Let x and y be any two arbitrary states, and $S_\alpha = \{1, 2, \ldots, \alpha\}$, $S_\beta = \{\alpha + 1, \ldots, \alpha + \beta\}$. If $x \in S_\alpha$ and $y \in S_\beta$, then $x \xrightarrow{p} y$. If $x \in S_\beta$ and $y \in S_\alpha$, then $x \xrightarrow{q} y$. If $x, y \in S_\alpha$ (resp. S_β), then $x \xrightarrow{p} z \xrightarrow{q} y$, for some $z \in S_\beta$, (resp. $x \xrightarrow{q} z \xrightarrow{p} y$, for some $z \in S_\alpha$). Hence $x \to y$ and, similarly, $y \to x$. Therefore, the chain, being irreducible and with finite state space, is a recurrent chain.

EXAMPLE 3. Let X be an MC on $S = \{0, 1, \ldots, 6\}$ with transition matrix

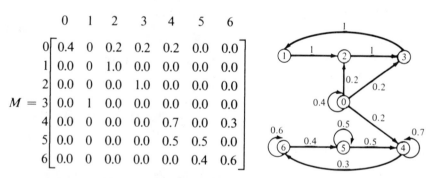

$$
M = \begin{array}{c} \\ 0 \\ 1 \\ 2 \\ 3 \\ 4 \\ 5 \\ 6 \end{array}
\begin{array}{c}
\begin{array}{ccccccc} 0 & 1 & 2 & 3 & 4 & 5 & 6 \end{array} \\
\left[\begin{array}{ccccccc}
0.4 & 0 & 0.2 & 0.2 & 0.2 & 0.0 & 0.0 \\
0.0 & 0 & 1.0 & 0.0 & 0.0 & 0.0 & 0.0 \\
0.0 & 0 & 0.0 & 1.0 & 0.0 & 0.0 & 0.0 \\
0.0 & 1 & 0.0 & 0.0 & 0.0 & 0.0 & 0.0 \\
0.0 & 0 & 0.0 & 0.0 & 0.7 & 0.0 & 0.3 \\
0.0 & 0 & 0.0 & 0.0 & 0.5 & 0.5 & 0.0 \\
0.0 & 0 & 0.0 & 0.0 & 0.0 & 0.4 & 0.6
\end{array}\right]
\end{array}
$$

Determine the recurrent, and the transient states.

From the directed graph it is clear that the state 0 leads to every state in S, whereas no state $x \neq 0$ leads to 0. Also $C_1 = \{1, 2, 3\}$ and $C_2 = \{4, 5, 6\}$ are irreducible closed sets of states. Hence 0 is transient and the rest of the states are recurrent.

EXAMPLE 4. *Birth–Death Chain*. Returning to Example 3.3.4 (3), we have seen that

$$P_x\{t_0 < t_b\} = \sum_{u=x}^{b-1} a_u \Big/ \sum_{u=0}^{b-1} a_u, \qquad (3.5.8)$$

where $a_u = (q_1 q_2 \cdots q_u / p_1 p_2 \cdots p_u)$ and $a_0 = 1$. Consider this chain $X = \{X_n, n \geqslant 0\}$ on $S = \{0, 1, 2, \ldots\}$ with $0 < p_x$, $x \geqslant 0$ and $0 < q_x$, $x \geqslant 1$, so that the chain becomes irreducible. Let us find conditions under which X becomes a recurrent chain. Toward this let $X_0 = 1$, AS, and $A_n = \{\omega \in \Omega : t_0 < t_n\}$, $n > 1$. Then, clearly, $A_n \uparrow \{t_0 < \infty\}$, and hence

$$P_1\{t_0 < \infty\} = \lim_{n \to \infty} P_1\{A_n\} = \lim_{n \to \infty} \left\{ \sum_{k=1}^{n-1} a_k \Big/ \sum_{k=0}^{n-1} a_k \right\}$$
$$= 1 - \left[\sum_{k=0}^{\infty} a_k \right]^{-1}. \qquad (3.5.9)$$

Since the chain is irreducible, if state 0 is recurrent, the chain is then recurrent. Hence we find conditions that will render the state 0 recurrent. To find a necessary condition, let 0 be recurrent. Then from Theorems 3.5.2 and 3.5.5 we see that $P_1\{t_0 < \infty\} = f^*(1,0) = 1$. Hence from (3.5.9) we get

$$\sum_{k=0}^{\infty} a_k = \sum_{k=0}^{\infty} \frac{q_1 \cdots q_k}{p_1 \cdots p_k} = \infty. \qquad (3.5.10)$$

Let us check whether (3.5.10) is a sufficient condition for the recurrence of state 0. From (3.5.10) and (3.5.9) we get $f^*(1,0) = 1$. Therefore, from the first-step decomposition,

$$f^*(0,0) = p(0,0) + p(0,1)f^*(1,0) = r_0 + p_0 = 1,$$

and hence 0 is recurrent. Thus we have established that *an irreducible birth–death chain on* $\{0, 1, 2, \ldots\}$ *is recurrent if and only if the following relation holds*:

$$\sum_{k=1}^{\infty} \left[\frac{q_1 \cdots q_k}{p_1 \cdots p_k} \right] = \infty. \qquad (3.5.11)$$

To further illustrate this example, first let $q_k = p_k k^2 / (k+1)^2$, $k \geqslant 1$. Then the LHS of (3.5.11) is

$$\frac{1}{2^2} + \frac{1}{3^2} + \frac{1}{4^2} + \cdots + = \sum_{k=2}^{\infty} k^{-2} < \infty \qquad \text{(why?)}.$$

In other words, the chain is transient. Next let $q_k \geqslant p_k$, $k \geqslant 1$. Then the series on the LHS of (3.5.11) is minorized by the divergent series $\Sigma_{k=1}^\infty 1$, and consequently the corresponding chain is recurrent.

EXAMPLE 5. *Random Walk.* In Section 2.4 we showed that a one-dimensional unrestricted simple random walk is recurrent iff it is a symmetric walk. Let us consider the d-dimensional unrestricted lattice walk $X = \{X_n, n \geqslant 0\}$ [see Example 3.1.5 (2)]. Here we present a criterion for the recurrence of the random walk X. Let J_n denote the nth jump. Then $X_n = X_0 + J_1 + \cdots + J_n$ where the jumps J_n are independent and identically distributed, say, according to $p(\mathbf{x}) = P\{J_n = \mathbf{x}\}$, where $\mathbf{x} \in I^d$, the d-dimensional lattice, and $n \geqslant 1$. Let $\phi(s)$ be the common characteristic function of the RVs J_n:

$$\phi(\mathbf{s}) = \sum_{\mathbf{x} \in I^d} p(\mathbf{x}) \exp\left\{ i \sum_{k=1}^d x_k s_k \right\}, \qquad \mathbf{s} = (s_1, \ldots, s_d) \in R^d.$$

Let $X_0 = \mathbf{O}$. Then $p^n(\mathbf{O}, \mathbf{x}) = P_{\mathbf{O}}\{X_n = \mathbf{x}\}$, and the characteristic function of X_n is

$$\phi^n(\mathbf{s}) = \sum_{\mathbf{x} \in I^d} p^n(\mathbf{O}, \mathbf{x}) \exp\left\{ i \sum_{k=1}^d x_k s_k \right\}.$$

From this it follows that

$$p^n(\mathbf{O}, \mathbf{x}) = \frac{1}{(2\pi)^d} \int_C \phi^n(\mathbf{s}) \exp\left\{ -i \sum_{k=1}^d x_k s_k \right\} d\mathbf{s},$$

where C is the cube with side 2π and center at $\mathbf{O}$. According to Theorem 3.5.4, X will be a transient walk if and only if $\mu(\mathbf{O}, \mathbf{O}) = \Sigma_{n \geqslant 0} p^n(\mathbf{O}, \mathbf{O}) < \infty$. Noting that $\mu(\mathbf{O}, \mathbf{O}) = \lim_{z \to 1-0} \Sigma_{n \geqslant 0} p^n(\mathbf{O}, \mathbf{O}) z^n$ (by Abel's lemma), we see that, for $0 \leqslant z < 1$,

$$\mu(\mathbf{O}, \mathbf{O}) = \lim_{z \to 1-0} \sum_{n \geqslant 0} [2\pi]^{-d} z^n \int_C \phi^n(\mathbf{s}) \, d\mathbf{s}$$

$$= \lim_{z \to 1-0} [2\pi]^{-d} \int_C [1 - z\phi(\mathbf{s})]^{-1} \, d\mathbf{s}$$

$$= \lim_{z \to 1-0} [2\pi]^{-d} \int_C \Re[1 - z\phi(\mathbf{s})]^{-1} \, d\mathbf{s},$$

where $0 \leqslant z < 1$ and $\Re$ denotes the real part. From this we have the following criterion.

Theorem 3.5.12. *A (temporally homogeneous) d-dimensional unrestricted lattice walk is transient if and only if*

$$\lim_{z \to 1-0} \int_C \Re[1 - z\phi(\mathbf{s})]^{-1} \, d\mathbf{s} < \infty.$$

Theorem 3.5.13. *Let X be an unrestricted d-dimensional lattice walk such that* $\mu = E[J_1] = \Sigma_{x \in I^d} xp(x)$ *exists (absolutely), that is,* $E[|J_1|] < \infty$. *If* $\mu \neq 0$, *then the random walk X is transient. Consequently, a one-dimensional lattice walk is recurrent if and only if* $\mu = 0$.

PROOF. Let $A_n = \{\omega: |n^{-1}X_n - \mu| > |\mu|/2\}$, $n \geqslant 1$, where $|\cdot|$ is the standard Euclidean norm in R^d. Since X_n is the sum of IID RVs, it follows from the strong law of large numbers that $P_O\{\limsup_{n \to \infty} A_n\} = 0$. Noting that the event $\{X_n = 0\}$ implies (i.e., $\subset$) the event A_n, we get $P_O\{\limsup_{n \to \infty} X_n = O\} = 0$. Therefore, the state O and hence the RW are transient.

Example **3.5.14.** We illustrate here the second statement in Theorem 3.5.5. It states that if x is a recurrent state and the particle starts its motion from some state $u \neq x$, it is possible that the particle may never hit x. Consider an RW on $S = \{\cdots, -1, 0, 1, \ldots\}$ with the following type of jumps. If the particle is in state $x \in S$, then $P\{x \to x + 4\} = \frac{1}{2} = P\{x \to x - 4\}$. That is, $P\{J = 4\} = \frac{1}{2} = P\{J = -4\}$. Therefore, $\mu = E[J] = 0$ and by Theorem 3.5.17, the RW is recurrent (i.e., every state is recurrent). Let $X_0 \equiv 1$. Then, with probability one, the particle never hits 0. This is due to the fact that the particle performs its walk only on a subset of S that depends on the initial state.

3.6. Stationary Distribution

Consider a Markov particle beginning its motion from an arbitrary state x. In studying its subsequent motion, we would like to ask what could be said about the motion after an elapse of a large number of steps. In physical terms, this is a question about the asymptotic *stability* of the motion; that is, we want to look at the limiting *steady-state* distribution—irrespective of the initial position of the particle.

Let $p_n(x) = P\{X_n = x\}$, $x \in S$, $n \geqslant 0$. Then $p_n(x) \geqslant 0$ and $\Sigma_{x \in S} p_n(x) = 1$, $n \geqslant 0$. Also

$$p_{n+m}(x) = \sum_{y \in S} p_n(y)p^m(y, x), \qquad x \in S, \quad n, m \geqslant 0. \qquad (3.6.1)$$

If the absolute distributions $p_n(x)$ are independent of n, say

$$p(x) = P\{X_n = x\}, \qquad x \in S, \quad n \geqslant 0,$$

then the probability distribution $\{p(x), x \in S\}$ is called the *steady-state distribution* of the MC $\{X_n, n \geqslant 0\}$. From (3.6.1) we have

$$p(x) = \sum_{u \in S} p(u)p^n(u, x), \qquad x \in S, \quad n \geqslant 0. \qquad (3.6.2)$$

Now it is natural to expect that the existence of a steady-state distribution is related to a nontrivial and nonnegative solution of

$$\pi(x) = \sum_{u \in S} \pi(u)p(u, x), \qquad x \in S. \qquad (3.6.3)$$

Definition 3.6.1. If $\pi(x) \geqslant 0$ for all $x \in S$, and $\Sigma_{x \in S} \pi(x) = 1$ and satisfies relation (3.6.3), $\{\pi(x), x \in S\}$ is then called a *stationary distribution*.

We are also interested in the behavior $\lim_{n \to \infty} p^n(x, y)$, which is partially related to the existence of stationary distributions.

We have seen earlier that the steady-state distribution of an MC is a stationary distribution. If a stationary distribution $\pi(x)$ is the initial distribution $p_0(x)$ of the MC $\{X_n\}$, it is easy to see that $\pi(x) = p_0(x)$ is the steady-state distribution of $\{X_n\}$. Now the physical relevance of stationarity or the steady state is clear. Consider a large number N of particles performing independent Markov motions on the same state space S. If $\{\pi(x)\}$ is the steady-state distribution, then at any time n the expected number of particles in state x is $N\pi(x)$, a constant. Thus observing the system as a whole, we see that a state of macroscopic equilibrium is maintained even though individual particles might spend a disproportionately large part of their time in a particular (large) subset of S.

Definition 3.6.2. A state $x \in S$ is called a *periodic state* if its period $d(x)$ defined by

$$d(x) = \mathrm{GCD}\{n \geqslant 1: p^n(x, x) > 0\} \qquad (3.6.4)$$

is greater than 1 (GCD = greatest common divisor). If $d(x) = 1$, then x is called *aperiodic*. A positive recurrent aperiodic state is called *ergodic*.

Theorem 3.6.3. *Let $\{X_n, n \geqslant 0\}$ be an MC with state space S and $x \in S$ be a periodic state. Then:*

(i)   *If $x \leftrightarrow y$, then y is also periodic and $d(x) = d(y)$.*
(ii) *The state x is an aperiodic state of the MC $\{X_{nd(x)}, n \geqslant 0\}$ whose one-step transition matrix is $[p^{d(x)}(x, y)]$.*
(iii) *An irreducible MC is aperiodic if $p(x, x) > 0$ for some $x \in S$.*

PROOF. Exercise. □

Theorem 3.6.4. *Let* $X = \{X_n\}$ *be an irreducible aperiodic* MC.

(i) *If* X *is a positive recurrent chain, then*

$$\lim_{n\to\infty} p^n(x,y) = \pi(y) > 0, \qquad x, y \in S, \qquad (3.6.5)$$

where:

$$\text{(a) } \pi(x) = \frac{1}{\mu(x)}, \qquad (3.6.6)$$

the reciprocal of the mean entrance time, (b) $\{\pi(x), x \in S\}$ *is a probability distribution that is uniquely determined by the system of equations* (3.6.3), (c) $\{\pi(x)\}$ *is the only stationary distribution of the* MC X, *and*

$$\text{(d) } \lim_{n\to\infty} p_n(x) = \pi(x), \qquad x \in S. \qquad (3.6.7)$$

(ii) *If the states are null recurrent or transient, then:*

$$\text{(a) } \lim_{n\to\infty} p^n(x,y) = 0; \qquad \text{(b) } \lim_{n\to\infty} p_n(x) = 0 \qquad (3.6.8)$$

for all $x, y \in S$, *and no stationary distribution exists.*

Now let X *be an irreducible positive recurrent periodic chain with period* d. *Then for each pair* $x, y \in S$ *there is an integer* k, $0 \leqslant k < d$ *such that*

$$\text{(a) } \lim_{n\to\infty} p^{k+nd}(x,y) = d\pi(y), \qquad (3.6.9)$$

and

$$\text{(b) } p^m(x,y) = 0 \qquad \text{unless } m = Nd + k, \text{ for some } N. \quad (3.6.10)$$

PROOF. The proof is divided into several steps.

Step 1. Let X be an irreducible positive recurrent MC with a stationary distribution $\pi(x)$. Then $\pi(x)$ is given by relation (3.6.6).

By induction on (3.6.3), we see that $\pi(x)$ satisfies (3.6.2). From (3.6.2) we have, rewriting $\mu^n(x,y)$ as $\mu(n, x, y)$,

$$\pi(x) = \frac{1}{n}\sum_{i=1}^n \sum_u \pi(u)p^i(u,x) = \sum_{u \in S} \pi(u)n^{-1}\mu(n, u, x).$$

Taking limits on both sides as $n \to \infty$, we get from Theorem 3.5.9 (iii),

$$\pi(x) = [\mu(x)]^{-1} \sum_u \pi(u) = \frac{1}{\mu(x)}.$$

Step 2. The function $\pi(x)$ given by (3.6.6) is a stationary distribution.

Case i. *Let S be a finite state space.* Recalling again that $\lim_{n\to\infty} \mu(n, u, x)/n = 1/\mu(x)$, and since $1 = \Sigma_x p^i(u, x)$ for all i, we get

$$1 = \frac{1}{n} \sum_{i=1}^{n} \sum_{x} p^i(u, x) = \sum_{x} \frac{\mu(n, u, x)}{n} \to \sum_{x} \left[\frac{1}{\mu(x)}\right].$$

Hence $\{1/\mu(x)\}$ is a probability distribution. Similarly, summing the Chapman–Kolmogorov relation $p^{i+1}(u, y) = \Sigma_x p^i(u, x) p(x, y)$ over $i = 1, \ldots, n$ and dividing by n, we get

$$\sum_{x} n^{-1} \mu(n, u, x) p(x, y) = n^{-1}[\mu(n + 1, u, y) - p(u, y)]. \qquad (3.6.11)$$

Letting $n \to \infty$ in this relation, we see that

$$\sum_{x} \frac{p(x, y)}{\mu(x)} = \mu(y). \qquad (3.6.12)$$

Case ii. *Let S be a countably infinite state space:* Let S_0 be an arbitrary finite subset of S. Then

$$\sum_{x \in S_0} n^{-1} \mu(n, u, x) \leqslant 1, \quad u \in S, \qquad \text{and hence} \sum_{x \in S_0} [\mu(x)]^{-1} \leqslant 1.$$

Since S_0 is arbitrary, we get

$$\sum_{x \in S} \frac{1}{\mu(x)} \leqslant 1. \qquad (3.6.13)$$

Similarly, from (3.6.11),

$$\sum_{u \in S} \frac{p(u, x)}{\mu(u)} \leqslant [\mu(x)]^{-1}, \quad x \in S. \qquad (3.6.14)$$

We want equality to hold in (3.6.13). Suppose, on the contrary, that we have only the strict inequality. Then, summing on x in (3.6.14),

$$\sum_{x} [\mu(x)]^{-1} > \sum_{x} \sum_{u} \frac{p(u, x)}{\mu(u)}$$

$$= \sum_{u} [\mu(u)]^{-1} \sum_{x} p(u, x) = \sum_{u} [\mu(u)]^{-1},$$

a contradiction. Hence $\{[\mu(x)]^{-1}\}$ satisfies (3.6.12). Now set $(1/\alpha) = \Sigma_x 1/\mu(x)$. Then from (3.6.12), $\pi(x) = \alpha/\mu(x)$. But, by Step 1, $\alpha = 1$. This proves Step 2.

Step 3. Let X be an ergodic chain ($=$ irreducible positive recurrent aperiodic chain). We claim that for every pair $x, y \in S$, there is an integer $n_0 > 0$ such that $p^n(x, y) > 0$ for all $n \geqslant n_0$.

First fix an $u \in S$ and let $A = \{n \geqslant 1: p^n(u, u) > 0\}$. Since X is aperiodic, GCD $A = 1$. From $p^{m+n}(u, u) \geqslant p^m(u, u)p^n(u, u)$, it follows that, if $m, n \in A$, then $(m + n) \in A$. From these two properties of the set A, we can find an integer $N > 0$ such that $n \in A$ for all $n \geqslant N$, that is, $p^n(u, u) > 0, n \geqslant N$. Now let $x, y \in S$ be an arbitrary pair. From the irreducibility of the chain, we can find integers $K, M > 0$ such that $p^K(x, u) > 0$ and $p^M(u, y) > 0$, so that $p^{K+n+M}(x, y) \geqslant p^K(x, u)p^n(u, u)p^M(u, x) > 0$, for all $n \geqslant N$. This proves Step 3.

Next introduce an MC $\{(X_n, Y_n), n \geqslant 0\}$ on the state space $S^2 = S \times S$ with one-step transition matrix defined by $q((u, v), (x, y)) = p(u, x)p(v, y)$ so that $\{X_n, n \geqslant 0\}$ and $\{Y_n, n \geqslant 0\}$ are each an MC with transition $p(\cdot, \cdot)$. The transitions of X_n and Y_n are chosen independently of each other.

Step 4. $Z = (X_n, Y_n)$ is an ergodic chain on S^2.

Let $(u, v), (x, y) \in S^2$. By Step 3, there exists an integer $N_0 > 0$ such that, for $n \geqslant N_0$, $p^n(u, x) > 0$ and $p^n(v, y) > 0$ and, consequently, $q^n((u, v), (x, y)) = p^n(u, x)p^n(v, y) > 0$. This proves that Z is irreducible and aperiodic.

Now let $\pi(x)$ be the stationary distribution of X. Define $\pi(u, v) = \pi(u)\pi(v)$. Then π is a stationary distribution, because:

$$\pi(u, v) = \pi(u)\pi(v) = \left\{ \sum_x \pi(x)p(x, u) \right\} \left\{ \sum_y \pi(y)p(y, v) \right\}$$

$$= \sum_x \sum_y \pi(x)\pi(y)p(x, u)p(y, v)$$

$$= \sum_{(x, y) \in S^2} \pi(x, y)q((x, y), (u, v)).$$

By Steps 1 and 2, Z is a positive recurrent chain. This proves Step 4.

Let D denote the diagonal of S^2: $D = \{(x, y) \in S^2 : x = y\}$. Let t_D be the hitting time of D for the MC Z. Since Z is a recurrent chain, it is easy to see that $P\{t_D < \infty\} = 1$. Since the chains X_n and Y_n are indistinguishable after t_D, it is clear that

$$P\{X_n = x, t_D \leqslant n\} = P\{Y_n = x, t_D \leqslant n\}, \qquad x \in S. \qquad (3.6.15)$$

Step 5. We shall now show that

$$\lim_{n \to \infty} [P\{X_n = x\} - P\{Y_n = x\}] = 0, \qquad x \in S. \qquad (3.6.16)$$

From (3.6.15) we have:

$$P\{X_n = x\} = P\{X_n = x, t_D \leqslant n\} + P\{X_n = x, t_D > n\}$$
$$= P\{Y_n = x, t_D \leqslant n\} + P\{X_n = x, t_D > n\} \quad (3.6.17)$$
$$\leqslant P\{Y_n = x\} + P\{t_D > n\}$$

Similarly, relation (3.6.17) holds with X_n and Y_n interchanged. Hence, for $n \geqslant 1$, $|P\{X_n = x\} - P\{Y_n = x\}| \leqslant P\{t_D > n\}$. Noting that $P\{t_D < \infty\}$ $= 1$, and passing to limit in this inequality, we obtain (3.6.16).

Step 6. Relation (3.6.5) holds: Let $P\{X_0 = u\} = 1$, for an arbitrarily fixed $u \in S$, and $\pi(y)$ be the initial distribution of $\{Y_n\}$. Then

$$P\{X_n = x\} = p^n(u, x) \quad \text{and} \quad P\{Y_n = x\} = \pi(x), \quad x \in S.$$

From this and Step 5 we see that

$$\lim_{n \to \infty} \{p^n(u, x) - \pi(x)\} = \lim_{n \to \infty} \{P\{X_n = x\} - P\{Y_n = x\}\} = 0.$$

To complete the proof of Theorem 3.6.4 (i), it remains to show that (3.6.7) holds. From (3.6.1) we get $p_n(x) = \Sigma_y p_0(y)p^n(y, x)$. Now from what we have seen so far [especially relation (3.6.5)], we get (3.6.7).

Step 7. Proof of Theorem 3.6.4 (ii). Let x be transient. From the criteria for transience (Theorem 3.5.4), $p^n(x, x) \to 0$ as $n \to \infty$. Let x be null recurrent. Then from Theorem 3.5.9 (v) we have $n^{-1}\mu(n, x, x) \to 0$, as $n \to \infty$. Consequently, this holds if x is either transient or null recurrent. Then, as observed from the proof of Step 1, $\pi(x) = \lim_{n \to \infty} \Sigma_u \pi(u)n^{-1}\mu(n, u, x) = 0$. This implies that $p_n(x) \to 0$ as $n \to \infty$. From (3.6.2) it is clear that no stationary distribution exists.

It remains to establish the periodic case. Let X be an irreducible positive recurrent periodic chain with period $d > 1$.

Step 8. Relations (3.6.9) and (3.6.10) hold:

First observe that Theorem 3.6.4 (i) can be restricted to an ergodic class C. Also, from Theorem 3.6.3 (ii) the chain $Y_n = X_{nd}$, $n \geqslant 0$, is aperiodic. If the chain X_n and hence the chain Y_n start from an $x \in S$, then X_n returns to x, for the first time at kd, for some $k > 0$. Therefore, the expected return time to x for the Y chain is $d^{-1}\mu(x)$, where $\mu(x)$ is the expected return time to x for the X chain. Let $q(\cdot, \cdot)$ denote the transition probability of Y. It follows, from what we saw in Theorem 3.6.4 (i), that

$$\lim_{n \to \infty} p^{nd}(x, x) = \lim_{n \to \infty} q^n(x, x) = d[\mu(x)]^{-1} = d\pi(x). \quad (3.6.18)$$

First we establish (3.6.10). Let, for any given pair $x, y \in S$, $M = \min\{n: p^n(x,y) > 0\}$. If $m > 0$ is an integer such that $p^m(y,x) > 0$, then $p^{M+m}(y,y) \geqslant p^m(y,x)p^M(x,y) > 0$, and hence $M + m$ is an integral multiple of d. Same is the case for any n with $p^n(x,y) > 0$. Therefore, $n - M = Kd$ for some $K \geqslant 0$. Hence we can find an integer N such that $M = Nd + k$, $0 \leqslant k < d$ and $p^n(x,y) = 0$ unless $n = Nd + k$. This proves (3.6.10).

To complete the proof of Step 8 and hence the theorem, it remains to establish (3.6.9). Now from (3.6.10) and (3.3.3) we have

$$p^{nd+k}(x,y) = \sum_{i=0}^{n} P_x\{t_y = id + k\} p^{(n-i)d}(y,y). \tag{3.6.19}$$

If $i > n$, redefine $p^{(n-i)d}(y,y) = 0$. Then, passing to the limit in (3.6.19) as $n \to \infty$, and using (3.6.18),

$$\lim_{n\to\infty} p^{nd+k}(x,y) = \sum_{i=0}^{\infty} P_x\{t_y = id + k\} \lim_{n\to\infty} p^{(n-i)d}(y,y)$$

$$= d\pi(y) \sum_{i=0}^{\infty} P_x\{t_y = id + k\}$$

$$= d\pi(y)P_x\{t_y < \infty\} = d\pi(y).$$

This completes the proof. $\qquad\square$

For further criteria for the ergodicity of an MC, we refer the interested reader to Foster (1953).

Examples 3.6.5

EXAMPLE 1. Let $X = \{X_n, n \geqslant 0\}$ be an MC on $S = \{1, \dots, 5\}$ with transition matrix

$$M = \begin{array}{c} \\ 1 \\ 2 \\ 3 \\ 4 \\ 5 \end{array} \begin{array}{ccccc} 1 & 2 & 3 & 4 & 5 \\ \begin{bmatrix} 0 & 0.5 & 0.5 & 0.0 & 0.0 \\ 0 & 0.0 & 0.0 & 0.2 & 0.8 \\ 0 & 0.0 & 0.0 & 0.4 & 0.6 \\ 1 & 0.0 & 0.0 & 0.0 & 0.0 \\ 1 & 0.0 & 0.0 & 0.0 & 0.0 \end{bmatrix} \end{array}.$$

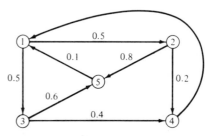

It follows from the directed graph or from the transitions $1 \to 2 \to 4 \to 1$, $1 \to 3 \to 4 \to 1$, $1 \to 3 \to 5 \to 1$ that X is an irreducible chain. Since S is finite, X is a recurrent chain. From the transitions shown from state 1 back to itself, it is clear that the period of state 1 is 3. As it is irreducible, it follows from Theorem 3.6.3 that X is periodic with period 3. To find the stationary distribution $\{\pi(x)\}$ of X, we use the defining relations of $\pi(x)$: $\pi(y) = \Sigma_x \pi(x)p(x,y)$. Then from matrix M,

$$\pi(1) = \pi(1)p(1,1) + \pi(2)p(2,1) + \pi(3)p(3,1) + \pi(4)p(4,1) + \pi(5)p(5,1)$$
$$= \pi(4) + \pi(5)$$

$$\pi(2) = 0.5\pi(1) = \pi(3), \quad \pi(4) = 0.2\pi(2) + 0.4\pi(3),$$

$$\pi(5) = 0.8\pi(2) + 0.6\pi(3).$$

Solving this set of equations along with

$$\pi(1) + \pi(2) + \pi(3) + \pi(4) + \pi(5) = 1,$$

(since $\{\pi(x)\}$ is a probability distribution), we obtain

$$\pi = \{\tfrac{1}{3}, \tfrac{1}{6}, \tfrac{1}{6}, \tfrac{1}{10}, \tfrac{7}{30}\} = \{\pi(x), x \in S\}.$$

EXAMPLE 2. *Ehrenfests Chain.* From the discussion of Example 3.1.9 (6), the transition matrix M of the Ehrenfests chain $X = \{X_n, n \geqslant 0\}$ is given by

	0	1	2	3	$\cdots$	$2N-2$	$2N-1$	$2N$
0	0	1	0	0	$\cdots$	0	0	0
1	$\dfrac{1}{2N}$	0	$\dfrac{2N-1}{2N}$	0	$\cdots$	0	0	0
2	0	$\dfrac{2}{2N}$	0	$\dfrac{2N-2}{2N}$	$\cdots$	0	0	0
.	.	.	.	.	$\cdots$	.	.	.
.	.	.	.	.	$\cdots$	.	.	.
.	.	.	.	.	$\cdots$	.	.	.
$2N-1$	0	0	0	0	$\cdots$	$\dfrac{2N-1}{2N}$	0	$\dfrac{1}{2N}$
$2N$	0	0	0	0	$\cdots$	0	1	0

Clearly, X is an irreducible recurrent chain. If the chain is in state x at a certain time, then it takes even number of steps to return to x. Also,

$$p^2(x,x) = \left(1 - \frac{x}{2N}\right)\frac{x+1}{2N} + \frac{x}{2N}\left(1 - \frac{x-1}{2N}\right) > 0.$$

Therefore, x is a periodic state, and hence X is a periodic chain, of period 2. Next let us compute the stationary distribution of the Ehrenfests chain. Using the defining relations (3.6.3) we obtain the set of equations

$$\pi(x) = \pi(x-1)\frac{2N-x+1}{2N} + \pi(x+1)\frac{x+1}{2N}, \qquad i \leqslant x \leqslant 2N-1,$$

$$\pi(0) = \frac{\pi(1)}{2N} \quad \text{and} \quad \pi(2N) = \frac{\pi(2N-1)}{2N}.$$

Solving these equations recursively, we see that

$$\pi(x) = \binom{2N}{x}\pi(0).$$

Using this with the fact that $\{\pi(x), x \in S\}$ is a probability distribution, we get

$$1 = \pi(0) \sum_{x=0}^{2N} \binom{2N}{x} = 2^{2N}\pi(0).$$

Hence the stationary distribution of the Ehrenfests chain is given by the binomial distribution $B(2N,\frac{1}{2})$:

$$\pi(x) = \binom{2N}{x}2^{-2N}, \qquad x = 0, 1, \ldots, 2N. \tag{3.6.20}$$

While introducing the Ehrenfests chain we remarked that following the work of Boltzmann and Gibbs an irreconcilable situation arose regarding the recurrence and the thermodynamical irreversibility of a conservative dynamical system. Now we can address to this quandary. We have seen that the Ehrenfests chain is an irreducible recurrent chain. From (3.6.20) and Theorem 3.6.4 it follows that the expected return time to a state x is

$$\mu(x) = 2^{2N}\frac{x!(2N-x)!}{(2N)!} < \infty. \tag{3.6.21}$$

It also follows from (3.6.20) that, regardless of the initial composition of urn I, after a long time the probability of finding a specified number of balls in urn I is nearly the same as if the $2N$ balls had been distributed randomly (i.e., each ball having probability $\frac{1}{2}$ of being in urn I). For large N we see that, by passing to normal approximation to the binomial if necessary, we can certainly find about half of the particle in each one of the urns. So it looks as if we are moving toward the equilibrium state N. This can be expected even by looking at the transition matrix M. Farther the particle away from the state N, larger is the probability of moving toward the equilibrium. This is the so-called diffusion with central force.

To reconcile between the recurrence and equilibrium, we have to appeal to the mean return time. First we note that, from (3.6.21),

$$\frac{\mu(N + x)}{\mu(N)} \simeq \left(1 + \frac{x}{N}\right)^x. \tag{3.6.22}$$

This shows how rapidly the expected return time increases as the chain moves away from the equilibrium position N. State 0 is the farthest from state N. Note that the expected return time is 2^{2N}. Therefore, for large N, it is no wonder the process looks irreversible.

EXAMPLE 3. Let $X = \{X_n, n \geqslant 0\}$ be the MC introduced in Example 3.5.11 (1). We have seen that X is an irreducible recurrent chain on $S = \{0, 1, 2, \ldots\}$. Let us compute the stationary distribution of X. Using Definition 3.6.1,

$$\mu(0) = q[\mu(0) + \mu(1) + \cdots] = q,$$

$$\mu(x) = p\mu(x - 1) = p^2\mu(x - 2) = \cdots = p^x\mu(0) = qp^x, \qquad x \geqslant 1.$$

Hence $\{\mu(x), x \geqslant 0\} = \{qp^x, x \geqslant 0\}$.

EXAMPLE 4. Let $X = \{X_n, n \geqslant 0\}$ be the MC introduced in Example 3.5.11 (2). This is an irreducible chain on a finite state space $S = \{1, 2, \ldots, \alpha, \alpha + 1, \ldots, \alpha + \beta\}$. We compute the stationary distribution $\{\mu(x), x \in S\}$ of X. If $1 \leqslant x \leqslant \alpha$, then

$$\mu(x) = \sum_{u \in S} \mu(u)p(u, x) = \alpha^{-1} \sum_{u=\alpha+1}^{\alpha+\beta} \mu(u),$$

so that

$$\mu(1) = \cdots = \mu(\alpha) = \alpha^{-1}\left[1 - \sum_{u=1}^{\alpha} \mu(u)\right].$$

For $\alpha + 1 \leqslant y \leqslant \alpha + \beta$, $\mu(y) = \beta^{-1}\sum_{u=1}^{\alpha}\mu(u)$. Therefore,

$$\mu(x) = \begin{cases} \alpha^{-1} - \mu(1) & \text{if } 1 \leqslant x \leqslant \alpha \\ \alpha\beta^{-1}\mu(1) & \text{if } \alpha + 1 \leqslant x \leqslant \alpha + \beta. \end{cases}$$

Let $x = 1$. Then $\mu(1) = \alpha^{-1} - \mu(1)$ and hence $\mu(1) = [2\alpha]^{-1}$. Consequently, $\mu(x) = (2\alpha)^{-1}$ for $1 \leqslant x \leqslant \alpha$, and $\mu(x) = (2\beta)^{-1}$ for $\alpha + 1 \leqslant x \leqslant \alpha + \beta$.

EXAMPLE 5. *Birth–Death Chain*. Let $X = \{X_n\}$ be a birth–death chain on $S = \{0, 1, 2, \ldots\}$ with transition probabilities, $p(x, x + 1) = p_x$, $(x \geqslant 0)$, $p(x, x) = r_x$, $(x \geqslant 0)$, and $p(x, x - 1) = q_x$, $(x \geqslant 1)$, where $p_x, q_x > 0$, and $p_x + q_x + r_x = 1$. Then X is an irreducible MC. Define

$$a_0 = 1, \quad \text{and} \quad a_x = \frac{p_0 \cdots p_{x-1}}{q_1 \cdots q_x}, \, x \geqslant 1.$$

We claim that X *has a stationary distribution if and only if* $\sum_{x=0}^{\infty} a_x < \infty$. From Definition 3.6.1, we have

$$\pi(0) = \pi(0)r_0 + \pi(1)q_1,$$

$$\pi(x) = \pi(x - 1)p_{x-1} + \pi(x)r_x + \pi(x + 1)q_{x+1}, \quad x \geqslant 1.$$

Because $p_x + q_x + r_x = 1$,

$$q_1 \pi(1) - p_0 \pi(0) = 0$$

$$q_{x+1} \pi(x + 1) - p_x \pi(x) = q_x \pi(x) - p_{x-1} \pi(x - 1), \quad x \geqslant 1,$$

so that, by induction, we get the recursive relation

$$\pi(x) = \frac{p_{x-1} \pi(x - 1)}{q_x}, \quad x \geqslant 0.$$

$$\therefore \pi(x) = \frac{p_{x-1}}{q_x} \pi(x - 1) = \cdots = \frac{p_0 \cdots p_{x-1}}{q_1 \cdots q_x} \pi(0) = a_x \pi(0).$$

Since $\{\pi(x)\}$ is a probability distribution,

$$1 = \sum_x \pi(x) = \pi(0) \sum_x a_x, \quad \text{or} \quad \pi(0) = \left[\sum_x a_x \right]^{-1}.$$

Consequently,

$$\pi(x) = a_x \Big/ \sum_u a_u, \quad x \geqslant 0.$$

Now it is clear that the stationary distribution exists if and only if $\sum a_x < \infty$.

As a special case of this example, let $p_0 = 1, p_x = p > 0, q_x = q > 0$, $(x \geqslant 1)$, and $p + q = 1$. Then

$$\sum_x a_x = \sum_x \frac{p^{x-1}}{q^x} = p \sum_x \left[\frac{p}{(1 - p)} \right]^x.$$

If $p < \frac{1}{2}$, then $0 < p/(1 - p) < 1$ and the geometric series converges. Letting $p < \frac{1}{2}$, we see that

$$\pi(0) = \frac{1 - 2p}{2(1 - p)}; \quad \pi(x) = \frac{(1 - 2p)p^{x-1}}{2(1 - p)^{x+1}}, \quad x \geqslant 1.$$

EXAMPLE 6. Find the stationary distribution concentrated on each of the irreducible closed sets of the MC whose transition matrix is given by

$$
\begin{array}{c@{\quad}ccccccc}
 & 0 & 1 & 2 & 3 & 4 & 5 & 6 \\
\begin{array}{c}0\\1\\2\\3\\4\\5\\6\end{array} &
\left[\begin{array}{ccccccc}
0.1 & 0.1 & 0.2 & 0.2 & 0.4 & 0.0 & 0.0 \\
0.0 & 0.0 & 0.5 & 0.5 & 0.0 & 0.0 & 0.0 \\
0.0 & 0.0 & 0.0 & 1.0 & 0.0 & 0.0 & 0.0 \\
0.0 & 1.0 & 0.0 & 0.0 & 0.0 & 0.0 & 0.0 \\
0.0 & 0.0 & 0.0 & 0.0 & 0.5 & 0.5 & 0.0 \\
0.0 & 0.0 & 0.0 & 0.0 & 0.5 & 0.0 & 0.5 \\
0.0 & 0.0 & 0.0 & 0.0 & 0.0 & 0.5 & 0.5
\end{array}\right]
\end{array}
$$

From the directed graph of transitions we see that the two irreducible closed sets are $C_1 = \{1, 2, 3\}$ and $C_2 = \{4, 5, 6\}$. On C_1, the distribution $\{\pi(x)\}$ satisfies

$$\pi(1) = \pi(3), \quad \pi(2) = 0.5\pi(1), \quad \pi(3) = 0.5[\pi(1) + \pi(2)],$$

from which the stationary distribution π_1 corresponding to class C_1 is given by $\pi_1 = \{0, \frac{2}{5}, \frac{1}{5}, \frac{2}{5}, 0, 0, 0\}$. Similarly, the stationary distribution corresponding to class C_2 is given by $\pi_2 = \{0, 0, 0, 0, \frac{1}{3}, \frac{1}{3}, \frac{1}{3}\}$.

EXAMPLE 7. *Predator-Encounter Chain*. For the predator-encounter chain we have shown [see equations (3.2.4) and (3.2.5)] that

$$P\{X_n = \mathrm{I}\} = \frac{q}{p+q} + (1-p-q)^n \left[p_0(\mathrm{I}) - \frac{q}{p+q}\right],$$

$$P\{X_n = \mathrm{II}\} = \frac{p}{p+q} + (1-p-q)^n \left[p_0(\mathrm{II}) - \frac{p}{p+q}\right].$$

This chain is clearly an irreducible aperiodic recurrent chain. Let us find the stationary distribution of this chain. Let $0 < p, q < 1$. Then $|1 - p - q| < 1$. Letting $n \to \infty$ in the preceding probabilities, we get

$$\lim_{n \to \infty} P\{X_n = \mathrm{I}\} = \frac{q}{p+q}; \quad \lim_{n \to \infty} P\{X_n = \mathrm{II}\} = \frac{p}{p+q}.$$

So, from Theorem 3.6.4 (i) [see relation (3.6.7)], the stationary distribution $\{\pi(x), x = \mathrm{I}, \mathrm{II}\}$ is given by

$$\left\{\pi(\mathrm{I}) = \frac{q}{p+q}, \quad \pi(\mathrm{II}) = \frac{p}{p+q}\right\}.$$

3.7. Branching Chain

As pointed out in Section 3.2, study of the branching chain began with Galton's work on the survival of family names. One can now find a wide range of applications of these chains in physics and biology. For example, consider the study of neutron transport where we are concerned with the prediction of neutron population in a reactor. A neutron, as a result of collision with a nucleus, is either scattered, absorbed, or multiplied by the process of fission. Since the neutrons move in a bounded region, there is an upper bound on the time for birth of a neutron counted in any generation. We obtain a branching chain as follows. Let $X_0 = 1$. This neutron, when it collides with a nucleus, splits the nucleus and the resulting fission produces a random number of new neutrons, thereby forming the first generation. Each of these neutrons, moving independently of each other, may hit some other nucleus and produce more neutrons. This process continues forming different generations of a neutron branching chain. If X_n denotes the number of neutrons in the nth generation, then $\{X_n, n \geq 0\}$ is a branching chain.

A similar situation arises in the study of electron multipliers, which is a device that amplifies a weak current of electrons. In the path of electrons one sets up a series of plates. As an electron hits the first plate, it generates a random number of electrons. These secondary electrons hit the next plate and produce additional electrons. The process continues. If X_n denotes the number of electrons emitted from the nth plate, then clearly $\{X_n, n \geq 0\}$ is a branching chain.

Definition 3.7.1. Let $J_n, n \geq 1$, be a sequence of IID RVs with values in $S = \{0, 1, 2, \ldots\}$ and the common distribution given by $p_k = P\{J = k\}, k \geq 0$. An MC $\{X_n, n \geq 0\}$ is called a *branching chain* with state space S if its transition probabilities are given by

$$p(x, y) = P\{X_n = y \mid X_{n-1} = x\} = P\{J_1 + \cdots + J_x = y\}, \quad (3.7.1)$$

for $x \geq 1$ and $y \geq 0$, and $p(0, 0) = 1$.

Theorem 3.7.2. (Watson). *Let* $f(s) = \sum_{i=0}^{\infty} p_i s^i, |s| \leq 1$, *be the probability generating function of* J_1, *and define inductively*

$$f_0(s) = s, \quad f_1(s) = f(s), \quad f_n(s) = f(f_{n-1}(s)), \quad n \geq 1. \quad (3.7.2)$$

Then $f_n(s)$ *is the generating function of* $X_n, n \geq 0$.

PROOF. Let $g_n(s)$ be the generating function of X_n. Clearly, $g_0(s) \equiv s$ and $g_1(s) = f(s)$. Now

$$
\begin{aligned}
g_{n+1}(s) &= \sum_{x \geqslant 0} P\{X_{n+1} = x\} s^x \\
&= \sum_{x \geqslant 0} \sum_{y \geqslant 0} P\{X_{n+1} = x | X_n = y\} P\{X_n = y\} s^x \\
&= \sum_{x \geqslant 0} s^x \sum_{y \geqslant 0} P\{X_n = y\} P\{J_1 + \cdots + J_y = x\} \\
&= \sum_{y \geqslant 0} P\{X_n = y\} \sum_{x \geqslant 0} P\{J_1 + \cdots + J_y = x\} s^x \\
&= \sum_{y \geqslant 0} P\{X_n = y\}[f(s)]^y, \qquad \text{since the RVs } J_k \text{ are IID,} \\
&= g_n(f(s)) = g_n(g(s)).
\end{aligned}
$$

Since $g_0 \equiv f_0$ and $g_1 \equiv f_1$, it follows by induction that $g_n \equiv f_n$. This completes the proof. $\qquad \square$

By iteration one can show that $f_{n+m}(s) = f_n(f_m(s))$. In particular, $f_{n+1}(s) = f_n(f(s))$.

Example 3.7.3. Let $m = E[X_1] < \infty$ and $\sigma^2 = \mathrm{var}(X_1) < \infty$. Show that

$$
E[X_n] = m^n; \quad \mathrm{var}(X_n) = \begin{cases} \sigma^2 m^{n-1}(m^n - 1)/(m-1) & \text{if } m \neq 1 \\ n\sigma^2 & \text{if } m = 1 \end{cases}.
$$

Since $E[X_n] = f_n'(1)$, we differentiate the relation $f_{n+1}(s) = f_n(f(s))$ and set $s = 1$:

$$
f_{n+1}'(1) = f_n'(f(1))f'(1) = f_n'(1)f'(1).
$$

By iteration:

$$
f_{n+1}'(1) = f_{n-1}'(1)[f'(1)]^2 = \cdots = f_1'(1)[f'(1)]^n = [f'(1)]^{n+1}.
$$

Since $f'(1) = E[X_1] = m$, we get $E[X_n] = m^n$.

Next note that $\mathrm{var}(X_n) = f_n''(1) + f_n'(1) - [f_n'(1)]^2$. Now

$$
f_{n+1}''(1) = f''(1)[f_n'(1)]^2 + f'(1)f_n''(1).
$$

But

$$
f'(1) = m \qquad \text{and} \qquad f''(1) = \sigma^2 + m^2 - m,
$$

and hence

$$f''_{n+1}(1) = (\sigma^2 + m^2 - m)m^{2n} + mf''_n(1).$$

Set $a = \sigma^2 + m^2 - m$. Then by induction

$$f''_{n+1}(1) = a(m^{2n} + m^{2n-1}) + m^2 f''_{n-1}(1) = \cdots$$
$$= a(m^{2n} + m^{2n-1} + \cdots + m^n),$$

from which

$$\operatorname{var}(X_{n+1}) = (\sigma^2 + m^2 - m)\left(\sum_{k=n}^{2n} m^k\right) + m^{n+1} - m^{2n+2}$$
$$= \sigma^2 m^n (m^{n+1} - 1)/(m - 1) \qquad \text{if } m \neq 1$$

$$\operatorname{var}(X_{n+1}) = (n+1)\sigma^2 \text{ if } m = 1.$$

Theorem 3.7.4. *Let* $X_n, n \geqslant 0$, *be a branching chain such that* $p_1 = P\{J = 1\} \neq 1$. *Then all nonzero states* $x(\neq 0)$ *are transient, and*

$$\lim_{n\to\infty} P\{X_n = x\} = 0, \; x \neq 0.$$

PROOF. First note that the state $x = 0$ is the absorbing state of the chain and hence is positive recurrent. Let $x \neq 0$. We claim that $f^*(x,x) < 1$ so that x is transient. Now

$$f^*(x,x) = P_x\{X_n = x \text{ for some } n \geqslant 1\}.$$

If $p_0 = 0$, then $f^*(x,x) = p^k < 1$. If $p_0 > 0$, then $f^*(x,x) = 1 - p(x,0) = 1 - p_0^k < 1$. This proves the claim. Since all the states $x \neq 0$ are transient,

$$P\{X_n = x\} = p^n(1,x) \to 0 \qquad \text{as } n \to \infty,$$

by Theorem 3.5.4.

$\square$

Before discussing the existence of a stationary distribution for X_n, let us study the *probability of extinction* of a population. Let q denote this probability; then $q = P\{\cup_{n\geqslant 1}\{X_n = 0\}\}$. Since 0 is an absorbing state,

$$q = P\{\bigcup_{n \geqslant 1}\{X_n = 0\}\} \qquad = \lim_{n \to \infty} P\{\bigcup_{k=1}^{n}\{X_k = 0\}\}$$

$$= \lim_{n \to \infty} P\{X_n = 0\} \qquad = \lim_{n \to \infty} f_n(0).$$

Note that $f_n(0)$ is a nondecreasing function of n and hence $0 \leqslant f_0(0)$ $\leqslant f_1(0) \leqslant \cdots \leqslant q = \lim_{n \to \infty} f_n(0) \leqslant 1$. Since $f_{n+1}(0) = f(f_n(0))$, we get $q = f(q)$; that is, q is a fixed point of $f(s)$.

Let $m \leqslant 1$. Then $f(0) > 0$ and $f'(s) \leqslant f'(1) = m \leqslant 1$, for $0 \leqslant s \leqslant 1$. Using the mean-value theorem to express $f(s)$ in terms of $f(1)$, we obtain $f(s) > s$ for $0 \leqslant s < 1$. Hence if $m \leqslant 1$, then $q = 1$.

Let $m > 1$. The mean-value theorem then implies that $f(s) < s$ for all $s \in (1 - \epsilon, 1)$, $\epsilon > 0$ sufficiently small. But $f(0) \geqslant 0$. Thus $f(s)$ has at least one fixed point in $[0, 1)$. Here we get the uniqueness due to the strict convexity of f, for if there were two fixed points, say, s and σ with $0 \leqslant s < \sigma < 1$, then by Rolle's theorem there exist a and b, $s < a < \sigma$ $< b < 1$, such that $f'(a) = 1 = f'(b)$, which is impossible due to strict convexity of f.

Now we claim that $q < 1$. If $q = 1$, then $f_n(0) \to 1$ as $n \to \infty$, and from what we saw above, we would have

$$f_{n+1}(0) = f(f_n(0)) < f_n(0),$$

contradicting the monotone increasing nature of the sequence $f_n(0)$.

In 1874 Watson established that q is the unique fixed point of $f(s)$ but failed to observe that $q < 1$ if $m > 1$. Steffensen gave a complete solution for the extinction problem. Summarizing the preceding arguments, we obtain Theorem 3.7.5.

Theorem 3.7.5. (H. Watson, J. Steffensen). *If $m = E[X_1] \leqslant 1$, then q $= 1$. If $m > 1$, then q is the unique nonnegative solution in $[0, 1)$ of $s = f(s)$.*

In view of Theorem 3.7.4, it is easy to see that the branching chain has a unique stationary distribution $(1, 0, 0, 0, \ldots)$ concentrated at 0 if $p_0 > 0$. Instead of requiring a stationary distribution, if we seek for a stationary measure, we have the following theorem due to Harris.

Theorem 3.7.6. *If $\{X_n\}$ is a branching chain with $0 < p_0 < 1$, then it has a stationary measure $\{\pi_x\}$ such that $\Sigma_{x \geqslant 1} \pi_x = \infty$. Let $\pi(\theta) = \Sigma_{x \geqslant 1} \pi_x \theta^x$, the generating function of π_x, and q the extinction probability of $\{X_n\}$. Then $\pi(\theta)$ is analytic in $|\theta| < q$ and satisfies the functional equation $\pi(f(\theta))$ $= 1 + \pi(\theta)$, $|\theta| < q$, provided that $\pi(p_0) = 1$.*

Examples 3.7.7

EXAMPLE 1. Consider a branching chain with $p_0 = \frac{1}{4}$, $p_1 = \frac{1}{2}$, $p_2 = \frac{3}{16}$, $p_3 = \frac{1}{16}$, and $p_k = 0$ for $k \geqslant 4$. Find the probability q of extinction.
First note that

$$m = EX_1 = 0 \cdot \frac{1}{4} + 1 \cdot \frac{1}{2} + 2 \cdot \frac{3}{16} + 3 \cdot \frac{1}{16} = \frac{17}{16} > 1.$$

Hence by Theorem 3.7.5, $q < 1$ and is the fixed point of $s = f(s)$. Now $f(s) = \frac{1}{4} + \frac{1}{2}s + \frac{3}{16}s^2 + \frac{1}{16}s^3$. Let us solve $f(s) = s$.

$$f(s) = s \Rightarrow s^3 + 3s^2 - 8s + 4 = 0$$

$$\text{or} \quad (s - 1)(s^2 + 4s - 4) = 0$$

so that $s = 1$, $s = -2 - 2\sqrt{2}$, or $s = -2 + 2\sqrt{2}$, and hence the extinction probability $q = 2(\sqrt{2} - 1)$.

EXAMPLE 2. Let $f(s) = (as^2 + bs + c)$ with $a > 0$, $b > 0$, $c > 0$, and $f(1) = 1$. If the probability of extinction q is such that $0 < q < 1$, show that $q = (c/a)$.
By assumption $q \in (0, 1)$. So we are looking for the unique fixed point of $f(s)$.

$$f(s) = s \Rightarrow as^2 + (b - 1)s + c = 0.$$

If $f(1) = 1$, $(s - 1)$ is a factor of $as^2 + (b - 1)s + c$. Then

$$0 = as^2 + (b - 1)s + c = (s - 1)[as + (a + b - 1)] = 0$$

and consequently $c = 1 - a - b$. Now

$$q = \frac{1 - a - b}{a} = \frac{c}{a}.$$

Exercises

1. Consider a sequence of independent tosses of a fair die. Let X_n denote the maximum of the outcomes in the first n throws. Find the transition probability matrix of this MC $\{X_n, n \geqslant 1\}$.

2. Consider a sequence of independent tosses of a pair of fair coins. Let X_n denote the number of heads in n tosses. Find the state space and the transition matrix of this MC $\{X_n, n \geqslant 1\}$.

3. Markov Chains

3. Consider two urns and $2N$ balls, of which N are red and N are green. Start an experiment with N balls in each urn. At each trial choose a ball at random from each urn and interchange the balls in the opposite urns. If X_0 is the initial number of red balls in urn I and X_n, $n \geqslant 1$, is the number of red balls in urn I at the conclusion of the nth trial, find the transition matrix of the MC $\{X_n, n \geqslant 0\}$.

4. Consider an RW on $0, 1, \ldots, N$ with 0 as a reflecting barrier. Let N be a sticky boundary in the following sense: the particle that is stuck at N is released and thrown to state $N - 2$ two time units after it hits N. Find the transition matrix of such an RW.

5. Find M^2, the two-step transition matrix, in each of the above Exercises 1–4.

6. A Markov particle moves on points 0, 1, and 2 arranged in a circle in the clockwise direction. A step in the clockwise direction occurs with probability p, $0 < p < 1$, and a step in the counter-clockwise direction occurs with probability $1 - p$. Prove that

$$M^n = \begin{bmatrix} p_{1n} & p_{2n} & p_{3n} \\ p_{3n} & p_{1n} & p_{2n} \\ p_{2n} & p_{3n} & p_{1n} \end{bmatrix}$$

where $p_{1n} + \omega p_{2n} + \omega^2 p_{3n} = (1 - p + p\omega)^n$, with ω denoting the primitive cube root of unity.

7. Consider a sequence of independent tosses of a biased coin having probability p for heads. For $n \geqslant 2$, $X_n = 0$ or 1 accordingly as the $(n - 1)$st and nth tosses, both resulted in heads. Prove that $\{X_n\}$ is not an MC.

8. Determine the classes and classify the states into transient and recurrent states for an MC with transition matrix:

(a)
$$\begin{bmatrix} 0.0 & 1.0 & 0.0 & 0.0 & 0.0 \\ 1.0 & 0.0 & 0.0 & 0.0 & 0.0 \\ 0.1 & 0.1 & 0.3 & 0.5 & 0.0 \\ 0.3 & 0.7 & 0.0 & 0.0 & 0.0 \\ 0.3 & 0.4 & 0.2 & 0.0 & 0.1 \end{bmatrix}$$

(b)
$$\begin{bmatrix} 0.2 & 0.0 & 0.0 & 0.8 \\ 1.0 & 0.0 & 0.0 & 0.0 \\ 0.0 & 1.0 & 0.0 & 0.0 \\ 0.0 & 0.0 & 1.0 & 0.0 \end{bmatrix}$$

(c)
$$\begin{bmatrix} 0.1 & 0.9 & 0.0 & 0.0 & 0.0 & 0.0 \\ 1.0 & 0.0 & 0.0 & 0.0 & 0.0 & 0.0 \\ 1.0 & 0.0 & 0.0 & 0.0 & 0.0 & 0.0 \\ 0.0 & 0.2 & 0.8 & 0.0 & 0.0 & 0.0 \\ 0.3 & 0.0 & 0.0 & 0.0 & 0.3 & 0.4 \\ 0.0 & 0.0 & 0.0 & 0.7 & 0.3 & 0.0 \end{bmatrix}$$

(d)
$$\begin{bmatrix} 0.4 & 0.5 & 0.1 & 0.0 & 0.0 \\ 0.0 & 0.3 & 0.2 & 0.5 & 0.0 \\ 0.0 & 1.0 & 0.0 & 0.0 & 0.0 \\ 1.0 & 0.0 & 0.0 & 0.0 & 0.0 \\ 0.3 & 0.7 & 0.0 & 0.0 & 0.0 \end{bmatrix}.$$

9. Let x and y be two states of an MC such that $f^*(x,y) = 1 = f^*(y,x)$. Then show that x and y are recurrent states.

10. Consider a simple symmetric RW on the three-dimensional lattice. Find a set A in this lattice such that $f^*((0,0,0),A) = 1$.

11. Let $\{Y_n\}$, $n \geqslant 1$, be a sequence of IID RVs with the common probability distribution $\{p_k\}$, $k = 0, \pm1, \pm2, \dots$. Set $X_n = Y_1 + \cdots + Y_n$. Let $E|Y| < \infty$ and $E[Y_n] = m$ for all n. Show that the MC $\{X_n\}$, $n \geqslant 1$, is recurrent if and only if $m = 0$.

12. Let $\{a_k\}$, $k = 0, 1, \dots$, be a positive sequence and $\{X_n\}$ an MC, the transition probability matrix of which is given by $p(k,0) = a_k$ and $p(k, k+1) = 1 - a_k$, $0 < a_k < 1$. Prove that this MC is transient if and only if $a_0 + a_1 + \cdots < \infty$.

13. Consider an irreducible birth–death chain on $\{0, 1, \dots\}$ such that $p_x \leqslant q_x$, $x \geqslant 1$. Show that the chain is recurrent.

14. Consider the birth–death chain on $S = \{0, 1, \dots\}$ such that $p_x = (x + 2)/2(x + 1)$ and $q_x = x/2(x + 1)$, $x \geqslant 0$. Show that this chain is transient. Also compute $P_x\{t_a < t_b\}$ for $a < x < b$ and $f^*(x,0)$, $x > 0$.

15. Consider an MC with N states. (a) If $p(m; x,y) > 0$, show that $0 < m \leqslant N - 1$. (b) If x is a recurrent state, show that one can find an a, $0 < a < 1$, such that for $n > N$, $P_x\{t_x > n\} \leqslant a^n$.

16. Let x be a recurrent state of an MC $\{X_n, n \geqslant 0\}$. Show that

$$\lim_{N \to \infty} P_x\{X_m \neq x, n + 1 \leqslant m \leqslant n + N\} = 0.$$

Prove also that the convergence shown above is uniform in n if x is positive recurrent.

17. Let π_x, $x \in S$, be a stationary distribution of an MC and $x, y \in S$ be two arbitrary states such that $\pi(x) > 0$ and $x \to y$. Show that $\pi(y) > 0$.

18. The transition probabilities of an MC on $S = \{0, 1, \dots\}$ are given by $p(x,0) = (x + 1)/(x + 2)$ and $p(x, x + 1) = 1/(x + 2)$. Determine whether this MC is positive recurrent, null recurrent, or transient. If it is positive recurrent, find its stationary distribution.

19. Now do Exercise 18 with $p(x,0) = 1/(x + 2)$ and $p(x, x + 1) = (x + 1)/(x + 2)$.

20. Following are the transition matrices of some MCs. In each case find the stationary distribution concentrated on irreducible closed sets. Also find the mean first passage time $m(x,y) = \Sigma_{n \geqslant 1} nf^n(x,y)$ for all communicating recurrent states x and y.

(a)–(d) The transition matrices in Exercise 8.

$$
\text{(e)} \begin{bmatrix}
1.0 & 0.0 & 0.0 & 0.0 & 0.0 \\
0.3 & 0.4 & 0.3 & 0.0 & 0.0 \\
0.0 & 0.3 & 0.7 & 0.0 & 0.0 \\
0.0 & 0.0 & 0.2 & 0.0 & 0.8 \\
0.0 & 0.0 & 0.0 & 0.0 & 1.0
\end{bmatrix}
\qquad
\text{(f)} \begin{bmatrix}
0.1 & 0.1 & 0.1 & 0.3 & 0.4 \\
0.0 & 0.7 & 0.3 & 0.0 & 0.0 \\
0.0 & 0.0 & 0.1 & 0.2 & 0.7 \\
0.3 & 0.1 & 0.1 & 0.4 & 0.1 \\
0.0 & 0.0 & 0.3 & 0.3 & 0.4
\end{bmatrix}
$$

21. Find the stationary distribution of the MC described in Exercise 3.

22. Let $\{X_n\}$ be a birth–death chain on $S = \{0, 1, \ldots\}$ such that $p_0 = 1, p_x = p > 0, x \geqslant 1$, and $q_x = q = 1 - p, x \geqslant$. Find the condition under which $\{X_n\}$ possesses a stationary distribution and find that distribution.

23. Consider a branching chain $\{X_n\}, n \geqslant 0$, with generating function $f(x) = (1 - b - c)/(1 - c) + (bs/1 - cs), 0 < c < b + c < 1$, with $(1 - b - c) > c(1 - c)$, and $X_0 = 1$. Show that

$$
\lim_{n \to \infty} P\{X_n = x \,|\, X_n > 0\} = (1 - r^{-1})r^{1-x},
$$

where $r = (1 - b - c)/c(1 - c)$.

24. Let $\{X_n\}$ be a branching chain with $p_k = p(1 - p)^k, k \geqslant 0$, where $0 < p < 1$. If q is the probability of extinction, then show that $q = 1$ if $p \geqslant \frac{1}{2}$ and $q = p/(1 - p)$ if $p < \frac{1}{2}$.

25. Let $\{X_n\}, n \geqslant 0$, be a branching chain with $X_0 = 1$. Show that

$$
P\{\bar{X}_n > N \text{ for some } 0 \leqslant n \leqslant m \,|\, X_m = 0\} \leqslant [P\{X_m = 0\}]^N.
$$

26. Initially start a blood culture with one red cell. At the end of a unit time the red cell dies and is replaced either by: (a) two red cells, (b) one red and one white or (c) two white cells with probabilities $\frac{1}{4}, \frac{2}{3}$, and $\frac{1}{12}$, respectively. Subsequently, each red cell reproduces this way and each white cell dies at the end of one unit of time without reproducing. Show that the probability of extinction of the entire culture is $\frac{1}{3}$.

27. Consider a society in which each man has exactly two children. Each child is a boy or a girl with equal probability $(\frac{1}{2})$. Let $\{X_n\}$ be a branching chain representing the number of males in the nth generation. Show that the male line is sure to extinct. Show that the probability of extinction of the male line is $q = \sqrt{5} - 2$ if each man has three children instead of two.

4

Poisson Processes

4.1. Definitions and Examples

Poisson processes are found to yield accurate models in several applications from such diverse areas as physics, geology, biology, nuclear medicine, anthropology, astronomy, and geography. In this chapter we study certain basic properties of this process, not only because it arises very naturally in several applications, but also as it is a very simple continuous-time stochastic process and a prototype of more general jump Markov processes and processes with independent increments. Poisson processes arise in situations where one is interested in the total number of occurrences of a "specified type" of event up to a time point $t \geqslant 0$, such as the number of successive occurrences of events of the following type: arrival of telephone calls, emission of α-particles by radioactive substances (each monotonically increasing with time), flaws in a material, stars in space, and skulls in an ancient burial ground (each monotonically increasing with d-dimensional volume, $d = 2$ or 3). Such processes are actually known as counting processes (and the Poisson processes are defined by additional restrictions).

Let $(\Omega, \mathcal{C}, P)$ be the basic probability space on which the RVs $X(t, \cdot)$, $t \geqslant 0$, of the stochastic process $X(t, \omega)$ are defined. A continuous-time stochastic process $\{X(t), t \geqslant 0\}$ with values in the state space $S = \{0, 1, 2, \ldots\}$ is called a *counting process* if $X(t)$, for any t, represents the total number of "events" that have occurred during the time period $[0, t]$.

Definition 4.1.1. A stochastic process $X(t)$, $t \geqslant 0$, is said to have *independent increments* if for all $n \geqslant 1$ and time points $0 \leqslant t_0 < t_1 < \cdots < t_n$

87

the increments $X(t_0) - X(0)$, $X(t_1) - X(t_0)$, ..., $X(t_n) - X(t_{n-1})$ are stochastically independent RVs. A process $X(t)$ is said to possess *stationary increments* if the distributions of the increments $X(t) - X(s)$, $s < t$, depend only on the length $(t - s)$ of the time interval $[s, t]$ over which we have taken the individual increment.

Recall that a RV $X: \Omega \to \{0, 1, 2, ...\}$ is called a Poisson RV with rate $\lambda > 0$, if the probability distribution of X is given by $P\{X = x\} = e^{-\lambda}\lambda^x/x!$, $x = 0, 1, 2, ...$

Definition 4.1.2. A counting process $\{X(t), t \geqslant 0\}$ is said to be a *Poisson process* with rate $\lambda > 0$ if: (1) $X(0) = 0$, (2) $X(t)$ is a process with independent increments, and (3) the number of events in any interval of length t is Poisson distributed with rate λt, that is, for all $s, t \geqslant 0$,

$$P\{X(t + s) - X(s) = x\} = e^{-\lambda t}(\lambda t)^x/x!, \qquad x = 0, 1, 2, ... \quad (4.1.1)$$

We see below that this definition of a Poisson process is equivalent to the following one.

Definition 4.1.3. A counting process $\{X(t), t \geqslant 0\}$ is said to be a *Poisson process* with rate $\lambda > 0$ if: (1) $X(0) = 0$, (2) $X(t)$ is a process with independent and stationary increments, and (3) relations (4.1.2) and (4.1.3) hold:

$$P\{X(t + h) - X(t) = 1\} = \lambda h + o(h), \qquad (4.1.2)$$

$$P\{X(t + h) - X(t) \geqslant 2\} = o(h), \qquad (4.1.3)$$

where a function $f(x)$ is said to be of order $o(h)$ if $\lim_{h \to 0} f(h)/h = 0$.

Next we give several counting processes that are modeled and studied as Poisson processes.

Examples 4.1.4

EXAMPLE 1. *Radioactive Decay.* Due to decay, certain radioactive substances emit γ-photons at a rate varying with the amount of the substance present. Let $X(t)$ represent the number of γ-photons reaching a registering counter during the time interval $[0, t]$. By using Definition 4.1.3, $\{X(t)\}$ can be modeled as a Poisson process.

EXAMPLE 2. *Electron Emissions.* Let the heating element in a vacuum tube be energized at time $t = 0$. Then an electron, after being emitted from a cathode, travels to the anode. One obtains a Poisson process, as in Example 4.1.4 (1), by taking $X(t)$ to represent the number of electrons emitted during the time $0 \leqslant s \leqslant t$.

A similar process arises in *optical detection.* Consider a photodetector (e.g., an avalanche photodiode) having a photoemissive surface. An optical field incident on the photodetector results into the emission of a stream of photoelectrons at the detector output. The number $X(t)$ of emissions during $0 \leqslant s \leqslant t$ can be modeled as a Poisson process (see Karp, et al. 1970).

EXAMPLE 3. *Neuron Spike Activity.* Let us insert a microelectrode into an exposed auditory nerve fiber. [This is a standard procedure in auditory electrophysiology; see Siebert (1970).] In response to an acoustic pressure stimulus applied to the outer ear, this microelectrode detects the electrical activity in the nerve in the form of randomly occurring spike activity. A Poisson process can be used to model this spike activity by taking $X(t)$ to represent the number of spike discharges until time t.

EXAMPLE 4. *Spatial Distribution of Random Points.* Models of Poisson process $X(t)$ where t is a space variable arise in disciplines such as astronomy, bacteriology, and ecology. In ecology, one assumes that the number of animal litters in a plot follows a Poisson process with intensity proportional to the area t of the plot. Along the same lines, consider the photograph of a Petri plate with bacterial colonies that one can observe under a microscope (these are visible as dark spots). The bacterial count on the Petri plate is modeled as a Poisson process with intensity proportional to the area. Next, consider the distribution in space of the centers of gravity of stars. This spacial distribution of stars is modeled as a Poisson process with the intensity proportional to the volume t. A model similar to the Petri plate in bacteriology arises in electron microscopy. Consider a microscopic object whose image has to be obtained. In electron microscopy, a flood of electrons illuminate this object, and the electrons are collected in a scintillation crystal. The distribution of the electron impacts in the crystal can be modeled as a Poisson process with the intensity depending on the area t.

EXAMPLE 5. *Nuclear Medicine.* Radioactive tracers are widely used in nuclear medicine. They provide (relatively) noninvasive diagnostic procedures for clinical medicine. The radioactive pharmaceutical tracer is also used as a tool in the research of physiological phenomena such as blood flow rate and lung ventilation. In the static study, one is interested in obtaining an image of the spatial distribution of the tracer, whereas in the dynamical study, one obtains quantitative information about the parameters relating to physiological phenomena. In order to visualize internal organs a technique used in nuclear medicine is to ingest into them a radioactive tracer and then monitor the time course of radioactive emissions by an external scintillation detector. The data so collected is a series of discrete events associated with the emissions of light pulses in the crystal of the detector. To a first approximation the series of light pulses can be modeled as a Poisson process $X(t)$. Here t varies in a region of the product $R_+ \times R^2$, where $s \in R_+$ and $\mathbf{x} \in R^2$ correspond, respectively, to the observation interval $[0, s]$ and to the surface of a scintillation crystal, so that $X(t)$ is a time–space process. See Jacques (1968) and other articles in Wagner (1968).

Theorem 4.1.5. *The two ways (Definitions* 4.1.2 *and* 4.1.3) *of defining Poisson processes are equivalent.*

PROOF.   First we show that Definition 4.1.2 implies Definition 4.1.3. From relation (4.1.1) it follows that the process $X(t)$ has stationary increments. To see that (4.1.2) holds, we use (4.1.1):

$$P\{X(t + h) - X(t) = 1\} = \frac{e^{-\lambda h}(\lambda h)^1}{1!}$$

$$= \lambda h \left[\frac{\sum\limits_{n=0}^{\infty} (-\lambda h)^n}{n!} \right]$$

$$= \lambda h[1 - \lambda h + o(h)] = \lambda h + o(h).$$

Similarly,

$$P\{X(t + h) - X(t) \geqslant 2\} = \sum_{n \geqslant 2} \frac{e^{-\lambda h}(\lambda h)^n}{n!} = o(h).$$

Next we show that Definition 4.1.3 implies Definition 4.1.2. Let $f(t, \cdot)$ be the probability distribution of $X(t, \omega)$. We obtain (4.1.1) through solving some simple linear differential equations:

$$f(t + h, 0) = P\{X(t + h) = 0\}$$
$$= P\{X(t) = 0, X(t + h) - X(t) = 0\}$$
$$= P\{X(t) = 0\} P\{X(t + h) - X(t) = 0\} \quad \text{(why?)}$$
$$= f(t, 0)[1 - P\{X(t + h) - X(t) \geqslant 1\}]$$
$$= f(t, 0)[1 - \lambda h + o(h)] \quad \text{(why?)},$$

and thus

$$h^{-1}[f(t + h, 0) - f(t, 0)] = -\lambda f(t, 0) + h^{-1}o(h).$$

Letting $h \to 0$, we get $f'(t, 0) = -\lambda f(t, 0)$, which gives $f(t, 0) = ce^{-\lambda t}$. But $f(0, 0) = P\{X(0) = 0\} = 1$; thus

$$f(t, 0) = e^{-\lambda t}. \tag{4.1.4}$$

Next, for $n \geqslant 1$,

$$f(t + h, n) = P\{X(t + h) = n\}$$

$$= P\left\{ \bigcup_{k=0}^{n} \{X(t) = n - k\} \cap \{X(t + h) - X(t) = k\} \right\}$$

$$= P\{X(t) = n, X(t + h) - X(t) = 0\}$$

$$+ P\{X(t) = n - 1, X(t + h) - X(t) = 1\}$$

$$+ \sum_{k=2}^{n} P\{X(t) = n - k, X(t + h) - X(t) = k\}$$

$$= f(t, n)f(h, 0) + f(t, n - 1)f(h, 1) + o(h)$$

$$= (1 - \lambda h)f(t, n) + \lambda h f(t, n - 1) + o(h),$$

from which

$$h^{-1}[f(t + h, n) - f(t, n)] = -\lambda f(t, n) + \lambda f(t, n - 1) + h^{-1}o(h).$$

Letting $h \to 0$ in this expression, we get

$$f'(t, n) = -\lambda f(t, n) + \lambda f(t, n - 1)$$

or, equivalently,

$$\frac{d}{dt}[e^{\lambda t}f(t, n)] = \lambda e^{\lambda t}f(t, n - 1). \tag{4.1.5}$$

Let $n = 1$ in (4.1.5) and use (4.1.4). Then

$$\frac{d}{dt}[e^{\lambda t}f(t, 1)] = \lambda e^{\lambda t}e^{-\lambda t} = \lambda,$$

or

$$f(t, 1) = (\lambda t + c)e^{-\lambda t}.$$

Since $f(0, 1) = 0$, $c = 0$ and $f(t, 1) = \lambda t e^{-\lambda t}$. We claim that $f(t, n) = (e^{-\lambda t}(\lambda t)^n/n!)$. This is true, as seen above, for $n = 1$. Let it also be true

91

for $n = m$. Then from (4.1.5), $(d/dt)[e^{\lambda t}f(t, m + 1)] = \lambda e^{\lambda t}e^{\lambda t}(\lambda t)^{m}/m!$ $= \lambda^{m+1}(t^{m}/m!)$, and by integration, $e^{\lambda t}f(t, m + 1) = c + (\lambda t)^{m+1}/(m + 1)!$. Since $f(0, m + 1) = 0$, $c = 0$, and we obtain our claim. This completes the proof. $\qquad\square$

4.2. Basic Properties of Poisson Processes

In this section we find the distributions of several time characteristics associated with a Poisson process $X = \{X(t), t \geqslant 0\}$. These distributions are of practical importance. Let $X(t)$, $t \geqslant 0$, denote the number of spike discharges, up to time t, modeled to a first approximation as a Poisson process. Consider the following diagrams:

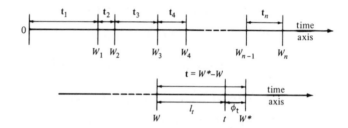

Here W_1, W_2, $\ldots$, denote the occurrence times of the first, second, $\ldots$, spike discharges, respectively, and $\mathbf{t}_n$, $n \geqslant 1$, denotes the time elapsed between the $(n - 1)$th and nth discharges. In general, W_n (resp. $\mathbf{t}_n$), $n \geqslant 1$, is called the nth *occurrence time* or the *waiting time* for the nth event (resp. nth *interarrival time*). *Waiting time* is the terminology used most often (especially in queuing theory) for the time characteristics W_n, $n \geqslant 1$, but we prefer to call them *occurrence times*. Let t be an arbitrarily fixed time with the constraint that $t \neq W_n$ for every $n \geqslant 1$. Then ϕ_t denotes the time elapsed beginning at t until the next occurrence of a spike discharge. Similarly, l_t denotes the time elapsed since the last spike discharge up to time t.

We show below that the interarrival times $\mathbf{t}_n$, $n \geqslant 1$, form a sequence of IID exponential RVs. This is a characteristic property of Poisson processes. The importance of this result lies in the fact that it provides a method of testing the hypothesis that a sequence of events randomly occurring in time are events of Poisson type. This property is useful in practice for simulating Poisson processes in Monte Carlo studies of situations that are too complicated for a direct analytical investigation. Here one proceeds as follows. Using standard algorithms, generate a sequence $\{U_n\}$ of IID RVs uniformly distributed on $[0, 1]$. Transform these variables into the sequence $\{t_n\}$ of IID exponential RVs by the transformation $t_n = -\lambda^{-1} \ln(U_n)$. Assigning these t_n values as the successive interarrival times, one can simulate a Poisson process.

Recall that a continuous RV X with density function given by

$$f(x) = \begin{cases} \lambda e^{-\lambda x}, & x \geq 0 \\ 0, & x < 0 \end{cases}$$

is called an *exponential* RV *with parameter* $\lambda > 0$. An important property of an exponential RV is that it is memoryless. A RV X is said to be *memoryless* if

$$P\{X > s + t \,|\, X > t\} = P\{X > s\} \qquad \text{for all } s, t \geq 0.$$

To understand this property, let X denote the lifetime of a light bulb. If the bulb is burning at time t, the distribution of the remainder of its lifetime is the same as the original lifetime distribution (i.e., the light bulb does not remember that it has already been in use for a time t). From the definitive assumptions that a Poisson process has stationary and independent increments, it follows that the process starts from scratch at any point in time; that is, a Poisson process has the Markov property and has no memory. In the light of this property it is no surprise that the interarrival times are exponentially distributed.

Theorem 4.2.1. *Every stochastic process $X(t)$, $t \geq 0$, with independent increments, and hence a Poisson process, has the Markov property; that is, for all $x \in R$ and $0 \leq t_0 < t_1 < \cdots < t_n < \infty$, we have*

$$P\{X(t_n) < x \,|\, X(t_i), 0 \leq i < n\} = P\{X(t_n) < x \,|\, X(t_{n-1})\}. \qquad (4.2.1)$$

A SKETCH OF THE PROOF. First observe that the collection of all events generated (i.e., the σ-algebra generated) by the RVs $\{X(t_i), 0 \leq i \leq n - 1\}$ coincides with the collection of those events generated by the increments $X(t_0) - X(0)$, $X(t_1) - X(t_0)$, $\ldots, X(t_{n-1}) - X(t_{n-2})$. Let $\mathcal{B}$ denote this collection (σ-algebra), and $B \in \mathcal{B}$. Then it is not difficult to see that

$$E[I_B P\{X(t_n) < x \,|\, X(t_{n-1})\}] = P\{B \cap \{X(t_n) < x\}\}, \qquad (4.2.2)$$

which is equivalent to (4.2.1). This proves the theorem. $\qquad \square$

Theorem 4.2.2. *Let $X = \{X(t), t \geq 0\}$ be a Poisson process with intensity parameter λ, and let $\{t_n, n \geq 1\}$ be the corresponding sequence of successive interarrival times. Then the RVs t_n, $n \geq 1$, are IID obeying an exponential density with mean λ^{-1}.*

PROOF. First we find the probability law of t_1. The event $\{t_1 > \tau\}$ occurs if and only if no Poisson event has occurred in the interval $[0, \tau]$. Hence for $\tau \geqslant 0$,

$$P\{t_1 > \tau\} = P\{X(\tau) = 0\} = e^{-\lambda \tau},$$

so that t_1 is exponentially distributed with mean $1/\lambda$. Next

$$P\{t_2 > \tau\} = E\left[E\left[I_{\{t_2 > \tau\}}\Big| t_1 = \sigma\right]\right] = E[P\{t_2 > \tau | t_1 = \sigma\}].$$

But

$$
\begin{aligned}
P\{t_2 > \tau | t_1 = \sigma\} &= P\{\text{no events in } (\sigma, \sigma + \tau] | t_1 = \sigma\} \\
&= P\{X(\sigma + \tau) - X(\sigma) = 0\} = P\{X(\tau) - X(0) = 0\} \\
&= e^{-\lambda t},
\end{aligned}
$$

since X has independent and stationary increments. In general, for $n > 1$ and $\tau, \sigma_1, \ldots, \sigma_{n-1} \geqslant 0$,

$$
\begin{aligned}
&P\{t_n > \tau | t_1 = \sigma_1, \ldots, t_{n-1} = \sigma_{n-1}\} \\
&= P\{X(\tau + \sigma_1 + \cdots + \sigma_{n-1}) - X(\sigma_1 + \cdots + \sigma_{n-1}) = 0\} \\
&= P\{X(\tau) = 0\} = e^{-\lambda t}.
\end{aligned}
$$

This completes the proof. $\qquad\qquad\square$

Theorem 4.2.3. *Let $\{W_n\}$, $n \geqslant 1$, be the sequence of successive occurrence times ($=$ waiting times) associated with a Poisson process $X(t)$, $t \geqslant 0$. Then the probability density $f_{W_n}(t)$ of the nth occurrence time W_n, $(n \geqslant 1)$, is the Γ-density*

$$f_{W_n}(t) = \lambda e^{-\lambda t} \frac{(\lambda t)^{n-1}}{(n-1)!}, \qquad t \geqslant 0. \tag{4.2.3}$$

PROOF. We need the probability of the event $\{W_n \leqslant t\}$ to obtain the distribution function F_{W_n} of W_n. But the event $\{W_n \leqslant t\}$ occurs if and only if the event $\{X(t) \geqslant n\}$ occurs. Hence

$$F_{W_n}(t) = P\{W_n \leqslant t\} = P\{X(t) \geqslant n\} = \sum_{m \geqslant n} e^{-\lambda t} \frac{(\lambda t)^m}{m!}.$$

Differentiating this with respect to t, we get

$$f_{W_n}(t) = \sum_{m \geqslant n} \lambda e^{-\lambda t} \frac{(\lambda t)^{m-1}}{(m-1)!} - \sum_{m \geqslant n} \lambda e^{-\lambda t} \frac{(\lambda t)^m}{m!}$$

$$= \lambda e^{-\lambda t} \frac{(\lambda t)^{n-1}}{(n-1)!},$$

and this completes the proof.   □

Example 4.2.4 Consider a mechanical device in which "shocks" occur according to a Poisson process and that fails when a total of K shocks occur. Find the density function for the lifetime T of the device.

This problem is simply a rewording of Theorem 4.2.3, and hence the required density is

$$f_T(t) = \lambda e^{-t} \frac{(\lambda t)^{K-1}}{(K-1)!}.$$

Theorem 4.2.5. *The distribution F_ϕ of ϕ_t is independent of t and is given by*

$$F_\phi(s) = P\{\phi_t \leqslant s\} = 1 - e^{-\lambda s}, \qquad s \geqslant 0. \qquad (4.2.4)$$

PROOF. Define $g(s) = P\{\phi_t > s\}$. Let τ be the first time point, after time t, at which a Poisson event occurs. Then the two events $\{\phi_t > s\}$ and $\{\tau > t + s\}$ are equivalent. Therefore, for $h \downarrow 0$,

$$g(s + h) = P\{\phi_t > s + h\} = P\{\tau > t + s + h\}$$

$$= P\{\tau > t + s \text{ and no event in } [t + s, t + s + h]\}$$

$$= P\{\phi_t > s\} P\{X(t + s + h) - X(t + s) = 0\}$$

$$= g(s)[1 - \lambda h + o(h)], \qquad (4.2.5)$$

where we obtain the third equality due to the fact that the event $\{\phi_t > s\}$ $= \{\phi_t \leqslant s\}^c$ refers to what happens during or prior to time $t + s$ and is independent of the increment $X(t + s + h) - X(t + s)$. From (4.2.5) we get

$$h^{-1}[g(s + h) - g(s)] = -\lambda g(s) + \frac{o(h)}{h}.$$

Letting $h \to 0$, we get the initial-value problem

$$g'(s) = -\lambda g(s), \qquad g(0) = P\{\phi_t > 0\} = 1.$$

This yields $\ln g(s) = -\lambda s + c$, or $g(s) = Ke^{-\lambda s}$, and by $g(0) = 1$,

$$g(s) = e^{-\lambda s} \quad \text{or} \quad F_\phi(s) = P\{\phi_t \leqslant s\} = 1 - e^{-\lambda s}, \quad s \geqslant 0. \quad \square$$

Theorem 4.2.6. *The distribution function of l_t has an atom at t; that is,* $P\{l_t = t\} > 0$, *and* $P\{l_t \leqslant s\} = 1 - e^{-\lambda s}, 0 \leqslant s < t.$

PROOF. If no event has occurred in $[0, t)$, then:

$$P\{l_t = t\} = P\{\mathbf{t}_1 > t\} = e^{-\lambda t} > 0.$$

Next, if at least one event has occurred in $[0, t)$, then

$$P\{l_t > s\} = P\{X(t) - X(t - s) = 0\} = e^{-\lambda s}.$$

Hence the theorem holds. $\qquad\qquad\square$

Theorem 4.2.7 shows how the uniform distributions are associated with a Poisson process, and this explains why Poisson processes are often called "random" or "completely random" processes. First consider the situation where it is given that exactly one Poisson event has occurred by time τ and we have to determine the distribution of time $\mathbf{t}$ at which the event occurred. Since the Poisson process possesses independent and stationary increments, it is intuitively clear that each interval in $[0, \tau]$ of equal length should have the same probability of containing the event, so that $\mathbf{t}$ is uniformly distributed in $[0, \tau]$. More precisely, for $s \leqslant t$,

$$
\begin{aligned}
P\{\mathbf{t} < s | X(t) = 1\} &= [P\{X(t) = 1\}]^{-1} P\{\mathbf{t} < s, X(t) = 1\} \\
&= [\lambda t e^{-\lambda t}]^{-1} P\{X(s) = 1, X(t) - X(s) = 0\} \\
&= [\lambda t e^{-\lambda t}]^{-1} \lambda s e^{-\lambda s} e^{-\lambda(t-s)} = \frac{s}{t}.
\end{aligned}
$$

In general we have the following theorem.

Theorem 4.2.7. *Let $X(t), t \geqslant 0$ be a Poisson process. If it is given that exactly n Poisson events have occurred in $[0, \tau]$, then the n successive occurrence times $W_1 < W_2 < \cdots < W_n$ are distributed according to the distribution of the order statistics $U_{(1)}, \ldots, U_{(n)}$ of n independent RVs $U_1, \ldots, U_n$ uniformly distributed on $[0, \tau]$, where $U_{(1)}(\omega) = \min\{U_1(\omega), \ldots, U_n(\omega)\}$, $U_{(k)}(\omega) = k$th smallest in $\{U_1(\omega), \ldots, U_n(\omega)\}, 1 < k < n - 1$, and $U_{(n)}(\omega) = \max\{U_1(\omega), \ldots, U_n(\omega)\}$.*

PROOF. The joint density function $g_n(u_1, \ldots, u_n)$ of $(U_1, \ldots, U_n)$ is given by

$$g_n(u_1, \ldots, u_n) = \begin{cases} \tau^{-n} & \text{if } 0 \leqslant u_1, \ldots, u_n \leqslant \tau \\ 0 & \text{otherwise} \end{cases},$$

and the corresponding joint density $g_{(n)}(x_1, \ldots, x_n)$ of $(U_{(1)}, \ldots, U_{(n)})$ is given by

$$g_{(n)}(x_1, \ldots, x_n) = \begin{cases} n! \, \tau^{-n} & \text{if } 0 \leqslant x_1 \leqslant \cdots \leqslant x_n \leqslant \tau \\ 0 & \text{otherwise} \end{cases}.$$

Now let it be given that exactly n Poisson events have occurred in $[0, \tau]$. Also let $\{[x_k, x_k + h_k], 1 \leqslant k \leqslant n\}$ be a sequence of n nonoverlapping intervals in $[0, \tau]$. Then

$$P\{x_1 \leqslant W_1 \leqslant x_1 + h_1, \ldots, x_n \leqslant W_n \leqslant x_n + h_n | X(\tau) = n\}$$
$$= [e^{-\lambda \tau}(\lambda \tau)^n / n!]^{-1} \lambda h_1 e^{-\lambda h_1} \cdots \lambda h_n \exp[-\lambda(\tau - h_1 - \cdots - h_n)]$$
$$= n! \, \tau^{-n} h_1 \cdots h_n. \tag{4.2.6}$$

But the conditional probability in (4.2.6) is approximately equal to $g_{(n)}(x_1, \ldots, x_n) h_1 \cdots h_n$. This observation completes the proof. $\qquad \square$

Examples 4.2.8

EXAMPLE 1. If it is given that n Poisson events have occurred in $[0, \tau]$ and $0 < s < \tau$, find $P\{X(s) = k | X(\tau) = n\}$ for $0 < k < n$.

$$P\{X(s) = k | X(\tau) = n\} = [P\{X(\tau) = n\}]^{-1} P\{X(s) = k, X(\tau) = n\}$$
$$= [e^{-\lambda \tau}(\lambda \tau)^n / n!]^{-1} P\{X(s) = k, X(\tau) - X(s) = n - k\}$$
$$= n! \, (\lambda \tau)^{-n} e^{\lambda \tau} e^{-\lambda s} \frac{(\lambda s)^k}{k!} e^{-\lambda(\tau - s)} \frac{[\lambda(\tau - s)]^{n-k}}{(n - k)!}$$
$$= \binom{n}{k} \left(\frac{s}{\tau}\right)^k \left(1 - \frac{s}{\tau}\right)^{n-k},$$

which is a binomial distribution with parameters n and s/τ.

EXAMPLE 2. If it is given that n Poisson events have occurred by time τ, find the density function of the time T_k of the occurrence of the kth event, $k < n$.

Set $H(s|n) = P\{T_k \leqslant s | X(\tau) = n\}$. We want to find

$$h(s|n) = \frac{d}{ds} H(s|n) = \lim_{h \to 0} h^{-1} P\{s \leqslant T_k \leqslant s + h | X(\tau) = n\}.$$

Now, as $h \to 0$,

$P\{s \leqslant T_k \leqslant s + h | X(\tau) = n\}$

$\quad = [P\{X(\tau) = n\}]^{-1} P\{s \leqslant T_k \leqslant s + h, X(\tau) = n\}$

$\quad = e^{\lambda\tau}(\lambda\tau)^{-n} n! \, P\{s \leqslant T_k \leqslant s + h, X(\tau) - X(s + h) = n - k\}$

$\quad = n! \, e^{\lambda\tau}(\lambda\tau)^{-n} P\{s \leqslant T_k \leqslant s + h\} P\{X(\tau) - X(s + h) = n - k\}. \quad \textbf{(4.2.7)}$

Dividing both sides of (4.2.7) by h and letting $h \to 0$, we get

$$h(s|n) = n! \, e^{\lambda\tau}(\lambda\tau)^{-n}[\text{absolute density of } T_k] P\{X(\tau) - X(s) = n - k\}$$

$$= n! \, e^{\lambda\tau}(\lambda\tau)^{-n} \lambda \frac{(\lambda s)^{k-1}}{(k-1)!} e^{-\lambda s} e^{-\lambda(\tau-s)} \frac{[\lambda(\tau - s)]^{n-k}}{(n-k)!},$$

from Theorem 4.2.3,

$$= \frac{n!}{(k-1)!(n-k)!} \frac{s^{k-1}}{\tau^k} \left(1 - \frac{s}{\tau}\right)^{n-k}.$$

Hence, given that $X(\tau) = n$, then $T_k (= W_k)$ follows Beta distribution.

EXAMPLE 3. Let $X(t)$ and $Y(t)$ be two independent Poisson processes with rates λ and μ, respectively. Let W and W^* be the times of two successive occurrences of Poisson events of the process $X(t)$; that is, $W < W^*$, $X(t) = X(W)$ for $W \leqslant t < W^*$, and $X(W^*) = X(W) + 1$. Define $N = Y(W^*) - Y(W)$. Find the probability distribution of N.

Set $t = W^* - W$. Then t is an interarrival time and $P\{t \leqslant s\} = 1 - e^{-\lambda s}$, by Theorem 4.2.2. Now

$$P\{N = k\} = P\{Y(W^*) - Y(W) = k\}$$

$$= E[P\{Y(W^*) - Y(W) = k | t = s\}]$$

$$= \int_0^\infty e^{-\mu s} \frac{(\mu s)^k}{k!} \lambda e^{-\lambda s} \, ds = \frac{\lambda\mu^k}{k!} \int_0^\infty s^k e^{-(\lambda+\mu)s} \, ds$$

$$= \frac{\lambda}{\lambda + \mu} \left(\frac{\mu}{\lambda + \mu}\right)^k.$$

4.3. Some Generalizations of the Poisson Process

Definition 4.3.1. A stochastic process $X: R_+ \times \Omega \to S = \{0, 1, 2, \ldots\}$ is called a *nonhomogeneous Poisson process* with intensity function $\lambda(t)$, $t \geqslant 0$, if: (1) $X(0) \equiv 0$, (2) X possesses independent increments, (3) $P\{X(t + h) - X(t) = 1\} = \lambda(t)h + o(h)$, and (4) $P\{X(t + h) - X(t) \geqslant 2\} = o(h)$.

Set $\Lambda(t) = \int_0^t \lambda(s)\, ds$; $\Lambda(t)$ is called the *mean value function* of the Poisson process $X(t)$.

Let $Y(t)$, $t \geqslant 0$, be a homogeneous Poisson process with intensity $\lambda(t) \equiv 1$. Define $X(t) = Y(\Lambda(t))$, $t \geqslant 0$. Then $X(t)$ is a nonhomogeneous Poisson process. Now let $X(t)$ be a nonhomogeneous Poisson process with intensity function $\lambda(t)$. The mean-value function $\Lambda(t)$ is continuous and nondecreasing. Let us define the inverse Λ^{-1}.

$$\Lambda^{-1}(u) = \min\{t: \Lambda(t) \geqslant u \text{ for } u > 0\}.$$

Now define a process $Y(t)$ by $Y(t) = X(\Lambda^{-1}(t))$, $t \geqslant 0$. Then $Y(t)$ is a homogeneous Poisson process with unit intensity. One important application of this observation is the simulation of a nonhomogeneous Poisson process. First simulate a homogeneous process as indicated in the second paragraph of Section 4.2, and then rescale the time to simulate a nonhomogeneous process.

In the examples cited in Section 4.1 we did not give explicitly the intensity functions of the processes. Actually, all those processes are better approximated by nonhomogeneous Poisson processes. Here we present some examples of possible intensity functions.

Examples 4.3.2

EXAMPLE 1. *Radioactive Decay.* Consider the emission of γ-photons by a radioactive source. This can be modeled to a close approximation by a nonhomogeneous Poisson process with intensity function $\lambda(t)$ given by $\lambda(t) = \alpha \exp[-\beta t]$, $t \geqslant 0$, $\alpha, \beta > 0$, where α is a parameter depending on the amount of the radioactive material and β^{-1} is the mean life of the source. The intensity functions of this form arise in nuclear medicine, nuclear physics, and geochronology (Evans 1965).

EXAMPLE 2. *Electron Emission in Optical Detection.* Emission of a stream of photoelectrons results when an optical field is incident on a photomultiplier. To a good approximation this can be modeled by a nonhomogeneous Poisson process. Let the incident light be characterized by the scalar electric field $e(t, x) = \sqrt{2}\, \mathrm{Re}\,\{E(t, x)\exp(2\pi i\phi t)\}$ at time t and position x. Here $E(t, x)$ is a complex space–time envelope that is slowly varying in comparison to the optical frequency ϕ. Noting that $|E(t, x)|^2$ gives the light intensity, the rate $\lambda(t)$ of electron emission is given by $\lambda(t) = (q/h\phi)\int_S |E(t, x)|^2\, dx + \lambda_0$, where q is the probability that a quantum of light energy is converted into an electron, h is Planck's constant, λ_0 is the rate of generation of extraneous electrons in the photodetector, and the integral is taken over the sensitive surface S of the detector.

EXAMPLE 3. *Neuron Spike Activity.* As seen in Example 4.1.4 (3), the spike discharges in an auditory nerve can be modeled by a Poisson process, a nonhomogeneous process for a better approximation. Here the intensity function can be taken in the form, (Siebert 1970)

$$\lambda(t) = \alpha \exp\{\beta \cos(2\pi\phi t + \gamma)\},$$

where the parameters α, β, and γ are connected with the physiological mechanisms involved in converting the pressure stimulus into electrical nerve activity and ϕ is the frequency of the applied pressure.

EXAMPLE 4. *Nuclear Medicine.* We have seen in Example 4.1.4 (5) that the emission of light pulses in the scintillation crystal can be modeled by a Poisson process. A form of the intensity function $\lambda(t)$ that is frequently used is

$$\lambda(t) = \alpha_0 + \sum_{k=1}^{n} \alpha_k \exp[-\alpha_{n+k}\, t], \qquad t \geqslant 0,\ \alpha_k > 0,\ 0 \leqslant k \leqslant 2n,$$

where the parameters α_k arise in clinical studies and $\lambda(t)$ describes the physiological transport phenomena. Note that the intensity decreases, as a function of time, which is partially due to the removal of the radioactive source from the field of view of the scintillation detector by blood flow (Sheppard 1962).

Theorem 4.3.3. *Let $X(t)$, $t \geqslant 0$, be a nonhomogeneous Poisson process with mean-value function $\Lambda(t)$. Then*

$$P\{X(t + s) - X(t) = n\} = e^{-[\Lambda(t+s)-\Lambda(t)]}\frac{[\Lambda(t + s) - \Lambda(t)]^n}{n!}, \qquad n \geqslant 0.$$

PROOF. Exercise (mimic the proof of Theorem 4.1.5). □

Definition 4.3.4. A stochastic process $\{Y(t), t \geqslant 0\}$ is called a *compound Poisson process* if it admits a representation of the form

$$Y(t) = \sum_{n=1}^{X(t)} \xi_n,$$

where $X(t)$ is a Poisson process and $\{\xi_n, n \geq 1\}$ is a sequence of IID RVs such that $\{X(t), t \geq 0\}$ and $\{\xi_n, n \geq 1\}$ are independent processes.

Examples 4.3.5

EXAMPLE 1. *Seismic Events.* Earthquakes and other seismic events can be observed on a worldwide basis with an array of seismometers. Let us observe the occurrences of earthquakes as a Poisson process $X(t)$, with t denoting the time. Each earthquake results in some damage ξ. Let us assume that the damages ξ_n caused by different earthquakes are independent of each other. Then $\sum_{n=1}^{X(t)} \xi_n$ gives the total damage caused by all earthquakes during the period $[0, t]$. Clearly, $Y(t)$ is a compound Poisson process.

EXAMPLE 2. *Total Claims on a Life-Insurance Company.* Let $W_1, W_2, \ldots$ denote the occurrence times of the death of the policyholders of a certain life-insurance company. Treating these times as the arrival times of insurance claims, the number of deaths can be treated as a Poisson process $X(t)$. Let ξ_n denote the amount that the policyholder dying at time W_n carries. The $Y(t)$ defined by $Y(t) = \sum_{n=1}^{X(t)} \xi_n$ and denoting the total amount of claims that the insurance company will have to pay during the time period $[0, t]$ is a compound Poisson process.

EXAMPLE 3. *Prey–Predator System.* Consider a parasite system preying on a host population. Each encounter of the predators with the prey results in the depletion of prey population by a certain amount ξ_n, and the ingestion of food results in an increase η_n in the predator system. Let $X(t)$ denote the Poisson process modeling the number of encounters. Define $Y(t) = \sum_{n=1}^{X(t)} \xi_n$ and $Z(t) = \sum_{n=1}^{X(t)} \eta_n$. Then $Y(t)$ and $Z(t)$ are compound Poisson processes denoting the cumulative depletion of the prey and cumulative increase in the predator population during $[0, t]$, respectively.

EXAMPLE 4. *Optical Communication System.* In optical communication system, a simple device for sensing optical position of the light beam arriving at a receiver consists of a photodetector having a photoemissive surface that is divided into four quadrants. Photoelectron emissions are observed in each quadrant to sense the position of an incident optical field. Let us assume that the intensity of the light falling on the device follows a circularly symmetric Gaussian density γ. Let us also assume that the photoelectron emissions in each quadrant form a Poisson process with intensity $\lambda = k \int_S \gamma$, where S is the sensitive surface of the detector, and that the four quadrants operate independently of each other. Then the photoelectron emissions for the entire sensor can be treated as a compound Poisson process.

Theorem 4.3.6. *Let* $Y(t) = \sum_{n=1}^{X(t)} \xi_n$ *be a compound Poisson process. Then:*

(i) *The process Y has independent increments.*

(ii) *The characteristic function* $\phi_{Y(t)}$ *of* $Y(t)$ *is given by* $\phi_{Y(t)}(u)$ $= \exp\{\lambda t(\phi_\xi(u) - 1)\}$, *where* ϕ_ξ *is the common characteristic function of the RVs* ξ_n *and* λ *is the rate of occurrence of Poisson events.*

(iii) *If* $E[\xi^2] < \infty$, *then* $E[Y(t)] = \lambda t E[\xi]$ *and* $\mathrm{var}(Y(t)) = \lambda t E[\xi^2]$.

PROOF. (i) Let $0 \leqslant t_0 < t_1 < \cdots < t_m$. Then

$$Y(t_0) = \sum_{n=1}^{X(t_0)} \xi_n, \qquad Y(t_k) - Y(t_{k-1}) = \sum_{n=X(t_{k-1})+1}^{X(t_k)} \xi_n, \qquad k = 1, \ldots, m.$$

From the basic assumptions on $\{X(t)\}$ and $\{\xi_n\}$, it is not difficult to see (we omit the details) that $Y(t)$ possesses independent increments.

(ii) $\phi_{Y(t)}(u) = \sum_{n=0}^{\infty} E[e^{iuY(t)} | X(t+s) - X(s) = n] P[X(t+s) - X(s) = n]$

$$= \sum_{n \geqslant 0} [\phi_\xi(u)]^n \frac{e^{-\lambda t}(\lambda t)^n}{n!}$$

$$= e^{-\lambda t} \exp\{\lambda t \phi_\xi(u)\}$$

(iii) Using (ii) one can establish (iii), or proceed directly as follows. Note that $E[Y(t)] = E[E[Y(t)|X(t)]]$. Now

$$E[Y(t)|X(t) = n] = E\left[\sum_{k=1}^{X(t)} \xi_k | X(t) = n\right]$$

$$= E\left[\sum_{k=1}^{n} \xi_k | X(t) = n\right] = E\left[\sum_{k=1}^{n} \xi_k\right] = nE[\xi],$$

where we have used the independence of $\{\xi_n\}$ and $\{X(t)\}$ in the third step and used identical distribution assumption on all ξ_n in the fourth step. Therefore,

$$E[Y(t)] = E[E[Y(t)|X(t)]] = E[X(t)E[\xi]] = \lambda t E[\xi].$$

Next, note that

$$\mathrm{var}(Y(t)) = E[\mathrm{var}(Y(t)|X(t))] + \mathrm{var}(E[Y(t)|X(t)]).$$

Now

$$\text{var}(Y(t)|X(t) = n) = \text{var}\left(\sum_{k=1}^{X(t)} \xi_k \,\middle|\, X(t) = n\right) = \text{var}\left(\sum_{k=1}^{n} \xi_k\right)$$

$$= n \, \text{var}(\xi),$$

and consequently, $\text{var}(Y(t)|X(t)) = X(t)\text{var}(\xi)$. Therefore,

$$\text{var}(Y(t)) = E[X(t)\text{var}(\xi)] + \text{var}[X(t)E[\xi]]$$

$$= \lambda t \, \text{var}(\xi) + \lambda t (E[\xi])^2 = \lambda t[\text{var}(\xi) + (E[\xi])^2]$$

$$= \lambda t E[\xi^2].$$ $\square$

Exercises

1. Consider a Poisson process of spatial distribution of points in R^3 with mean rate $\lambda = 1$. With center at each of these points construct spheres of radius r. Let X denote the number of these spheres that contain the origin. Show that X is a Poisson RV with mean $4\pi r^3/3$.

2. Let $X(t)$ be a Poisson process in R^3 with mean rate λ and let D the distance from the origin to the nearest point of the process. Show that the density function of D is given by $4\pi\lambda r^2 \exp(-4\pi\lambda r^3/3)$.

3. Consider the processes $\{l_t\}$ and $\{\phi_t\}$ defined in Section 4.2 (see the figure there). Show that: (a) l_t and ϕ_t are independent and (b) the distribution F of $l_t + \phi_t$ is given by

$$F(x) = \begin{cases} 0 & \text{for } x < 0 \\ 1 - e^{-\lambda x}(1 + \lambda x) & \text{for } 0 \leqslant x < t \\ 1 - e^{-\lambda x}(1 + \lambda t) & \text{for } t \leqslant x < \infty \end{cases}.$$

4. Customers arrive at a checkout counter at a Poisson mean rate of 12 per hour. What is the probability that the interarrival times is: (a) greater than 10 min, (b) less than 5 min, (c) greater than 5 min, and (d) 2–8 min?

5. Certain nuclear particles arrive at a counter according to a Poisson process with mean rate λ. Each such particle gives rise to a pulse of unit duration and is registered by the counter if no other pulse is present. Show the probability that a particle is counted in the interval $[t, t + 1]$, $t \geqslant 1$, is given by $\lambda e^{-\lambda}$.

6. Let $X(t)$ be a Poisson process on $[0, \infty)$ with mean rate λ, and let T be an exponential RV with mean ν and independent of the process. Let N be the number of Poisson events in $[0, T]$. Show that the probability distribution of N is given by $p_n = \nu\lambda^n/(\lambda + \nu)^{n+1}$, $n \geqslant 0$.

7. Consider a Poisson process $X(t)$ with mean rate λ. With the kth event associate a RV Y_k such that the RVs Y_k are independent and exponentially distributed with mean ν. Let $Z(t) = \Sigma_{k < t} Y_k$. Show that the density function of $Z(t)$ is given by

4. Poisson Processes

$$\sqrt{\frac{\lambda \nu t}{x}}\, e^{-\lambda t - \nu x} I_1(2\sqrt{\lambda \nu t x}\,), \qquad x > 0,$$

where $I_1(\cdot)$ is a Bessel function.

8. For a Poisson process with mean rate λ, let W_1 be the random time of the occurrence of first Poisson event. Show that, for $0 \leqslant s \leqslant t$ and $n \geqslant 1$,

$$P\{W_1 \leqslant s | X(t) = n\} = 1 - (1 - st^{-1})^n.$$

9. Let $X(t)$ be a Poisson process subject to the constraint that following each occurrence of the Poisson event there is a "dead period" of length T in which no event can occur. If $N(t)$ is the number of events occurring in $[0, t)$, show that

$$P\{N(t) < x\} = \sum_{n=0}^{x-1} e^{-\lambda(t - xT)} \frac{[\lambda(t - xT)]^n}{n!}.$$

10. Let $\{X(t)\}$ be a Poisson process with mean rate λ, W the time until the occurrence of the first event, and $N(W/a)$ the number of events in the next W/a units of time. Show that

$$E\left[WN\left(\frac{W}{a}\right)\right] = \frac{2}{\lambda a}, \quad E\left[\left\{WN\left(\frac{W}{a}\right)\right\}^2\right] = \frac{6a + 24}{\lambda^2 a^2}.$$

11. Let $X(t)$ and $Y(t)$ be two independent Poisson processes with mean rates λ and ν, respectively. Let $X(0) = n$, $Y(0) = m$, and $N > m, n$. Show that the probability for the Y process to reach N earlier than the X process can reach N is given by

$$\sum_{k=0}^{N-n-1} \binom{N - m + k - 1}{k} p^k q^{N-m}, \qquad p = \frac{\lambda}{\lambda + \mu}, \; q = 1 - p.$$

12. Consider a telephone exchange with an indefinitely large number of switches. Each call lasts a random amount of time, obeying an exponential distribution with parameter μ. Let the incoming calls form a queue with parameter λ, and let $X(t)$ denote the number of busy lines at time t with $X(0) = 0$. Set $p(t, n) = P\{X(t) = n\}$. Show that:

$$(a) p(t, n) = \frac{1}{n!}\left\{\frac{\lambda}{\mu}(1 - e^{-\mu t})\right\}^n \exp\left\{-\frac{\lambda}{\mu}(1 - e^{-\mu t})\right\},$$

$$(b) \lim_{t \to \infty} p(t, n) = \frac{1}{n!}\left(\frac{\lambda}{\mu}\right)^n e^{-\lambda/\mu}.$$

13. Electrons flow from a heated cathode with rate λ. The flight times for different electrons are independent RVs following a common distribution $F(x)$. Let $X(t)$ denote the number of electrons between the electrodes of the electronic tube at time t. Set $p(t, n) = P\{X(t) = n\}$. Show that $p(t, n) = [\Lambda(t)]^n e^{-\Lambda(t)}/n!$, where $\Lambda(t) = \lambda \int_0^t [1 - F(x)] dx$.

5

Purely Discontinuous Markov Processes

While introducing the Poisson process we stated that such processes form a prototype of more general jump Markov processes. Our purpose now is to abstract and extend the basic characteristics of Poisson processes. Let the time parameter $t \in R_+ = [0, \infty)$ and the state space S be an at most countable set. The processes that we study in this chapter can be described as follows. A physical system starts its evolution from a state $x_0 \in S$. It stays there a random length of time $\mathbf{t}_1$ and then jumps to another state x_1, where it is a possibility that the system stays at x_0 forever. After jumping to x_1, the system remains there for a random length of time $\mathbf{t}_2$ and then jumps to a state $x_2 \neq x_1$. As before, it is possible that $\mathbf{t}_2 = \infty$. The evolution repeats this process. It is easy to perceive of systems that will make an infinite number of jumps in a finite time; such important systems do exist. The time of nth jump is $\mathbf{t}_1 + \cdots + \mathbf{t}_n = W_n$. So as not to allow an infinite number of jumps in a finite time, we have to require that $W_n \to \infty$ with probability one, as $n \to \infty$. The jump stochastic processes satisfying this condition are called *pure-jump processes* or *purely discontinuous processes*. Having described the jump motion of the system, we next have to look for basic mathematical quantities that will succinctly describe the process. The Poisson events occur with intensity λ (in the homogeneous case) or $\lambda(t)$ (in the nonhomogeneous case), and the system moves by unit steps in the forward direction. In extending these ideas to the jump process case, we allow arbitrary jumps and define an intensity function $q(t, x)$ and relative jump probabilities $Q(t, x, y)$. Of course, we also need the transition probabilities $p(s, x, t, y)$ of transition from state x at time s to state y at time t. Now, without further ado let us get to the mathematical framework.

5. Purely Discontinuous Markov Processes

5.1. Kolmogorov–Feller Equations

Let $\{X(t,\omega), t \in R_+, \omega \in \Omega\}$ be a stochastic process with state space $S = R$. By $\mathscr{B}(R)$ we denote the σ-algebra of Borel subsets of the real line R.

Definition 5.1.1. A stochastic process $X: R_+ \times \Omega \to R$ is called a *Markov process* if, for $0 \leqslant t_0 < t_1 < \cdots < t_n < s < t$, any $x_0, \ldots, x_n, x \in R$ and $B \in \mathscr{B}(R)$,

$$P\{X(t) \in B | X(t_k) = x_k, \quad 0 \leqslant k \leqslant n, \quad X(s) = x\}$$
$$= P\{X(t) \in B | X(s) = x\}. \tag{5.1.1}$$

The function $p(\cdot, \cdot; \cdot, \cdot): R_+ \times R \times R_+ \times \mathscr{B}(R) \to [0, 1]$ defined by

$$p(s, x; t, B) = P\{X(t) \in B | X(s) = x\} \tag{5.1.2}$$

is called the *transition function* of the process $X(t)$. Markov processes are also called the *processes without memory* or *after effect*.

It is clear that $\int_R p(s, x; t, dy) = 1$. Let the function $p(s, x; t, B)$ be continuous in s, t and x. Let $s < t < u$. Since the system passes from state x at time s to an intermediate state at time t and from y at time t to B at time u, we obtain the Chapman–Kolmogorov relation

$$p(s, x; u, B) = \int p(s, x; t, dy) p(t, y; u, B). \tag{5.1.3}$$

From the continuity assumption on p we obtain

$$\lim_{t \to s} p(s, x; t, B) = I_B(x) = \lim_{s \to t} p(s, x; t, B). \tag{5.1.4}$$

Definition 5.1.2. A function $q(t, x)$, $t \in R_+$, $x \in R$, is called the *intensity function* of the Markov process if $q(t, x)\, dt + o(dt)$ is the probability that $X(t)$ will undergo a random change in the infinitesimal interval $(t, t + dt)$ when $X(t) = x$. The conditional probability $Q(t, x, B)$ of $X(t)$ taking a value in B at time $(t + dt)$ given that $X(t) = x$ and that X has undergone a change in $(t, t + dt)$ is called the *relative transition function* of the Markov process $X(t)$.

It is clear that $q(t, x) \geqslant 0$ for all t and x, $0 \leqslant Q(t, x, B) \leqslant 1$ for all t, x, and B, $\int Q(t, x, dy) = 1$ for all t and x, and $Q(t, x, B) = 0$ if $x \in B$.

Definition 5.1.3. A Markov process $\{X(t)\}$ is called a *purely discontinuous Markov process* if in an arbitrary time interval $(t, t + dt)$ the system $X(t)$

undergoes a change with probability $q(t, x) dt + o(dt)$, remains unchanged with probability $1 - q(t, x) dt + o(dt)$, or undergoes more than one change with probability $o(dt)$.

Let $X(t)$ be a purely discontinuous Markov process with transition function $p(s, x; t, B)$, intensity function $q(t, x)$, and relative transition function $Q(t, x, B)$. Then it is clear that

$$p(s, x; t, B) = [1 - q(s, x)(t - s)]I_B(x) + (t - s)q(s, x)Q(s, x, B) + o(t - s).$$
(5.1.5)

Theorem 5.1.4. (Feller 1940). *The transition probability function $p(s, x; t, B)$ of a purely discontinuous Markov process $X(t)$ satisfies the following integrodifferential equations*:

(i) *Kolmogorov–Feller backward equation*:

$$\frac{\partial p(s, x; t, B)}{\partial s} = q(s, x)[\, p(s, x; t, B)$$
$$- \int p(s, y; t, B)Q(s, x, dy)].$$
(5.1.6)

(ii) *Kolmogorov–Feller forward equation*:

$$\frac{\partial p(s, x; t, B)}{\partial t} = - \int_B q(t, y)p(s, x; t, dy)$$
$$+ \int q(t, y)Q(t, y, B)p(s, x; t, dy).$$
(5.1.7)

PROOF. From Chapman–Kolmogorov equation (5.1.3) and relation (5.1.5), we obtain:

$$p(s, x; t, B) = \int p(s, x; s + \Delta s, dy)p(s + \Delta s, y; t, B)$$
$$= \int p(s + \Delta s, y; t, B)[1 - q(s, x)\Delta s + o(\Delta s)]I_{dy}(x)$$
$$+ \int p(s + \Delta s, y; t, B)[q(s, x)\Delta s + o(\Delta s)]Q(s, x, dy)$$
$$= [1 - q(s, x)\Delta s]p(s + \Delta s, x; t, B)$$
$$+ \Delta s q(s, x) \int p(s + \Delta s, y; t, B)Q(s, x, dy) + o(\Delta s).$$

This gives

5. Purely Discontinuous Markov Processes

$$\frac{1}{\Delta s}[p(s + \Delta s, x; t, B) - p(s, x; t, B)]$$

$$= q(s, x)p(s + \Delta s, x; t, B) - q(s, x) \int p(s + \Delta s, y; t, B)Q(s, x, dy)$$

$$+ \frac{o(\Delta s)}{\Delta s}.$$

Letting $\Delta s \to 0$ in this relation, we get the backward equation (5.1.6). Next

$$p(s, x; t + \Delta t, B) = \int p(s, x; t, dy)p(t, y; t + \Delta t, B)$$

$$= \int [1 - q(t, y)\Delta t]I_B(y)p(s, x; t, dy)$$

$$+ \Delta t \int q(t, y)Q(t, y, B)p(s, x; t, dy) + o(\Delta t)$$

$$= p(s, x; t, B) - \Delta t \int_B q(t, y)p(s, x; t, dy)$$

$$+ \Delta t \int q(t, y)Q(t, y, B)p(s, x; t, dy) + o(\Delta t).$$

Subtracting the term $p(s, x; t, B)$ on both sides, dividing by Δt throughout, and taking limit as $\Delta t \to 0$, we obtain the forward equation (5.1.7). □

Equation (5.1.6) [resp. equation (5.1.7)] is called the *backward equation* (resp. the *forward equation*) since it involves the differentiation with respect to the earlier time s (resp. the later time t). Now we specialize the state space S to an at most countable set. We take the power set $\mathbb{S}$ of S in place of $\mathcal{B}(R)$. To fix our ideas, let us take $S = \{0, 1, 2, \ldots\}$ or $S = \{\cdots, -1, 0, 1, \ldots\}$. Now the transition functions p and Q take the forms $p(s, x; t, y)$ and $Q(t, x, y)$, and equations (5.1.3), (5.1.5)–(5.1.7) take the forms

$$p(s, x; u, y) = \sum_{z \in S} p(s, x; t, z)p(t, z; u, y), \tag{5.1.8}$$

$$p(s, x; t, y) = [1 - q(s, x)(t - s)]\delta_{xy}$$
$$+ (t - s)q(s, x)Q(s, x, y) + o(t - s), \tag{5.1.9}$$

$$\frac{\partial}{\partial s}p(s, x; t, y) = q(s, x)\{p(s, x; t, y)\}$$
$$- \sum_{z \in S} Q(s, x, z)p(s, z; t, y) \tag{5.1.10}$$

$$\frac{\partial}{\partial t}p(s, x; t, y) = -q(t, y)p(s, x; t, y)$$
$$+ \sum_{z \in S} q(t, z)Q(t, z, y)p(s, x; t, z). \tag{5.1.11}$$

In the following sections we mainly study the temporally homogeneous case so that $p(s, x; t, y) = p(t - s, x, y)$, which gives the probability that the system moves from state x to state y in a time period of length $(t - s)$. Set

$$a(t, x, y) = \begin{cases} -q(t, x), & x = y \\ q(t, x)Q(t, x, y), & x \neq y. \end{cases} \tag{5.1.12}$$

The quantities $a(t, x, y)$ are called *infinitesimal parameters*, and the matrix $A(t) = [a(t, x, y)]$, $x, y \in S$, is called the *infinitesimal generator* of $X(t)$, $t \geqslant 0$. Most of the time we restrict ourselves to the time-independent functions $q(x) \equiv q(t, x)$ and $Q(x, y) \equiv Q(t, x, y)$. Kolmogorov's backward and forward equations become

$$\frac{\partial}{\partial t} p(t, x, y) = \sum_z a(x, z) p(t, z, y), \qquad t \geqslant 0, \tag{5.1.13}$$

$$\frac{\partial}{\partial t} p(t, x, y) = \sum_z p(t, x, z) a(z, y), \qquad t \geqslant 0, \tag{5.1.14}$$

respectively.

Examples 5.1.5

EXAMPLE 1. Find the probability $p_0(s, x, t)$ that a system found in state x at time s will undergo no change until time t.

The absence of any change in the position of the system during the period (s, t) can occur in two mutually exclusive ways. Either the system undergoes no change until time t and makes a transition in the time interval $(t, t + \Delta t)$, or there is no transition until time $t + \Delta t$. Therefore

$$p_0(s, x, t) = p_0(s, x, t)[1 - p_0(t, x, t + \Delta t)] + p_0(s, x, t + \Delta t)$$

$$= p_0(s, x, t)[1 - \{1 - q(t, x) \Delta t\}] + p_0(s, x, t + \Delta t) + o(\Delta t)$$

so that

$$\frac{p_0(s, x, t + \Delta t) - p_0(s, x, t)}{\Delta t} = -p_0(s, x, t)q(t, x) + \frac{o(\Delta t)}{\Delta t}.$$

Letting $\Delta t \to 0$ in this expression, we get

$$\frac{\partial}{\partial t} p_0(s, x, t) = -p_0(s, x, t)q(t, x).$$

Solving this equation, we get $p_0(s, x, t) = \alpha \exp\{-\int_s^t q(u, x) du\}$. Noting that $p_0(t, x, t) = 1$, we obtain $\alpha = 1$ and hence

$$p_0(s, x, t) = \exp\left\{ -\int_s^t q(u, x) du \right\}. \tag{5.1.15}$$

109

EXAMPLE 2. Let $p_n(s, x, t)$, $n \geq 1$, denote the probability that the system found in state x at time s undergoes n transition in the time interval (s, t). Find a recurrence relation connecting the probabilities $p_n(s, x, t)$, $n \geq 1$.

Observe that n transitions can occur in the time interval (s, t) as follows. First, for $s < u < t$, there is no change until time u [occurring with probability $p_0(s, x, u)$] and a transition occurs in the time interval $(u, u + \Delta u)$ [with probability $p_1(u, x, u + \Delta u) = q(u, x)\Delta u + o(\Delta u)$]. The jump, whenever it occurs, takes the system to the state interval $(y, y + \Delta y) \subset R$ with probability $Q(u, x, y + \Delta y) - Q(u, x, y)$. During the remainder of the time $(u + \Delta u, t)$ the system undergoes $(n - 1)$ transitions with probability $p_{n-1}(u + \Delta u, y, t)$. Noting that u and y vary continuously on (s, t) and R, respectively, we then obtain that

$$
\begin{aligned}
p_n(s, x, t) &= \int_s^t \int_R p_0(s, x, u)q(u, x)p_{n-1}(u, y, t)Q(u, x, dy)\, du \\
&= \int_s^t p_0(s, x, u)q(u, x) \int_R p_{n-1}(u, y, t)Q(u, x, dy)\, du,
\end{aligned}
\tag{5.1.16}
$$

which is the recurrence relation we are looking for.

EXAMPLE 3. Let $X(t)$, $t \geq 0$, be a homogeneous purely discontinuous Markov process with time-independent intensity $q(x)$ and relative transition $Q(x, y)$. Show that the transition probability $p(t, x, y)$ satisfies the equation

$$
p(t, x, y) = e^{-q(x)t}\delta_{xy} + \int_0^t q(x)e^{-q(x)s}\left\{ \sum_{z \neq x} Q(x, z)p(t - s, z, y) \right\} ds.
\tag{5.1.17}
$$

Let $\mathbf{t}$ denote the random holding time until the first jump. Noting that the intensity $q(x)$ is independent of time, it follows from Example 1 that

$$
P_x\{\mathbf{t} \geq t\} = P_x\{\text{no jump until } t\} = e^{-q(x)t}.
\tag{5.1.18}
$$

That is, $\mathbf{t}$ is exponentially distributed. Now

$$
\begin{aligned}
p(t, x, y) &= P_x\{X(t) = y\} \\
&= P_x\{X(t) = y, \ \mathbf{t} > t\} + P_x\{X(t) = y, \ \mathbf{t} \leq t\} \\
&= e^{-q(x)t}\delta_{xy} + P_x\{X(t) = y, \ \mathbf{t} \leq t\}.
\end{aligned}
\tag{5.1.19}
$$

But

$$
P_x\{\mathbf{t} \leq t, X(t) = y\} = \sum_{z \neq x} P_x\{\mathbf{t} \leq t, X(\mathbf{t}) = z, \ X(t) = y\}.
$$

For a system starting at x, the event $\{\omega: \mathbf{t}(\omega) \leq t, X(\mathbf{t}(\omega), \omega) = z, X(t, \omega) = y\}$ occurs if and only if a first transition occurs at some time $s \leq t$ taking the system to state z, and then the system moves from state z to state y in the remaining time $(t - s)$. Therefore,

$$P_x\{t \leqslant t, X(t) = y\} = \int_0^t q(x)e^{-q(x)s}\left\{\sum_{z \neq x} Q(x,z)p(t - s, z, y)\right\}ds.$$

Substituting this in (5.1.19), we get the required integrodifferential equation (5.1.17).

5.2. Birth–Death Processes (A)

Let $X(t)$, $t \geqslant 0$, be a purely discontinuous Markov process with state space $S = \{0, 1, 2, \ldots\}$ and infinitesimal generator $A(t)$, which we take to be independent of time.

Definition 5.2.1. Let $b(x)$ and $d(x)$, $x \in S$, be nonnegative numbers with $d(0) = 0$. A purely discontinuous homogeneous Markov process $X(t)$ with state space S is said to be a *birth–death process* with infinitesimal *birth rate* $b(x)$ and *death rate* $d(x)$ if its infinitesimal generator is defined by

$$a(x,y) = \begin{cases} b(x) & \text{if } y = x + 1 \\ d(x) & \text{if } y = x - 1 \\ -(b(x) + d(x)) & \text{if } y = x \\ 0 & \text{if } |y - x| > 1 \end{cases} \tag{5.2.1}$$

We interpret $X(t)$ as the size of the population at time t. If $b(0) = 0$, there will be no birth, and the state 0 represents the extinction of the population. By assuming $b(0) > 0$, we can allow immigration to revive the population.

Definition 5.2.2. A birth–death process $X(t)$ is called a *pure birth process* if $d(x) = 0$, for all $x \in S$, and is called a *pure death process* if $b(x) = 0$, for all $x \in S$.

5.2.3. Backward and Forward Equations

From Kolmogorov equations (5.1.13) and (5.1.14) we obtain the corresponding equations for the birth–death processes. For a birth–death process the backward equations reduce to

$$p'(t, x, y) = d(x)p(t, x - 1, y) - [b(x) + d(x)]p(t, x, y) + b(x)p(t, x + 1, y)$$
$$p'(t, 0, y) = -b(0)p(t, 0, y) + b(0)p(t, 1, y), \qquad x \geqslant 1, y \geqslant 0, \tag{5.2.2}$$

and the forward equations take the form

$$p'(t,x,y) = b(y-1)p(t,x,y-1) - [b(y) + d(y)]p(t,x,y)$$
$$+ d(y+1)p(t,x,y+1), \tag{5.2.3}$$
$$p'(t,x,0) = -b(0)p(t,x,0) + d(1)p(t,x,1), \qquad x \geqslant 0, y \geqslant 1.$$

5.2.4. Pure Birth Processes

Let $X(t)$ be a pure birth process. The system moves only to the right since $d(x) = 0$ for all $x \geqslant 0$, and hence

$$p(t,x,y) = 0 \qquad \text{for all } y < x \text{ and } t \geqslant 0. \tag{5.2.4}$$

The forward equations (5.2.3) reduce to

$$p'(t,x,y) = b(y-1)p(t,x,y-1) - b(y)p(t,x,y). \tag{5.2.5}$$

From (5.2.4) and (5.2.5) we get, under the assumption that $b(0) = 0$,

$$p'(t,x,x) = -b(x)p(t,x,x),$$

which with the obvious conditions $p(0,x,y) = \delta_{xy}, y \geqslant x$, gives

$$p(t,x,x) = \exp[-b(x)t]. \tag{5.2.6}$$

[But this is also clear from Example 5.1.5 (1).] Since (5.2.5) is a linear differential equation, it is also easy to see that

$$p(t,x,y) = p(0,x,y)e^{-b(y)t} + \int_0^t b(y-1)e^{-b(y)(t-s)}p(s,x,y-1)\,ds$$
$$= b(y-1)\int_0^t e^{-b(y)(t-s)}p(s,x,y-1)\,ds, \qquad y > x. \tag{5.2.7}$$

In particular, the following recursive formula is useful:

$$p(t,x,x+1) = b(x)\int_0^t e^{-b(x+1)(t-s)}p(s,x,x)\,ds$$
$$= b(x)\int_0^t e^{-(t-s)b(x+1)}e^{-sb(x)}\,ds$$
$$= \begin{cases} \dfrac{b(x)[e^{-tb(x)} - e^{-tb(x+1)}]}{b(x+1) - b(x)} & \text{if } b(x) \neq b(x+1) \\ tb(x)e^{-tb(x)} & \text{if } b(x) = b(x+1) \end{cases}. \tag{5.2.8}$$

Better yet, we have the following theorem.

Theorem 5.2.5. (Feller). *Let $X(t)$, $t \geqslant 0$, be a pure birth process with distinct birth rates $b(x)$ and $b(0) \geqslant 0$. Then the forward equations have the unique solution given by*

$$p(t, x, y) = \begin{cases} \sum_{k=x}^{y} B(k, x, y) e^{-b(k)t} & y \geqslant x \\ 0 & y < x \end{cases}, \qquad (5.2.9)$$

where

$B(k, x, y)$

$$= \frac{b(x) b(x + 1) \cdots b(y - 1)}{[b(x) - b(k)] \cdots [b(k - 1) - b(k)][b(k + 1) - b(k)] \cdots [b(y) - b(k)]}.$$

$$(5.2.10)$$

PROOF. Let $\pi(\theta, x, y)$ denote the Laplace transform of $p(t, x, y)$, that is, $\pi(\theta, x, y) = \int_0^\infty e^{-\theta t} p(t, x, y) \, dt$, $\theta > 0$. First note that

$$\int_0^\infty e^{-\theta t} p'(t, x, y) \, dt = e^{-\theta t} p(t, x, y) \big|_0^\infty + \theta \int_0^\infty e^{-\theta t} p(t, x, y) \, dt$$

$$= -\delta_{xy} + \theta \pi(\theta, x, y). \qquad (5.2.11)$$

Using this in the forward equations (5.2.5), we obtain

$$-\delta_{xy} + \theta \pi(\theta, x, y) = b(y - 1) \pi(\theta, x, y - 1) - b(y) \pi(\theta, x, y), \qquad (5.2.12)$$

so that

$$[\theta + b(y)] \pi(\theta, x, y) = \delta_{xy} + b(y - 1) \pi(\theta, x, y - 1), \qquad y \geqslant 0. \quad (5.2.13)$$

First we note from (5.2.13) that

$$\pi(\theta, x, y) = 0 \qquad \text{for } y < x. \qquad (5.2.14)$$

Let $x = y$. Then from (5.2.14) we get

$$\pi(\theta, x, x) = \frac{1}{\theta + b(x)},$$

from which it follows that for $y = x + 1$,

$$\pi(\theta, x, x + 1) = \frac{b(x)}{[\theta + b(x)][\theta + b(x + 1)]}.$$

Proceeding successively, we obtain that

$$\pi(\theta, x, y) = \frac{b(x)b(x + 1) \cdots b(y - 1)}{[\theta + b(x)][\theta + b(x + 1)] \cdots [\theta + b(y)]}, \qquad y > x. \quad (5.2.15)$$

Resolving the RHS of (5.2.15) into partial fractions, we get

$$\pi(\theta, x, y) = \sum_{u=x}^{y} \frac{B(u, x, y)}{\theta + b(u)}, \qquad B(u, x, y) = \lim_{\theta \to -b(u)} [\theta + b(u)]\pi(\theta, x, y).$$

This gives (5.2.10). But the inverse Laplace transform of $[\theta + b(u)]^{-1}$ is given by $\exp[-tb(u)]$, $t \geqslant 0$, and hence we obtain (5.2.9). This completes the proof. □

While introducing a jump process we remarked that we do not want to allow the system to make infinite number of jumps in a finite time period. To clarify the situation, let us consider a deterministic birth process in which the birth rate (per person) is proportional to the size of the population. (Treat the population size as a continuous variable.) Now, $b(x) = bx$ for some constant $b > 0$. Therefore, the population of size x increases by an amount $bx^2 \Delta t$ in time Δt. Hence the growth of the population is given by $(dx/dt) = bx^2$, say, with initial condition $x(0) = 1$. Then $x(t) = 1/(1 - bt)$ and consequently,

$$\lim_{t \to 1/b} x(t) = \infty.$$

That is, the population size explodes in a finite time period. This type of growth of a population is called *divergent* or *explosive*. This suggests that a stochastic birth process might become explosive in some sense if the infinitesimal birth rates $b(x)$ were not properly controlled. That is, with a positive probability it might happen that there were an infinite number of transitions in a finite time. Thus the probability that only a finite number of transitions occur during time t is

$$\sum_{y \geqslant x} p(t, x, y) < 1,$$

and $1 - \sum p(t, x, y)$ is the probability of explosion. Therefore, it becomes important to find condition(s) under which the pure birth process will be nonexplosive. Toward this we have the following theorem.

Theorem 5.2.6. (Feller–Lundberg). *Let $X(t)$, $t \geqslant 0$, be a pure birth process satisfying the hypotheses of Theorem 5.2.5. Then, $\Sigma p(t, x, y) = 1$ if and only if*

$$\frac{1}{b(x)} + \frac{1}{b(x+1)} + \cdots = \infty. \tag{5.2.16}$$

If (5.2.16) holds, the nth jump time $W_n \to \infty$ with probability one, as $n \to \infty$.

PROOF. Define $S(t, x, y) = \Sigma_{u=0}^{y} p(t, x, u)$. From (5.2.4), $S(t, x, y) = \Sigma_{u=x}^{y} p(t, x, u)$. Using the forward equations (5.2.5), we obtain

$$S'(t, x, y) = \sum_{u=x}^{y} p'(t, x, u)$$

$$= \sum_{u=x}^{y} \{b(u-1)p(t, x, u-1) - b(u)p(t, x, u)\}$$

$$= -b(y)p(t, x, y). \tag{5.2.17}$$

Let $\sigma(\theta, x, y)$ denote the Laplace transform of $S(t, x, y)$. Noting that $S(0, x, y) = 1$, it follows from (5.2.17) that

$$\theta\sigma(\theta, x, y) - 1 = -b(y)\pi(\theta, x, y). \tag{5.2.18}$$

But from the expression (5.2.15) for $\pi(\theta, x, y)$ we see that

$$\lim_{y \to \infty} b(y)\pi(\theta, x, y) = \begin{cases} \prod_{u \geqslant x} \left[1 + \dfrac{\theta}{b(u)}\right]^{-1} & \text{if } \sum \dfrac{1}{b(u)} < \infty \\ 0 & \text{if } \sum \dfrac{1}{b(u)} = \infty \end{cases}$$

Hence: (i) $\lim_{y \to \infty} \theta\sigma(\theta, x, y) < 1$ if the series $\Sigma_{u \geqslant x}[b(u)]^{-1} < \infty$, or (ii) $\lim_{y \to \infty} \theta\sigma(\theta, x, y) = 1$ if $\Sigma_{u \geqslant x}[b(u)]^{-1} = \infty$. Consequently, $\Sigma_y p(t, x, y) = \lim_{y \to \infty} S(t, x, y) < 1$ or $= 1$ accordingly as the series $\Sigma_{u \geqslant x}[b(u)]^{-1}$ converges or diverges. This proves the first part of the theorem.

Let $X(0) = x$ and W_n be the nth jump time $W_n = t_1 + \cdots + t_n$, where t_k is the interarrival time between $(x + k - 1)$ and $(x + k)$th births. Since the times t_k are independent and exponentially distributed [see Example 5.1.5 (i) and establish this], the characteristic function $\psi_n(\theta)$ of W_n is

$$\psi_n(\theta) = \prod_{u=x}^{n-1} \left[1 - \frac{i\theta}{b(u)}\right]^{-1},$$

5. Purely Discontinuous Markov Processes

and clearly $\lim_{n\to\infty} \psi_n(\theta) = 0$ if $\Sigma[b(u)]^{-1} = \infty$. Therefore, as $n \to \infty$, $W_n \to \infty$ AS. This completes the proof. $\qquad\square$

Theorem 5.2.7. (Feller). *The solution (5.2.9)–(5.2.10) of the forward equations of the pure birth process also solves the backward equations*

$$p'(t,x,y) = b(x)p(t,x+1,y) - b(x)p(t,x,y), \qquad x \geqslant 0. \qquad (5.2.19)$$

This solution, denoted by $q(t,x,y)$, is minimal in the sense that, if $p(t,x,y)$ is any nonnegative solution of (5.2.19), then

$$q(t,x,y) \leqslant p(t,x,y). \qquad (5.2.20)$$

Moreover, (5.2.9)–(5.2.10) solves uniquely both the forward and backward equations if $\Sigma_{u\geqslant x}[b(u)]^{-1} = \infty$.

PROOF. Proceeding as we did in establishing (5.2.13), we obtain from the backward equations (5.2.19) that

$$[\theta + b(x)]\pi(\theta,x,y) = \delta_{xy} + b(x)\pi(\theta,x+1,y), \qquad x \geqslant 0. \quad (5.2.21)$$

Because of (5.2.4) we set $\pi(\theta,x,y) = 0 = Q(\theta,x,y)$, for $y < x$, where Q is the Laplace transform of q. Then, from (5.2.15), $\pi(\theta,x,y) = Q(\theta,x,y)$, for $y \geqslant x$, and consequently $q(t,x,y) = p(t,x,y)$ solves the backward equations (5.2.19).

Next, treating (5.2.19) as a linear differential equation in $p(t,x,y)$, we obtain

$$p(t,x,y) = e^{-b(x)t}\delta_{xy} + b(x)\int_0^t e^{-b(x)(t-s)}p(s,x+1,y)\,ds. \quad (5.2.22)$$

Now, (5.2.20) is obvious for all $x > y$. If $x = y$, then from (5.2.22) and (5.2.6) we see that $p(t,x,x) \geqslant e^{-b(x)t} = q(t,x,x)$. From this and (5.2.22), and since $q(t,x,y)$ satisfies (5.2.22), we obtain

$$p(t,x-1,x) \geqslant b(x-1)\int_0^t e^{-b(x-1)(t-s)}q(s,x,x)\,ds$$

$$= q(t,x-1,x).$$

Proceeding recursively, it follows that (5.2.20) holds for all $x \geqslant 0$.

Let $\Sigma[b(u)]^{-1} = \infty$. Then $\Sigma q(t,x,y) = 1$ by the Feller–Lundberg theorem, and it follows from (5.2.20) that

$$\sum_{y\geqslant 0} p(t,x,y) \geqslant \sum_{y\geqslant 0} q(t,x,y) \geqslant 1.$$

Therefore, $\Sigma_{y \geqslant 0} p(t, x, y) = 1$ and consequently $q(t, x, y) = p(t, x, y)$ for every y. This completes the proof. $\qquad\square$

5.2.8. Linear Birth Process

A pure birth process is called a *linear birth process* if the birth rates $b(x)$ are given by $b(x) = bx$, for all $x \geqslant 0$ and some positive constant b. Linear birth processes are also called *simple birth processes* or *Furry–Yule processes*. Note that

$$\sum_{u \geqslant x} [b(u)]^{-1} = b^{-1} \sum_{u \geqslant x} u^{-1} = \infty.$$

Applying Theorem 5.2.7, the transition probabilities $p(t, x, y)$ are uniquely determined by (5.2.9)–(5.2.10). But $p(t, x, y)$ can also be easily determined by treating the forward equation as a linear differential equation. From (5.2.8),

$$p(t, x, x + 1) = x e^{-btx}(1 - e^{-bt}). \qquad (5.2.23)$$

We claim that, for all $y \geqslant x$, we have

$$p(t, x, y) = \binom{y - 1}{y - x} e^{-btx}(1 - e^{-bt})^{y - x}. \qquad (5.2.24)$$

From (5.2.6) and (5.2.23) it follows that (5.2.24) holds for $y = x$ and $y = x + 1$. To proceed inductively, let (5.2.24) hold for $y = z \geqslant x$. Then from (5.2.7) we obtain

$$
\begin{aligned}
p(t, x, z + 1) &= bz \int_0^t e^{-b(t-s)(z+1)} p(s, x, z)\, ds \\[2mm]
&= bz \int_0^t e^{-b(t-s)(z+1)} \binom{z - 1}{z - x} e^{-bsx}(1 - e^{-bs})^{z - x}\, ds \\[2mm]
&= bz \binom{z - 1}{z - x} e^{-bt(z+1)} \int_0^t e^{bx}(e^{bs} - 1)^{z - x}\, ds \\[2mm]
&= \frac{z(z - 1)!}{(z - x)!\,(x - 1)!}\, e^{-bt(z+1)} \frac{(e^{bs} - 1)^{z - x + 1}}{(z - x + 1)} \bigg|_0^t \\[2mm]
&= \binom{z}{z + 1 - x} e^{-btx}(1 - e^{-bt})^{z + 1 - x}.
\end{aligned}
$$

Hence (5.2.24) holds for all $y \geqslant x$.

117

5. Purely Discontinuous Markov Processes

5.2.9. Pure Death Process

Let $X(t)$ be a pure death process. Since the system moves only to the left,

$$p(t, x, y) = 0 \qquad \text{for all } y > x. \tag{5.2.25}$$

From (5.2.3) the forward equations for a pure death process are

$$p'(t, x, x) = -d(x)p(t, x, x)$$
$$p'(t, x, y) = -d(y)p(t, x, y) + d(y + 1)p(t, x, y + 1), \qquad y \leqslant x - 1. \tag{5.2.26}$$

From (5.2.26) one obtains, as in the case of pure birth process, that

$$p(t, x, x) = e^{-d(x)t}$$
$$p(t, x, y) = d(y + 1) \int_0^t e^{-(t-s)d(y)} p(s, x, y + 1)\, ds, \qquad y \leqslant x - 1. \tag{5.2.27}$$

In particular,

$$
p(t, x, x - 1)
=
\begin{cases}
\dfrac{d(x)}{d(x - 1) - d(x)}[e^{-td(x)} - e^{-td(x-1)}] & \text{if } d(x) \neq d(x - 1) \\
td(x)e^{-td(x)} & \text{if } d(x) = d(x - 1)
\end{cases}. \tag{5.2.28}
$$

If $d(x) = dx$ for a constant $d > 0$, the process $X(t)$ is called a *simple* or *linear death process*. In this case it is easy to see that, for $0 \leqslant y \leqslant x$,

$$p(t, x, y) = \binom{x}{y} e^{-dty}(1 - e^{-dt})^{x-y}. \tag{5.2.29}$$

5.2.10. Poisson Process

We give a third definition of a Poisson process here. This can easily be seen to be equivalent to the earlier ones (see Definitions 4.1.2 and 4.1.3). A pure birth process $X(t)$ is called a *Poisson process* if the birth rate is given by $b(x) = b > 0, x \geqslant 0$.

Let $X(t)$ be a Poisson process. As a special type of birth process, $X(t)$ satisfies equation (5.2.4), and (5.2.6) becomes

$$p(t, x, x) = e^{-bt}, \qquad t \geqslant 0. \tag{5.2.30}$$

The forward equations reduce to

$$p'(t, x, y) = bp(t, x, y - 1) - bp(t, x, y), \qquad y \neq 0, t \geqslant 0,$$

and consequently

$$p(t, x, y) = e^{-bt}\delta_{xy} + b \int_0^t e^{-b(t-s)} p(s, x, y - 1) \, ds, \qquad t \geq 0.$$

In particular, we have from (5.2.4) and (5.2.30) that

$$p(t, x, x + 1) = b \int_0^t e^{-b(t-s)} e^{-bs} \, ds = bte^{-bt}.$$

As in the simple-birth-process case, we can now show by induction that, for $0 \leq x \leq y$ and $t \geq 0$,

$$p(t, x, y) = \frac{e^{-bt}(bt)^{y-x}}{(y - x)!}. \tag{5.2.31}$$

Relations (5.2.4) and (5.2.31) yield spatial homogeneity.

Now let $0 \leq s \leq t$ and n be a positive integer. Then:

$$P\{X(t) - X(s) = n\} = \sum_{i \in S} P\{X(s) = i, X(t) - X(s) = n\}$$

$$= \sum_i P\{X(s) = i\} P\{X(t) = i + n \,|\, X(s) = i\}$$

$$= \sum_i P\{X(s) = i\} p(t - s, 0, n)$$

(from (5.2.4) and (5.2.17))

$$= p(t - s, 0, n)$$

$$= e^{-b(t-s)}[b(t - s)]^n / n! \qquad \text{(by (5.2.31)).} \tag{5.2.32}$$

Let $0 \leq t_0 < t_1 < \cdots < t_n$ and $x_1, \ldots, x_n \in S$. We claim that the increments $X(t_1) - X(t_0), \ldots, X(t_n) - X(t_{n-1})$ are independent. [The stationarity of these increments follows from (5.2.32).]

$$P\{X(t_1) - X(t_0) = x_1, \ldots, X(t_n) - X(t_{n-1}) = x_n\}$$

$$= \sum_{x \in S} P\{X(t_0) = \dot{x}, X(t_1) - X(t_0) = x_1, \ldots, X(t_n) - X(t_{n-1}) = x_n\}$$

$$= \sum_{x \in S} P\{X(t_0) = x\} p(t_1 - t_0, 0, x_1) \cdots p(t_n - t_{n-1}, 0, x_n) \qquad \text{(why?)}$$

$$= p(t_1 - t_0, 0, x_1) \cdots p(t_n - t_{n-1}, 0, x_n)$$

$$= P\{X(t_1) - X(t_0) = x_1\} \cdots P\{X(t_n) - X(t_{n-1}) = x_n\}.$$

119

5.3. Birth–Death Processes (B)

In Section 5.2 we defined the birth–death processes by specifying the infinitesimal rates and solving the Kolmogorov equations for the transition probabilities. In modeling several biological and physical processes it is also customary to formulate the corresponding differential equation for $p(t, x)$ $= P\{X(t) = x\}$ and work with this equation. This section presents an introduction to this variant of the above technique.

Let $X(t)$ be a stochastic process on the state space $S = \{0, 1, 2, \ldots\}$. Under the following assumptions, $X(t)$ becomes a birth–death process:

1. Let $X(t) = x$. The probability of transition $x \to x + 1$ in the interval $(t, t + \Delta t)$ is $b(x)\Delta t + o(\Delta t)$.
2. The probability of transition $x \to x - 1$ in $(t, t + \Delta t)$ is $d(x)\Delta t + o(\Delta t)$.
3. The probability of transition $x \to y$ with $|y - x| > 1$ is $o(\Delta t)$.
4. The probability of no change is $1 - [b(x) + d(x)]\Delta t + o(\Delta t)$.

State 0 will be an absorbing state. Consider $p(t + \Delta t, x) = P\{X(t + \Delta t) = x\}$. The event $\{X(t + \Delta t) = x\}$ occurs in the following mutually exclusive ways: (a) there were $x - 1$ persons in the population at time t and a birth occurs in the interval $(t, t + \Delta t)$ with probability $b(x - 1)\Delta t + o(\Delta t)$; (b) there were already x persons at time t and no change occurs in $(t, t + \Delta t)$ with probability $1 - [b(x) + d(x)]\Delta t + o(\Delta t)$; (c) there were $x + 1$ persons at time t, with one death occurring in $(t, t + \Delta t)$ with probability $d(x + 1)\Delta t + o(\Delta t)$, or (d) more than one change occurs in $(t, t + \Delta t)$ with probability $o(\Delta t)$. Thus

$$p(t + \Delta t, x) = b(x - 1)p(t, x - 1)\Delta t + \{1 - [b(x) + d(x)]\Delta t\}p(t, x)$$
$$+ d(x + 1)p(t, x + 1)\Delta t + o(\Delta t). \tag{5.3.1}$$

Transforming $p(t, x)$ to the LHS, dividing throughout by Δt and taking limit as $\Delta t \to 0$, we obtain, for $x \geqslant 1$,

$$p'(t, x) = b(x - 1)p(t, x - 1) - [b(x) + d(x)]p(t, x) + d(x + 1)p(t, x + 1). \tag{5.3.2}$$

For $x = 0$,

$$p'(t, 0) = d(1)p(t, 1), \tag{5.3.3}$$

if 0 is an absorbing state; otherwise

$$p'(t, 0) = -b(0)p(t, 0) + d(1)p(t, 1). \tag{5.3.4}$$

Let $X(0) = x_0$. Then the initial condition used to solve these equations is

$$p(0, x) = \delta_{xx_0}. \tag{5.3.5}$$

Example 5.3.1 *Feller–Arley Process.* A birth–death process is called a *Feller–Arley process* if $b(x) = bx$ and $d(x) = dx$, where $b > 0, d > 0$. It is not difficult to solve equation (5.3.2) in this case. To proceed further one uses either the probability generating function or the Laplace transform. We use the former. Let

$$g(t, \xi) = \sum_{x \geq 0} p(t, x)\xi^x \quad \text{with } g(0, \xi) = \xi^{x_0}. \tag{5.3.6}$$

Then:

$$\frac{\partial g}{\partial \xi} = \sum_{x \geq 1} xp(t, x)\xi^{x-1}. \tag{5.3.7}$$

From (5.3.2) and (5.3.7), we obtain

$$\frac{\partial g}{\partial t} - (\xi - 1)(b\xi - d)\frac{\partial g}{\partial \xi} = 0. \tag{5.3.8}$$

Equation (5.3.8) is a special case of the more general linear equation

$$\sum_{k=1}^{n} h_k(\xi_1, \ldots, \xi_n, \eta)\frac{\partial \eta}{\partial \xi_k} = R(\xi_1, \ldots, \xi_n, \eta), \tag{5.3.9}$$

subject to suitable boundary conditions. To solve equation (5.3.9), one first forms the auxiliary equations

$$\frac{d\xi_1}{h_1} = \cdots = \frac{d\xi_n}{h_n} = \frac{d\eta}{R} \tag{5.3.10}$$

and finds the n independent solutions of (5.3.10):

$$\zeta_k(\xi_1, \ldots, \xi_n, \eta) = \text{constant}, \quad 1 \leq k \leq n. \tag{5.3.11}$$

Then the general solution of (5.3.9) is given by

$$\psi(\zeta_1, \ldots, \zeta_n) = 0. \tag{5.3.12}$$

Therefore, for the Feller–Arley process, equations (5.3.10) become

$$\frac{dt}{1} = \frac{d\xi}{-(\xi - 1)(b\xi - d)} = \frac{dg}{0}. \tag{5.3.13}$$

5. Purely Discontinuous Markov Processes

From $dg = 0$, $g = c_1$. From $dt = -d\xi/(\xi - 1)(b\xi - d)$, $c_2 = e^{(d-b)t}$ $\times (b\xi - d)/(\xi - 1)$. Eliminating one of the constants c_1 and c_2, we obtain

$$g(t, \xi) = \psi \left[\frac{b\xi - d}{\xi - 1} e^{-(d-b)t} \right], \tag{5.3.14}$$

where the function ψ is determined from $g(0, \xi) = \xi^{x_0}$. Now $\psi(\alpha) = [(d - \alpha)/(b - \alpha)]^{x_0}$, and consequently

$$g(t, \xi) = \left\{ \frac{d[e^{(b-d)t} - 1] + \xi[b - de^{(b-d)t}]}{[be^{(b-d)t} - d] - b\xi[e^{(b-d)t} - 1]} \right\}^{x_0}. \tag{5.3.15}$$

To simplify expression (5.3.15), define functions $\alpha = \alpha(t)$ and $\beta = \beta(t)$ by

$$\frac{\alpha}{d} = \frac{\beta}{b} = \frac{e^{(b-d)t} - 1}{be^{(b-d)t} - d}. \tag{5.3.16}$$

Using (5.3.16) in (5.3.15), we get

$$g(t, \xi) = \left\{ \frac{\alpha + (1 - \alpha - \beta)\xi}{1 - \beta\xi} \right\}^{x_0}. \tag{5.3.17}$$

Based on the geometric distribution (5.3.17), we can draw some simple properties of a Feller–Arley process.

Proposition 5.3.2. Let $X(t)$, $t \geqslant 0$, be a Feller–Arley process with initial population size $X(0) = 1$. Then

(i)
$$p(t, 0) = \begin{cases} bt/(1 + bt) & \text{if } b = d \\ [de^{(b-d)t} - 1]/[be^{(b-d)t} - d] & \text{if } b \neq d \end{cases}.$$

(ii)
$$p(t, x) = \begin{cases} (bt)^{x-1}/(1 + bt)^{x+1} & \text{if } b = d \\ (1 - \alpha)(1 - \beta)\beta^{x-1} & \text{if } b \neq d \end{cases}, \quad x \geqslant 1.$$

(iii)
$$E[X(t)] = \begin{cases} 1 & \text{if } b = d \\ e^{(b-d)t} & \text{if } b \neq d \end{cases}.$$

$$\lim_{t \to \infty} E[X(t)] = \begin{cases} 0 & \text{if } b < d \\ 1 & \text{if } b = d. \\ \infty & \text{if } b > d \end{cases}$$

$$
\text{(iv)} \qquad \text{var}[X(t)] = \begin{cases} 2bt & \text{if } b = d \\ \dfrac{b + d}{b - d} e^{(b-d)t} \{e^{(b-d)t} - 1\} & \text{if } b \neq d \end{cases}.
$$

(v) *Probability of eventual extinction of the population is*

$$
\lim_{t \to \infty} p(t, 0) = \begin{cases} 1 & \text{if } b \leqslant d \\ d/b & \text{if } b > d \end{cases}.
$$

Examples 5.3.3

EXAMPLE 1. *Radioactive Transformations.* In the theory of radioactive transformations a basic assumption is that the radioactive atoms are unstable and disintegrate stochastically (Rutherford et al. 1930). These disintegrations lead to atoms that are chemically and physically different from the original atom. Each of these new atoms is also unstable. By the emission of radioactive particles, these new atoms pass through a number of transformations on physical states $\{0, 1, 2, \ldots, N\}$. Let $b(k)$ denote the decay rate from state k to state $k + 1$. We can describe the radioactive transformation as a pure birth process. Denoting by $p(t, k)$ the probability that the system is in state k at time t and assuming that in the interval $(t, t + \Delta t)$ the probability of an atom in state k transforming to state $k + 1$ is $b(k) \Delta t + o(\Delta t)$, we obtain

$$
p(t + \Delta t, k) = b(k - 1)p(t, k - 1) \Delta t + [1 - b(k) \Delta t]p(t, k).
$$

From this we get the system of equations

$$
\frac{d}{dt}p(t, 0) = -b(0)p(t, 0),
$$

$$
\frac{d}{dt}p(t, k) = b(k - 1)p(t, k - 1) - b(k)p(t, k), \tag{5.3.18}
$$

for $1 \leqslant k \leqslant N$, where $b(N) = 0$. System (5.3.18) is easy to solve and is left as an exercise.

EXAMPLE 2. *Blood Clotting.* The kinetic theory of chemical reactions is the study of the time rates of chemical reactions and of the factors that influence these rates. In the stochastic theory of chemical kinetics one is interested in the probability distribution of the concentrations. Below we model the enzyme reaction of blood clotting as a pure birth process. In closing a cut, the gelation process of blood clotting is caused by an enzyme known as *fibrin*, which is formed from the proenzyme *fibrinogen*. In the presence of the agent *thrombin* the conversion occurs at a faster rate. Even though blood clotting occurs as a *waterfall sequence* and could be modeled as in the case of radioactive transformation, we prescribe it as follows. Consider a large number of samples, each of which initially containing only one fibrin molecule and a very large number of fibrinogen molecules. Let $X(t)$ be the concentration (the number of molecules per unit or constant volume) of the enzyme system at time t. Let b be the reaction rate in each sample. Then the probability that a conversion occurs in the interval $(t, t + \Delta t)$ is $b \Delta t + o(\Delta t)$. Let there be k fibrin molecules in the system at time t. Then it is easy to see that

$$p'(t, k) = -bkp(t, k) + b(k - 1)p(t, k - 1),$$

$$p(0, k) = \delta_{kx_0}, \quad \text{if } X(0) = x_0.$$

Successively solving these equations, we get

$$p(t, x) = \binom{x - 1}{x_0 - 1} e^{-btx_0}(1 - e^{-bt})^{x-x_0}, \quad x \geqslant x_0.$$

This is a negative binomial-type distribution. Hence

$$E[X(t)] = x_0 e^{bt}, \quad \text{and} \quad \text{var}(X(t)) = x_0 e^{bt}(e^{bt} - 1).$$

EXAMPLE 3. *Unimolecular Reaction.* An irreversible conversion of a reactant ρ into a product π is called an *unimolecular reaction* $\rho \xrightarrow{d} \pi$, where $d > 0$ is the reaction rate. Let $X(t)$ denote the concentration (the number of molecules per unit volume) of the reactant ρ at time t. Set $X(0) = n_0 > 0$. Since the reaction is irreversible, $X(t)$ is a death process. The basic assumptions on the unimolecular reaction are: (1) the probability of the reverse reaction $\pi \to \rho$ is zero, (2) the probability of more than one conversion in $(t, t + \Delta t)$ is $o(\Delta t)$, and (3) the probability of a single conversion of n molecules in $(t, t + \Delta t)$, when it is known that there were $(n - n_0)$ conversions in the $(0, t]$, is $dn \Delta t + o(\Delta t)$. Then the probabilities $p(t, n) = P\{X(t) = n\}, 0 \leqslant n \leqslant n_0$, clearly satisfy

$$p'(t, n) = d(n + 1)p(t, x + 1) - dn p(t, n).$$

Solving this system recursively with the initial condition $p(0, n) = \delta_{nn_0}$, we obtain

$$p(t, n) = \binom{n_0}{n} e^{-n_0 dt}(e^{dt} - 1)^{n_0-n}, \quad 0 \leqslant n \leqslant n_0.$$

[Compare this with equation (5.2.29).] Now

$$E[X(t)] = n_0 e^{-dt} \quad \text{and} \quad \text{var}(X(t)) = n_0 e^{-dt}(1 - e^{-dt}).$$

EXAMPLE 4. *A Stochastic Epidemic.* Consider a homogeneously mixing population of *susceptibles.* After a susceptible contracts an infectious disease, the disease develops within the *infected* person for a period of time called the *latent period.* Following the latent period, the *infective* discharges the infectious matter during an *infectious period,* thereby spreading the disease. After an infective shows the symptoms of disease he is removed from the population until he recovers or dies. The time period between the contraction of the infection and the appearance of symptoms is called the *incubation period.* For a detailed treatment of stochastic theory of epidemics, the students should consult Bailey (1957, 1963), Bartlett (1960), and Bharucha-Reid (1957, 1960).

Consider a homogeneously mixing population with a mild infection of the upper respiratory tract where the time period between the infection of any susceptible and his removal from the population is sufficiently long. Let the total size of the population be $(n + 1)$ such that there was one infective at time 0. Let $X(t)$ denote the number of susceptibles at time t. Then $X(0) = n$, and $X(t)$ is clearly a pure death process with a finite number of states. Let us assume that the probability of one new infection in the interval $(t, t + \Delta t)$ is $dx(n - x + 1)\Delta t + o(\Delta t)$ when it is known that there are x susceptibles at time t. Here d is the *contact rate.* If $p(t, x) = P\{X(t) = x\}$, then

$$p'(t, x) = (x + 1)(n - x)p(t, x + 1) - x(n - x + 1)p(t, x),$$
$$0 \leqslant x \leqslant n - 1,$$
$$p'(t, n) = -np(t, n), \qquad p(0, x) = \delta_{xn}. \tag{5.3.19}$$

The death process involved in here is a nonlinear process. A method of solving Equations (5.3.19) using Laplace transform can be found in Bailey (1957, 1963).

EXAMPLE 5. Some nonlinear death equations are not so difficult to solve. As an example, consider the following equations for a death process

$$p'(t, x) = d(x + 1)^2 p(t, x + 1) - dx^2 p(t, x), \qquad 0 < x < n$$
$$p'(t, n) = -dn^2 p(t, n),$$
$$p'(t, 0) = dp(t, 1), \qquad p(0, x) = \delta_{xn}. \tag{5.3.20}$$

5. Purely Discontinuous Markov Processes

From the second equation and the initial condition in (5.3.20), we obtain $p(t, n) = e^{-dn^2 t}$. Using this in the equation for $p(t, n - 1)$, we get

$$p'(t, n - 1) = dn^2 e^{-dn^2 t} - d(n - 1)^2 p(t, n - 1).$$

This equation, with $p(0, n - 1) = \delta_{n-1,n} = 0$, gives

$$p(t, n - 1) = \frac{dn^2}{d(n - 1)^2 - dn^2} \left\{ e^{-dn^2 t} - e^{-d(n-1)^2 t} \right\}.$$

We claim that

$$p(t, x) = c_x e^{-dx^2 t} + \sum_{i=0}^{n-x-1} \frac{\prod\limits_{j=x+1}^{n-i} dj^2 c_{n-i} e^{-d(n-i)^2 t}}{\prod\limits_{j=x}^{n-i-1} d[j^2 - (n - i)^2]} \tag{5.3.21}$$

for $1 \leqslant x \leqslant n$, where

$$c_n = 1, \quad \text{and} \quad c_x = -\sum_{i=0}^{n-x-1} \frac{\prod\limits_{j=x+1}^{n-i} dj^2 c_{n-i}}{\prod\limits_{j=x}^{n-i-j} d[j^2 - (n - i)^2]}. \tag{5.3.22}$$

From (5.3.20), $p'(t, x - 1) = dx^2 p(t, x) - d(x - 1)^2 p(t, x - 1)$. From (5.3.21) and (5.3.22), it follows that

$$p'(t, x - 1) + d(x - 1)^2 p(t, x - 1)$$

$$= dx^2 c_x e^{-dx^2 t} + dx^2 \sum_{i=0}^{n-x-1} \frac{c_{n-1} \prod\limits_{j=x+1}^{n-i} dj^2 e^{-d(n-i)^2 t}}{\prod\limits_{j=x}^{n-i-1} d[j^2 - (n - i)^2]}. \tag{5.3.23}$$

It is clear that $p(t, x - 1)$ will be of the form

$$p(t, x - 1) = \sum_{i=0}^{n-x+1} a_{n-i} e^{-d(n-i)^2 t}$$

so that

$$p'(t, x - 1) + d(x - 1)^2 p(t, x - 1)$$

$$= \sum_{i=0}^{n-x+1} a_{n-i} d[(x - 1)^2 - (n - i)^2] e^{-d(n-i)^2 t}.$$

Comparing this with equation (5.3.23), we have for $i = n - x + 1$ that $a_{x-1} d[(x-1)^2 - (x-1)^2] = 0$ and for $i = n - x$,

$$a_x d[(x-1)^2 - x^2] = dx^2 c_x, \quad \text{so that}$$

$$a_x = \frac{dx^2 c_x}{d[(x-1)^2 - x^2]}$$

$$a_{n-i} = \frac{c_{n-i} \prod\limits_{j=x+1}^{n-i} dj^2}{\prod\limits_{j=x}^{n-i-1} d[j^2 - (n-i)^2]}, \quad 0 < i < n - x.$$

Now it is easy to see that we obtain (5.3.21) and (5.3.22) for $(x - 1)$, thereby establishing them by induction in general. Therefore, the solution of (5.3.20) is given by

$$p(t, n) = e^{-dn^2 t},$$

$$p(t, x) = c_x e^{-dx^2 t} + \sum_{i=0}^{n-x-1} \frac{c_{n-i} e^{-d(n-i)^2 t} \prod\limits_{j=x+1}^{n-i} j^2}{\prod\limits_{j=x}^{n-i-1} [j^2 - (n-i)^2]}, \quad 0 < x < n,$$

$$p(t, 0) = 1 - c_1 e^{-dt} - \sum_{i=0}^{n-2} \frac{e^{-d(n-i)^2 t} [c_{n-i}/(n-i)^2] \prod\limits_{i=2}^{n-i} j^2}{\prod\limits_{j=1}^{n-i-1} [j^2 - (n-i)^2]},$$

where c_x is given by (5.3.22).

EXAMPLE 6. *A Nonhomogeneous Birth–Death Process.* Certain physical processes such as cascade showers in cosmic-ray theory and telephone traffic are modeled along birth–death processes with time-dependent infinitesimal rates. Let us consider the Feller–Arley process with the rates b and d depending on time t. In place of equation (5.3.8) we obtain

$$\frac{\partial g}{\partial t} = (\xi - 1)\{b(t)\xi - d(t)\}\frac{\partial g}{\partial \xi},$$

$$g(0, \xi) = \xi^{x_0}. \tag{5.3.24}$$

The corresponding auxiliary equations are

$$\frac{dt}{1} = \frac{d\xi}{(\xi - 1)(d - b\xi)} = \frac{dg}{0}. \tag{5.3.25}$$

From the first equality in (5.3.25), we obtain $(d\xi/dt) = (\xi - 1)(d - b\xi)$. Set $\eta = 1/(\xi - 1)$. Then

$$\frac{d\eta}{dt} = b(t) - [d(t) - b(t)]\eta. \tag{5.3.26}$$

This is a linear equation in η and

$$\eta e^{a(t)} = \int_0^t b(s)e^{a(s)}\,ds + c,$$

where $a(t) = \int_0^t [d(s) - b(s)]\,ds$. Therefore, transforming to the variable ξ and noting that $g = $ constant from (5.3.25), the general solution of (5.3.24) is

$$g(t,\xi) = \psi(e^{a(t)}(\xi - 1)^{-1} - \int_0^t b(s)e^{a(s)}\,ds). \tag{5.3.27}$$

But $\xi^{x_0} = g(0,\xi) = \psi((\xi - 1)^{-1})$. Set $\xi = 1 + \zeta^{-1}$. Then $\psi(\zeta) = [1 + \zeta^{-1}]^{x_0}$. Hence

$$g(t,\xi) = \{1 + [e^{a(t)}(\xi - 1)^{-1} - \int_0^t b(s)e^{a(s)}\,ds]^{-1}\}^{x_0}. \tag{5.3.28}$$

It is not difficult to see, from (5.3.28), that

$$p(t,x) = \sum_{u=0}^{x_0 \wedge x} \binom{x_0}{u}\binom{x_0 + x - u - 1}{x_0 - 1}[f(t)]^{x_0 - u}[h(t)]^{x-u}$$
$$\times [1 - f(t) - h(t)]^u,$$
$$p(t,0) = [f(t)]^{x_0}, \tag{5.3.29}$$

where

$$f(t) = 1 - \frac{1}{e^{a(t)} + F(t)}, \qquad h(t) = 1 - \frac{e^{a(t)}}{e^{a(t)} + F(t)},$$
$$F(t) = \int_0^t b(s)e^{a(s)}\,ds.$$

But it can be easily seen that

$$e^{a(t)} + F(t) = 1 + \int_0^t d(s)e^{a(s)}\,ds.$$

Using this, we see that the probability of extinction at time t is given by

$$p(t,0) = \left\{ \int_0^t d(s)e^{a(s)}\,ds \Big/ [1 + \int_0^t d(s)e^{a(s)}\,ds] \right\}^{x_0}.$$

Since $\lim_{t \to \infty} \int_0^t d(s)e^{a(s)} \, ds = \infty$, the eventual extinction is almost sure.

EXAMPLE 7. *A Nonlinear Birth–Death Process in Carcinogenesis.* In this example we consider the growth of tumor as a birth–death process. Let a tumor be initiated by carcinogenic action with normal cell. A *birth* in the population of cancer cells occurs either as the mutation of the normal cell or as the reproduction by the existing cancer cells. A *death* of a tumor cell occurs as a combination of nonimmunological and immunological elements. Death due to nonimmunological element occurs at a linear rate, whereas the immunological feedback yields a nonlinear rate. Therefore, the cumulative death rate is actually a quadratic rate, at least as a first approximation. Let $X(t)$ denote the number of tumor cells at time t. Under the condition that there are x cells at time t, the probability of a: (1) birth in $(t, t + \Delta t)$ due to mutation is $m(t) \Delta t + o(\Delta t)$, (2) birth in $(t, t + \Delta t)$ due to reproduction is $bx \Delta t + o(\Delta t)$, (3) death in $(t, t + \Delta t)$ due to a nonimmunological element is $dx \Delta t + o(\Delta t)$, (4) death in $(t, t + \Delta t)$ due to immunological response is $kx^2 \Delta t + o(\Delta t)$, and (5) no change in $(t, t + \Delta t)$ is $1 - [m(t) + bx + dx + kx^2] \Delta t + o(\Delta t)$, where $b, d, k > 0$, $b - d > k$, and $m(t) \geqslant 0$ for all t. Let $X(0) \equiv 1$. Now the forward equation takes the form, abusing the notation and taking $m(t) = 0$,

$$p'(t, x) = b(x - 1)p(t, x - 1) - (b + d + kx)xp(t, x)$$
$$+ [d + k(x + 1)](x + 1)p(t, x + 1), \qquad x \geqslant 1, \quad (5.3.30)$$
$$p'(t, 0) = (d + k)p(t, 1).$$

Several conventional linearization techniques to solve (5.3.30) fail in the sense that they lead to contradictory results. We present here a method due to Dubin (1976). The idea is to linearize the quadratic term in death rate as $x^2 = xX(t)$ [cf. (5.3.32)]. This would linearize the system at the expense of homogeneity. Example 6 now comes to our rescue. Let $y(t)$ denote the solution of equation $y'(t) = (b - d - ky)y$, which describes the deterministic growth of the tumor. Then

$$y(t) = \frac{(b - d)}{[(b - d - k)e^{-(b-d)t} + k]}. \qquad (5.3.31)$$

Now we replace the death rate

$$dx + kx^2 \qquad \text{by} \qquad dx + ky(t)x. \qquad (5.3.32)$$

Now appealing to Example 6, we obtain

5. Purely Discontinuous Markov Processes

$$a(t) = \int_0^t \left\{ d - b + \frac{k(b-d)}{[(b-d-k)e^{-(b-d)s} + k]} \right\} ds$$

$$= \log \left\{ \frac{[(b-d-k)e^{-(b-d)t} + k]}{(b-d)} \right\},$$

so that

$$F(t) = \int_0^t b(s)e^{a(s)} ds = (b-d)^{-1} \left\{ bkt + \frac{b(b-d-k)}{b-d}[1 - e^{-(b-d)t}] \right\}.$$

Hence

$$g(t,\xi) = 1 + (b-d) \left\{ \frac{(b-d-k)e^{-(b-d)t} + k}{\xi - 1} - bkt \right.$$

$$\left. - \frac{b(b-d-k)}{b-d}[1 - e^{-(b-d)t}] \right\}^{-1}.$$

The mean number $m(t)$ of cancer cells is

$$m(t) = \frac{\partial g}{\partial \xi} \Big|_{\xi=1} = (b-d)[(b-d-k)e^{-(b-d)t} + k]^{-1}.$$

The variance is given by

$$\sigma^2(t) = \left[\frac{\partial^2 g}{\partial \xi^2} + \frac{\partial g}{\partial \xi} - \left\{ \frac{\partial g}{\partial \xi} \right\}^2 \right] \Big|_{\xi=1}$$

$$= \frac{2bk[(b-d)t + 1] + (b-d)(b+d+k) - (b+d)(b-d-k)e^{-(b-d)t}}{[(b-d-k)e^{-(b-d)t} + k^2]^2}.$$

Also

$$p(t,0) = 1 - \{(b-d)^2[k(b-d)(bt+1)$$

$$+ (b-d-k)(b - de^{-(b-d)t})]^{-1}\}.$$

It is easy to see that

$$\lim_{t\to\infty} m(t) = \frac{b-d}{k}$$

$$\lim_{t\to\infty} \sigma^2(t) = \infty$$

$$\lim_{t\to\infty} p(t,0) = 1.$$

5.4. Recurrence and Ergodic Properties

In this section, which parallels Sections 3.5 and 3.6, we discuss the notions of recurrence, transience, and stationary distributions associated with a purely discontinuous homogeneous Markov process. Let $X(t)$, $t \geqslant 0$, be a purely discontinuous Markov process with an at most countable state space S. Suppose that the Markov system is starting from a nonabsorbing state x. Denoting the first jump time (from x) by $\mathbf{t}_1$, define the *first return time* to x by

$$\mathbf{t}_x(\omega) = \inf\{t \geqslant \mathbf{t}_1(\omega): X(t, \omega) = x\}$$

and *the first entrance time* from x into y, $x \neq y$, by

$$\mathbf{t}_{xy}(\omega) = \inf\{t \geqslant \mathbf{t}_1(\omega): X(t, \omega) = y\}.$$

Note that $\mathbf{t}_x = \mathbf{t}_{xx}$. Set $f^*(x, y) = P\{\mathbf{t}_{xy} < \infty\}$.

Definition 5.4.1.

1. A state x is called *recurrent* if $f^*(x, x) = 1$ and *transient* if $f^*(x, x) < 1$. If all the states in S are recurrent (resp. transient), the process X is called a *recurrent* (resp. *transient*) *process*.
2. Let $\mu(x) = E_x[\mathbf{t}_x]$ denote the *mean recurrence time* for the state x. A nonabsorbing state x is called *positive recurrent* (resp. *null recurrent*) if $\mu(x) < \infty$[resp. $\mu(x) = \infty$]. The process $X(t)$ is called a *positive recurrent process* (resp. *null recurrent process*) if all the states in S are positive recurrent (resp. null recurrent).
3. A class C of states is called *closed* if $p(t, x, y) = 0$ for every $x \in C$ and every $y \notin C$ and for all $t > 0$. A closed set C is called *minimal* if it contains no closed proper subclass. The process $X(t)$ is called *irreducible* if $f^*(x, y) > 0$ for all $x, y \in S$.

Throughout this section the following assumptions hold. The process $X(t)$, $t \geqslant 0$, is a purely discontinuous homogeneous Markov process with an at most countable state space S, intensity function $q(x)$, jump probabilities $Q(x, y)$, transition probabilities $p(t, x, y)$, and infinitesimal generator $\mathbf{A} = [a(x, y)]$, $(x, y \in S)$, such that $q = \sup_{x \in S} q(x) < \infty$ and $\lim_{t \to 0} p(t, x, y) = \delta_{xy}$. Let $\mathbf{Q}$ denote the stochastic matrix of jump probabilities $Q(x, y)$. By Y_n, $n \geqslant 0$, we denote the MC with $\mathbf{Q} = [Q(x, y)]$ as its transition probability matrix. Consider the matrix $\mathbf{M}$ defined by $\mathbf{M} = (I + q^{-1}\mathbf{A})$, where I is the identity matrix and $q = \sup_{x \in S} q(x)$. Denote the elements of $\mathbf{M}$ and $\mathbf{M}^n$ by $p(x, y)$ and $p^n(x, y)$ respectively, where

131

$\mathbf{M}^0 = I, n \geqslant 1$. First we note that, from Example 5.1.5 (3) and the Kolmogorov backward equation (5.1.14), that

$$a(x,y) = \frac{\partial}{\partial t} p(t,x,y)\big|_{t=0} \quad \text{and} \quad q(x) = -a(x,x) = \sum_{y \neq x} a(x,y). \quad (5.4.1)$$

From this we obtain that $p(x,y) \geqslant 0$ and $\Sigma_y p(x,y) = 1$, $x, y \in S$. Let Z_n, $n \geqslant 0$, denote the MC associated with the stochastic matrix $\mathbf{M} = [p(x,y)]$. The chains $Y = \{Y_n\}$ and $Z = \{Z_n\}$ are called the *embedded* $\mathbf{Q}$ *-chain* and the *embedded* $\mathbf{A}$ *-chain*, respectively. They are generally called *embedded chains*. Now define

$$p^*(t,x,y) = e^{-qt} \sum_{n \geqslant 0} \frac{q^n t^n p^n(x,y)}{n!}, \quad t \geqslant 0, \, x, y \in S. \quad (5.4.2)$$

Using this in the integrodifferential equation (5.1.17), it is not difficult to see that $p^*(t,x,y)$ solves that equation. Also

$$\sum_{y \in S} p^*(t,x,y) = e^{-qt} \sum_{n \leqslant 0} \frac{q^n t^n}{n!} \sum_{y \in S} p^n(x,y) = e^{-qt} \sum_{n \leqslant 0} \frac{q^n t^n}{n!} = 1.$$

Therefore, it follows, from the Kolmogorov–Feller theorem, that $p^*(t,x,y) = p(t,x,y)$, $t \geqslant 0$, $x, y \in S$, (we omitted several details here). The basic assumptions following Definition 5.4.1 are not repeated in the statements of our theorems; they are assumed to hold throughout this section.

Theorem 5.4.2. (i) *The limit* $\lim_{t \to \infty} p(t,x,x)$ *exists for all* $x \in S$.

(ii) *A state* x *is transient if and only if* $\int_0^\infty p(t,x,x) \, dt < \infty$.

PROOF. (i) Since $p(x,x) > 0$ for all $x \in S$, the embedded A-chain Z is an aperiodic chain. Then it follows from Theorem 3.6.4 (i)–(ii) that $p^n(x,x) \to \lambda(x)$, say. Then using (5.4.2) for $p(t,x,y)$, we obtain

$$p(t,x,x) - \lambda(x) = e^{-qt} \sum_{n=0}^{N} \frac{q^n t^n}{n!} (p^n(x,x) - \lambda(x))$$

$$+ e^{-qt} \sum_{n \geqslant N+1} \frac{q^n t^n}{n!} (p^n(x,x) - \lambda(x)). \quad (5.4.3)$$

We claim that $\lim_{t \to \infty} p(t,x,x) = \lambda(x)$. To see this, fix an $\epsilon > 0$ arbitrarily. Since $p^n(x,x) \to \lambda(x)$, choose a sufficiently large N such that $|p^n(x,x) - \lambda(x)| < \epsilon$ for all $n \geqslant N$. Therefore, the absolute value of the second term in the RHS of (5.4.3) can be made arbitrarily small (for all t). As $t \to \infty$, the first term goes to 0. This proves our claim and part (i) of the theorem.

(ii) Let $F_{xy}(t)$ be the distribution function of the first entrance time $\mathbf{t}_{xy}$, that is, $F_{xy}(t) = P\{\mathbf{t}_{xy} < t\}$. Then it follows as in Example 5.1.5 (3) that, for $t \geqslant 0$ and $x, y \in S$,

$$p(t, x, y) = e^{-q(x)t}\delta_{xy} + \int_0^t p(t - s, y, y)\, dF_{xy(s)}. \qquad (5.4.4)$$

Let $\pi(\theta, x, y)$ and $\phi(\theta, x, y)$ be the Laplace and Laplace–Stieltjes transforms of $p(t, x, y)$ and $F_{xy}(t)$, respectively. From (5.4.4) it follows that

$$\pi(\theta, x, x) = [\theta + q(x)]^{-1}[1 - \phi(\theta, x, x)]^{-1}, \qquad \theta > 0. \qquad (5.4.5)$$

An Abelian theorem for Laplace–Stieltjes transforms states that *if $F_{xy}(t)$ is of bounded variation in every finite interval, the transform $\phi(\theta, x, y)$ is convergent and* $\lim_{t\to\infty} F_{xy}(t)$ *exists, then*

$$\lim_{\theta \to 0} \phi(\theta, x, y) = \lim_{t\to\infty} F_{xy}(t). \qquad (5.4.6)$$

If x is transient, then, from (5.4.6) and $\lim_{t\to\infty} F_{xx}(t) = f^*(x, x) < 1$,

$$\lim_{\theta \to 0} \phi(\theta, x, x) = \lim_{t\to\infty} F_{xx}(t) < 1.$$

Now it follows from (5.4.5) that $\pi(\theta, x, x)$ approaches a finite limit as $\theta \downarrow 0$. Therefore, $\int_0^\infty p(t, x, x) < \infty$, as follows from an Abelian theorem for Laplace transforms, which states that if $p(t, x, x) \geqslant 0$ for $t \geqslant 0$ *and* $\pi(\theta, x, x)$ converges for $\theta > 0$, then

$$\lim_{\theta \to 0} \pi(\theta, x, x) = \int_0^\infty p(t, x, x)\, dt. \qquad (5.4.7)$$

To see the converse, let $\int_0^\infty p(t, x, x)\, dt < \infty$. Then another part of the Abelian theorem states that $\pi(\theta, x, x)$ is uniformly convergent for $\theta > 0$, and (5.4.7) holds. Therefore, it follows from (5.4.5) that $F_{xy}(\infty) = \lim_{t\to\infty} F_{xx}(t) < 1$ so that x is a transient state. This completes the proof. $\square$

Theorem 5.4.3. (i) *If x is a positive recurrent state, then* $\lim_{t\to\infty} p(t, x, x) = [q(x)\mu(x)]^{-1} > 0$, *where $\mu(x)$ is the mean recurrence time.*

(ii) *If x is a null recurrent state, then* $\lim_{t\to\infty} p(t, x, x) = 0$.

PROOF. Let x be a recurrent state. Then $\phi(\theta, x, x) \to 1$ as $\theta \downarrow 0$. From Theorem 5.4.2, $\lim_{t\to\infty} p(t, x, x)$ exists and hence from the Abelian theorem

$$\lim_{t \to \infty} p(t, x, x) = \lim_{\theta \downarrow 0} \theta \pi(\theta, x, x) = \frac{1}{q(x)} \lim_{\theta \downarrow 0} \frac{\theta}{1 - \phi(\theta, x, x)}$$

$$= \frac{-1}{q(x)} \lim_{\theta \downarrow 0} \left[\frac{d}{d\theta} \phi(\theta, x, x) \right]^{-1}$$

$$= \frac{1}{q(x)} \lim_{\theta \downarrow 0} \int_0^\infty t e^{-\theta t} dF_{xx}(t), \qquad \theta > 0$$

$$= \begin{cases} [q(x)\mu(x)]^{-1} & \text{if } \mu(x) = E_x[\mathbf{t}_x] < \infty \\ 0 & \text{if } \mu(x) = \infty \end{cases}$$

This proves both parts (i) and (ii). □

Theorem 5.4.4. *Let* $\{Y_n\}$ *and* $\{Z_n\}$, $n \geq 0$, *be the embedded chains associated with the purely discontinuous Markov process* $X(t)$, $t \geq 0$. *Then:*

(i) *If* x *is transient, recurrent, positive recurrent, or null recurrent for one of the processes* $X(t)$, Y_n, Z_n, *it is of the same type in all the three processes.*

(ii) *If* C *is a minimal closed set in one process, it is so in all three processes.*

PROOF. Let $p_{xy}(t, n)$ denote the probability that the system originating from x will be in state y at time t in exactly n transitions. As in Example 5.1.5 (1) it follows that

$$p_{xy}(t, 0) = \delta_{xy} e^{-q(x)t}, \qquad x, y \in S, t > 0.$$

Following arguments similar to that we used in Examples 5.1.5 (2) and (3), one can show that, for $n \geq 0$,

$$p_{xy}(t, n + 1) = \sum_{z \neq y} \int_0^t p_{xz}(s, n) a(z, y) e^{-q(y)(t-s)} ds. \tag{5.4.8}$$

Note that $p_{xy}(t, n + 1)$ is continuous in t, and

$$p(t, x, y) = \sum_{n \geq 0} p_{xy}(t, n), \tag{5.4.9}$$

which converges uniformly in every finite interval in t. This implies the continuity of $p(t, x, y)$. Now by iteration

$$\int_0^\infty p_{xy}(t, n) q(y) dt = \sum_{z \neq y} \frac{a(z, y)}{q(z)} \int_0^\infty p_{xz}(s, n) q(z) ds$$

$$= \cdots$$

$$= Q^n(x, y), \qquad n \geq 0, x, y \in S. \tag{5.4.10}$$

From (5.4.2), (5.4.9), and (5.4.10) we obtain that

$$q^{-1} \sum_{n \geqslant 0} p^n(x,y) = \int_0^\infty p(t,x,y)\,dt = \sum_{n \geqslant 0} \int_0^\infty p_{xy}(t,n)\,dt$$
$$= [q(y)]^{-1} \sum_{n \geqslant 0} Q^n(x,y), \tag{5.4.11}$$

where all the terms are finite or infinite simultaneously. Then, from Theorem 3.5.4 and Theorem 5.4.2, it follows by taking $x = y$ in (5.4.11) that x is recurrent or transient in all the processes $X(t)$, Y_n, Z_n if it is so in one of them. From Theorems 3.6.4 and 5.4.3 and the fact that $\lim_{t \to \infty} p(t, x, x) = \lim_{n \to \infty} p^n(x, x)$, it follows that x is positive recurrent to $X(t)$ if and only if it is so for Z_n. To prove (i), it remains to show that x is positive recurrent in Z_n if and only if it is so in Y_n.

Since part (ii) is clear let us assume that it holds. Let S be irreducible (for all three processes). Consider the systems

$$u(y) = \sum_{x \in S} u(x)p(x,y) \tag{5.4.12}$$

$$v(y) = \sum_{x \in S} v(x)Q(x,y) \tag{5.4.13}$$

in the unknowns $u(x)$ and $v(x)$, $x \in S$. Using the definitions of $\mathbf{M}$ and $\mathbf{Q}$, we see that systems (5.4.12) and (5.4.13) are equivalent to

$$q(y)u(y) = \sum_{x \neq y} u(x)a(x,y) \quad \text{and} \quad v(y) = \sum_{x \neq y} v(x)\frac{a(x,y)}{q(x)},$$

respectively. For simplicity, assume that all the states be aperiodic. Now it follows from Theorem 3.6.4 that if $u(x)$ and $v(x)$ are probability distributions, then

$$u(y) = \lim_{n \to \infty} p^n(x,y) \quad \text{and} \quad v(y) = \lim_{n \to \infty} Q^n(x,y).$$

Noting that $u(x)$ is a solution of (5.4.12) if and only if $q(x)u(x)$ solves (5.4.13), it follows that

$$q(y) \lim_{n \to \infty} p^n(x,y) = \lim_{n \to \infty} Q^n(x,y).$$

From this it follows that x is positive recurrent for Y if and only if it is so for Z.

Part (ii) follows from the observation that $p(t,x,y) > 0$ iff $p^n(x,y) > 0$ iff $a^n(x,y) > 0$. Hence the theorem is proved. □

135

Theorem 5.4.5. *Let $X(t)$ be an irreducible process. Then the following properties hold.*

(i) *If $X(t)$ is a positive recurrent process, the limit $\lim_{t \to \infty} p(t, x, y)$ exists and the limit $u(y)$, $y \in S$, say, is a probability distribution that is uniquely determined by*

$$q(y)u(y) = \sum_{x \neq y} u(x)a(x,y), \qquad y \in S.$$

Also $\{u(y), y \in S\}$ is the only stationary distribution of the Markov process, and

$$u(x) = \lim_{t \to \infty} p_t(x),$$

where $p_t(x) = P\{X(t) = x\}$.

(ii) *If $X(t)$ is null recurrent or transient, then*

$$\lim_{t \to \infty} p(t, x, y) = 0 = \lim_{t \to \infty} p_t(x), \qquad x, y \in S.$$

PROOF. We prove the theorem only for the aperiodic case. Applying Theorem 3.6.4 to the chain Z_n, which by hypothesis is positive recurrent, we obtain that

$$u(y) = \lim_{t \to \infty} p(t, x, y) - \lim_{n \to \infty} p^n(x, y) > 0$$

and that $\{u(x), x \in S\}$ is a probability distribution uniquely determined by the system

$$u(y) = \sum_{x} u(x)p(x, y),$$

which is equivalent to $q(y)u(y) = \sum_{x \neq y} u(x)a(x, y)$. Now using (5.4.2) it is easy to show that

$$u(y) = \sum_{x} u(x)p(t, x, y), \qquad t \geq 0, y \in S.$$

Note that $\{u(x), x \in S\}$ is the only stationary distribution of $X(t)$, $t \geq 0$, and that $u(x) = \lim_{t \to \infty} p_t(x)$ follows from

$$p_{t+s}(y) = \sum_{x \in S} p_s(x)p(t, x, y).$$

Part (ii) follows by arguing along similar lines. This proves the theorem. $\square$

Example **5.4.6**

EXAMPLE 1. *Birth–Death Process.* Let $X(t)$ be a birth–death process with state space $S = \{0, 1, 2, \ldots\}$. From Theorem 5.4.4 it follows that $X(t)$ is recurrent if and only if the embedded chain Y_n is recurrent. Therefore, applying Example 3.5.11 (4) to

$$
Q(x, y) = \begin{cases}
\dfrac{d(x)}{b(x) + d(x)} & \text{if } y = x - 1 \\[2ex]
\dfrac{b(x)}{b(x) + d(x)} & \text{if } y = x + 1 \\[2ex]
0 & \text{otherwise,}
\end{cases}
$$

we see that the birth–death process $X(t)$ is recurrent or transient according as the series $\Sigma_{x \geqslant 1}[d(1) \cdots d(x)/b(1) \cdots b(x)]$ is divergent or convergent. Set $a(x) = 1$ if $x = 0$ and $a(x) = [b(0) \cdots b(x - 1)/d(1) \cdots d(x)]$ for $x \geqslant 1$. Let $u(x)$ denote the stationary distribution of $X(t)$. Then

$$
u(1)d(1) - u(0)b(0) = 0
$$

$$
u(x + 1)d(x + 1) - u(x)b(x) = u(x)d(x) - u(x - 1)b(x - 1), \qquad x \geqslant 1.
$$

$$
= \cdots
$$

$$
= 0.
$$

Hence $u(x + 1) = b(x)u(x)/d(x + 1)$, $x \geqslant 0$, and consequently

$$
u(x) = \frac{b(0) \cdots b(x - 1)}{d(1) \cdots d(x - 1)} u(0) = a(x)u(0).
$$

Now it is clear that the stationary distribution exists (uniquely) if and only if $\Sigma_{x \geqslant 1}[b(0) \cdots b(x - 1)/d(1) \cdots d(x - 1)] < \infty$. If this series converges, then the stationary distribution is given by $u(x) = a(x)/\Sigma_{y \geqslant 0} a(y)$, $x \geqslant 0$.

Exercises

1. The contagion process is a birth process with rate $b(x) = (b + cx)$, where $x \geqslant 0$, $b > 0$, and $c > 0$. Show that the transition probabilities of $X(t)$, $t \geqslant 0$, are given by

$$
p(t; x, y) = e^{-(b + cx)t} \binom{-x - \dfrac{b}{c}}{y - x}(e^{-ct} - 1)^{y - x}, \qquad y \geqslant x.
$$

2. By solving the backward equation for a two-state birth–death process, show that the transition probabilities are given by

$$p(t; 0, 0) = \frac{d + be^{-(b+d)t}}{b + d}$$

$$p(t; 1, 0) = \frac{d - de^{-(b+d)t}}{b + d}$$

$$p(t; 0, 1) = \frac{b - be^{-(b+d)t}}{b + d}$$

$$p(t; 1, 1) = \frac{b + de^{-(b+d)t}}{b + d},$$

where $b = b(0)$ and $d = d(1)$. [Note that $b(1) = 0 = d(0)$.]

3. The Pólya process is a nonhomogeneous birth process having the probability $b(1 + cx)\Delta t/(1 + bct) + o(\Delta t)$ of a change in the interval $(t, t + \Delta t)$. Here $b, c \geqslant 0$. Let $p(t, x)$ be the probability that there are x individuals in the population at time t, and let $p(0, x) = \delta_{x0}$. Show that:

(a) $p(t, 0) = (1 + bct)^{-c^{-1}}$;

(b) $p(t, x) = \dfrac{(bt)^x}{x!}(1 + bct)^{-x-c^{-1}} \displaystyle\prod_{k=1}^{x-1}(1 + ck), \qquad x \geqslant 1$;

(c) $\displaystyle\sum_{x \geqslant 0} p(t, x) = 1$;

(d) $E[X(t)] = bt$;

(e) $\text{var}(X(t)) = bt(1 + bct)$;

(f) The Kolmogorov equations are given by

$$\frac{d}{dt}p(t, x, y) = -b\frac{1 + cy}{1 + bct}p(t, x, y) + b\frac{1 + c(y - 1)}{1 + bct}p(t, x, y - 1)$$

$$\frac{d}{dt}p(t, x, y) = -b\frac{1 + cx}{1 + bct}p(t, x, y) + b\frac{1 + c(x + 1)}{1 + bct}p(t, x + 1, y).$$

4. Let $X(t)$ be a birth–death process with rates $b(x) = b[N - x]$ and $d(x) = dx$, $0 \leqslant x \leqslant N$, and with $X(0) = 0$. Set $p = b[1 - e^{-(b+d)t}]/(b + d)$ and $q = 1 - p$. Show that

$$p(t, x) = P\{X(t) = x\} = \binom{N}{x}p^x q^{N-x}, \qquad 0 \leqslant x \leqslant N.$$

5. Investigate in detail the quadratic birth–death process with $b(x) = (bx^2 + c)$ and $d(x) = (dx^2 + e)$ (Karlin and McGregor 1958).

6. Solve the differential equations for a birth–death process with rates

$$b(x) = \begin{cases} b + cx, & x \geqslant 0 \\ b - c^* x, & x < 0 \end{cases}, \qquad d(x) = \begin{cases} d + ex, & x \geqslant 0 \\ d - e^* x, & x < 0 \end{cases}$$

and with initial condition $p(0, 1) = 1$. [*Hint:* Derive the equation for the generating function $g(t, \xi)$ of $p(t, x)$ and then apply the bilateral Laplace transform.]

7. Let $X(t)$ be a linear birth–death process $(b(x) = bx, d(x) = dx)$ with $X(0) = 1$. Show that

$$P\{x \text{ births before the first death}\} = \frac{b^x d}{(b + d)^{x+1}}.$$

8. Let $X(t)$ be a linear birth process with $X(0) = 1$. Let $N(t, a)$ be the number of individuals in the population at time t with age not exceeding a. Show that

$$P_1\{N(t, a) = x\} = e^{-bt}(1 - e^{-ba})^x (1 - e^{-ba} + e^{-bt})^{-x-1}.$$

9. Consider the stochastic epidemic described in Example 5.3.3. (4). Let T be the random time to complete the epidemic and S_x the time spent in state x. Assume that S_x is exponentially distributed with parameter $d(x) = dx(n - x + 1)$. Show that

$$E[e^{-\theta T}] = \prod_{x=1}^{n} \frac{dx(n - x + 1)}{[\theta + dx(n - x + 1)]}.$$

Show also that

$$E[T] \simeq \frac{2[\log(n + 1)]}{d(n + 1)},$$

as $n \to \infty$.

10. Let $X_1(t)$ and $X_2(t)$ be two independent linear birth processes with the same parameter b and with $X_1(0) = x_1$ and $X_2(0) = x_2$. Show that, for $x \geqslant x_1$ and $N \geqslant x_1 + x_2$,

$$P\{X_1(t) = x | X_1(t) + X_2(t) = N\}$$

$$= \binom{x - 1}{x_1 - 1}\binom{N - x - 1}{x_2 - 1} \Big/ \binom{N - 1}{x_1 + x_2 - 1}.$$

11. Let $X(t)$ be a linear birth–death process with rates b and d and denoting a fish population in a pond. Assuming, as a first approximation, that the pond has a constant carrying capacity K, let us set the birth rate as $b = 0$ whenever the population size exceeds K. Such a process is called a *birth–death process with carrying capacity* (MacArthur and Wilson 1967). Show that:

(a) for any initial population size, there is a nonzero probability of extinction in at most $(K + 1)$ steps;

(b) with probability one the population eventually becomes extinct;

(c) if $T_x = E_x$ [time to extinction], then

$$\text{(i)} \quad T_x = \frac{[bT_{x+1} + dT_{x-1} + 1]}{(b + d)},$$

$$\text{(ii)} \quad T_1 = b^{-1}\left[(b - d)T_K + \sum_1^K x^{-1}\right] + [d(K + 1)]^{-1},$$

$$\text{(iii)} \quad T_1 = b^{-1} \sum_{x=1}^K x^{-1}\left(\frac{d}{b}\right)^{-x} + \frac{(b/d)^K}{d(K + 1)}.$$

12. Let U and R denote the groups of underdeveloped and developed countries, respectively. Consider a process of population migration from U to R. Let $X(t)$ denote the number of individuals in developed countries at time t that have migrated from U. Assume that at time $t = 0$ there were u and r individuals in U and R, respectively. Given that x individuals have migrated in the time period $[0, t]$, let $p_x(t)\Delta t$ denote the probability that a single individual will migrate from U to R, and let $q_x(y, t)\Delta t$ denote the probability that y individuals will migrate from U to R in the interval $(t, t + \Delta t)$. Assuming that different individuals act independently of each other, show that

$$q_x(1, t) = p_x(t)(u - x)\Delta t + o(\Delta t)$$
$$q_x(0, t) = 1 - q_x(1, t).$$

Let $p_x(t) = f(x)$ such that $(u - y)f(y) \neq (u - z)f(z)$, $y \neq z$. In this case show that

$$p(t, 0) = e^{-b_0 t}, \qquad p(t, x) = a_x \sum_{k \geqslant 0} c_{xk}^{-1} e^{-b_k t}, \qquad x = 1, \ldots, n,$$

where

$$b_x = (u - x)f(x), \qquad a_x = \prod_{k=0}^{x-1} b_x, \qquad c_{xk} = \prod_{\substack{i=0 \\ i \neq j}}^x (b_i - b_j), \qquad j \leqslant x.$$

13. A liquid consists of molecules of types A and B that produce a molecule of type C on collision. Let $X(t)$, $Y(t)$, and $Z(t)$ denote the number of A, B, and C molecules in the liquid at time t, respectively. If $X(0) = x_0$ and $Y(0) = y_0$, then

$X(t) = x_0 - Z(t)$ and $Y(t) = y_0 - Z(t)$. Let c denote the rate of collision between molecules of types A and B and set $h(z) = (x_0 - z)(y_0 - z)$. Show that

$$P\{Z(t) = z\} = \sum_{k=0}^{z} e^{-ch(k)t} \prod_{k=0}^{x-1} h(k) \prod_{\substack{j \neq k \\ 0 \leq k \leq z}} [h(k) - h(j)]^{-1}$$

for $z = 0, 1, \ldots, (x_0 \wedge y_0)$. Show also that

$$E[Z(t)] = \frac{e^{-cx_0 t} - e^{-cy_0 t}}{x_0^{-1} e^{-c-x_0 t} - y_0^{-1} e^{-c-y_0 t}},$$

and that $E[Z(t)] \to x_0 \wedge y_0$, as $t \to \infty$.

14. In an epidemic model let S_n, I_n, and R_n denote the number of susceptibles, infectives, and removed at time n, respectively. Let p denote the probability of contact for each infective. Then $Q_n = (1 - p)^{I_n} = q^{I_n}$ is the probability of no contact. Note that $S_n + I_n + R_n = $ total population size. Let $p^n(S, I)$ denote the probability of having S susceptibles and I infectives at time n, and let $p(R|S)$ be the probability of R removals starting with S susceptibles.

(a) Show that

(i) $p^{n+1}(S_n, 0) = Q^{S_n}$,

(ii) $p^{n+1}(S_{n-1}, 1) = S_n Q_n^{S_{n-1}}(1 - Q_n)$,

(iii) $p^{n+1}(S_n - I_{n+1}, I_{n+1}) = \binom{S_n}{I_{n+1}} Q_n^{S_n - I_{n+1}}(1 - Q_n)^{I_{n+1}}$

(b) Show that

(i) $p(0|S) = Q_0^S$,

(ii) $p(R|S) = \binom{S}{R} p(R|R) Q_0^{S-R} q^{R(S-R)}$, $S < R$,

(iii) $p(S|S) = 1 - \sum_{R=0}^{S-1} p(R|S)$

(Ludwig 1975).

15. Let $d(x) = x$, $x \geq 0$, be the death rate of a birth–death process $X(t)$ on $S = \{0, 1, \ldots\}$. Show that $X(t)$ is null recurrent if $b(x) = x + 1$, $x \geq 0$, and is transient if $b(x) = x + 2$, $x \geq 0$.

16. Find the stationary distribution of the two-state birth–death process (see Exercise 2).

17. Consider a birth–death process on $S = \{0, 1, \ldots, N\}$ with rates $b(x) = b > 0$, and $d(x) = dx$, $d > 0$. Show that the stationary distribution of $X(t)$ is given by

$$\pi(x) = \frac{1}{x!}\left(\frac{b}{d}\right)^x \sum_{y=0}^{N} \frac{1}{y!}\left(\frac{b}{d}\right)^y.$$

6

Calculus with Stochastic Processes

6.1. Introduction

Consider a stochastic process $X = \{X(t, \omega)\}$, $t \in T$, $\omega \in \Omega$, with state space $S = R$, the real line. Most of what we do in this chapter holds also if S is the complex plane or a suitable metric space. The time parameter set T is taken as either $T = R$ or $T = \overline{a, b}$, a finite or infinite interval. The process X is an RV for every fixed $t \in T$. If we fix the sample or chance point ω, then $X_\omega(\cdot) = X(\cdot, \omega)$ is a function on T called a *sample path* or *trajectory*. It is also common to denote the sample path $X_\omega(t)$ by $\omega(t)$. The collection of all sample paths is called the *sample path space* of the process X. In general the sample path space is a subcollection of the space of all real-valued functions defined on T. One such space is obtained by considering all possible electrocardiogram outputs over a time interval $[a, b]$. A similar space is obtained by considering all possible seismograph recording over a time interval $[a, b]$. If $\{X(t, \omega)\}$ denotes the evolution of a physical process, then $X_\omega(t)$ represents a possible realization of the physical phenomenon. This analogy of a sample path suggests the necessity to study the analytic properties of the sample paths. First we study the sample properties such as the continuity, differentiation, and integration. For fixed time points s and t, the RVs $X(s, \cdot)$ and $X(t, \cdot)$ express, say, the random strength of the seismic events at the time points s and t. One would like to measure the "dissimilarity" or the "distance" between the two measurements $X(s)$ and $X(t)$. Such a notion helps us to define the continuity, differentiation, and integration in terms of this "distance." We also study such analytical properties of a stochastic process.

6. Calculus with Stochastic Processes

Before establishing certain sample path properties a remark is in order. This is about the concept of separability of a stochastic process. The technical details behind this notion are beyond the scope of this textbook. So we are content here with a brief remark about this concept. Let $X(t)$ be a real-valued stochastic process on T. We see later that subsets of Ω of the form

$$\{\omega: X(t, \omega) \in G \quad \text{for all } t \in I\},$$

where I is an open interval in T and G is a Borel subset of R, arise in many important problems. Since I is an uncountable set, one cannot in general assert that the sets of the preceding form are in the σ-algebra $\mathcal{C}$. That is, such collection of sample points ω may not form an event, and thus no probability can be assigned to it. This problem is more serious than it appears on the surface. To rectify this situation, Doob (1953) introduced the concept of separable stochastic processes.

Let $X(t)$ be a real-valued process on an infinite or finite interval T. The process $X(t)$, $t \in T$, is said to be *separable* if there is a countable set $C \subset T$, called *separant*, and a subset $N \subset \Omega$ with $P\{N\} = 0$ such that, for any open interval $I \subset T$ and an arbitrary closed set $F \subset R$, the difference

$$\{\omega: X(t, \omega) \in F \quad \text{for all } t \in I\} \backslash \{\omega: X(t, \omega) \in F \quad \text{for all } t \in C \cap I\}$$

is a subset of N. (Recall that the probability space $(\Omega, \mathcal{C}, P)$ is always assumed to be complete.)

It is a result of Doob (1953) that corresponding to any real-valued stochastic process $X(t)$, $t \in T$, one can always find an extended real-valued process $Y(t)$, $t \in T$, which is separable and equivalent to $X(t)$.

Based on Doob's result we can and do assume hereafter that every stochastic process $X(t)$ we work with is separable. This is a great technical convenience. To see the power of this, let us consider an illustration. In Section 6.2 we present conditions under which a given process $X(t)$ can be equivalent to a process $Y(t)$ almost all of whose sample paths are continuous. If $X(t)$ were a separable process, then this conclusion reduces to the claim that

$$P\{\omega: X_\omega(t) \quad \text{is continuous on } T\} = 1,$$

where $T = [a, b]$, say. That is we get continuity for $X(t)$ itself. Let C be the countable separant set that appears in the definition of a separable process. Clearly, $X(t) = Y(t)$ for all $t \in C$, with probability one, where $Y(t)$ is the process equivalent to $X(t)$ and almost all of whose sample paths are continuous. We want to show that $P\{X(t) = Y(t), t \in T\} = 1$. Fix t^*

$\in T$ and arbitrarily choose an $\epsilon > 0$. Then there exists a $\delta > 0$ such that $|Y(t) - Y(t^*)| < \epsilon$ provided $|t - t^*| < \delta$. Let $I = \{t \in T: |t - t^*| < \delta\}$. Then

$$Y(t^*) - \epsilon \leqslant \inf_{t \in I \cap C} Y(t) = \inf_{t \in I \cap C} X(t)$$

$$= \inf_{t \in I} X(t), \qquad \text{by separability,}$$

$$\leqslant X(t^*).$$

Letting $\epsilon \downarrow 0$, we have $Y(t^*) \leqslant X(t^*)$. Similarly, $X(t^*) \leqslant Y(t^*)$ and consequently $X(t^*) = Y(t^*)$. Since t^* is arbitrary, $X(t) = Y(t)$ for all $t \in T$. This proves that $P\{X(t) = Y(t), t \in T\} = 1$, since almost all sample paths of $Y(t)$ are continuous on T.

Let X be an RV with finite second moment, that is, $E[X^2] < \infty$. We do not distinguish between any two equivalent RVs. The collection of all (equivalence class of) RVs with finite second moment is denoted by L_2. For any two $X, Y \in L_2$, define

$$\|X - Y\| = \{E[(X - Y)^2]\}^{\frac{1}{2}}. \tag{6.1.1}$$

We claim that $\|X - Y\|$ defines a distance or metric on L_2. First we establish the Cauchy–Schwarz inequality that for $X, Y \in L_2$

$$E[|XY|] \leqslant \|X\|\|Y\| = \{E[X^2]E[Y^2]\}^{\frac{1}{2}} \tag{6.1.2}$$

From the classical inequality $|xy| \leqslant (x^2 + y^2)/2$ we get $E|XY| \leqslant E[(X^2 + Y^2)/2]$. Therefore, if $X, Y \in L_2$, then $E|XY| \leqslant (\|X\|^2 + \|Y\|^2)/2 < \infty$. Now let $x = X/\|X\|$ and $y = Y/\|Y\|$. Then

$$E[|xy|] = E\left[\frac{|XY|}{\|X\|\|Y\|}\right] \leqslant \frac{1}{2}\left\{E\left[\frac{X^2}{\|X\|^2}\right] + E\left[\frac{Y^2}{\|Y\|^2}\right]\right\} = 1,$$

and we obtain (6.1.2).

Next we show that the triangle inequality

$$\|X + Y\| \leqslant \|X\| + \|Y\|, \tag{6.1.3}$$

holds. Now

$$\|X + Y\|^2 = E[(X + Y)^2] = E[X^2] + 2E[XY] + E[Y^2]$$

$$\leqslant \|X\|^2 + 2E[|XY|] + \|Y\|^2$$

$$\leqslant \|X\|^2 + 2\|X\|\|Y\| + \|Y\|^2 \qquad \text{[by (6.1.2)]}$$

$$= (\|X\| + \|Y\|)^2.$$

This proves (6.1.3). Now it is clear that $\|\cdot\|$ defines a metric on L_2.

A stochastic process $X(t, \omega)$, $t \in T$, is called a *second-order process* if $E[(X(t))^2] < \infty$ for every $t \in T$. Important processes such as Gaussian and Wiener–Levy processes belong to this class. The *mean-square distance* $\|\cdot\|$ introduced above is used here to define the analytical notions of continuity and so on. Let $X(t)$ be a second-order process. The *mean function* of $X(t)$ is defined by $m(t) = E[X(t)]$, $t \in T$. The *covariance function* $K(s, t)$ of $X(t)$ is defined by

$$K_X(s, t) = \text{cov}(X(s), X(t)) = E[X(s)X(t)] - m_x(s)m_X(t). \quad (6.1.4)$$

Examples 6.1.1

EXAMPLE 1. Find the covariance function of the process $X(t)$ $= \Sigma_{k=1}^n [\xi_k \cos \lambda_k t + \eta_k \sin \lambda_k t]$, where λ_k are real constants, $\xi_1, \ldots, \xi_n$, $\eta_1, \ldots, \eta_n$ are independent random variables with mean zero and $\text{var}(\xi_k)$ $= \sigma_k^2 = \text{var}(\eta_k)$, $1 \leqslant k \leqslant n$.

$$m_X(t) = E[X(t)] = \sum_{k=1}^n \{E[\xi_k] \cos \lambda_k t + E[\eta_k] \sin \lambda_k t\} = 0.$$

By the independence and zero mean assumptions,

$$E[\xi_i \xi_j] = 0 = E[\eta_i \eta_j] \qquad \text{if} \qquad i \neq j,$$

$$E[\xi_i \eta_j] = 0 \qquad \text{for all} \qquad i, j.$$

Therefore

$$K_X(s, t) = E\left[\left\{ \sum_{j=1}^n (\xi_j \cos \lambda_j s + \eta_j \sin \lambda_j s) \right\} \left\{ \sum_{k=1}^n (\xi_k \cos \lambda_k t + \eta_k \sin \lambda_k t) \right\} \right]$$

$$= \sum_{k=1}^n \{E[\xi_k^2] \cos \lambda_k s \cos \lambda_k t + E[\eta_k^2] \sin \lambda_k s \sin \lambda_k t\}$$

$$= \sum_{k=1}^n \sigma_k^2 \cos \lambda_k (t - s).$$

EXAMPLE 2. Find the mean and covariance functions of a Poisson process $X(t)$ with parameter λ.

$$m_X(t) = E[X(t)] = \lambda t, \qquad t \in T.$$

Let $0 \leqslant s \leqslant t$. Then

$$
\begin{aligned}
E[X(s)X(t)] &= E[X(s)\{X(t) - X(s) + X(s)\}] \\
&= E[(\dot{X}(s))^2] + E[X(s)]E[X(t) - X(s)] \quad \text{(why?)} \\
&= \text{var}(X(s)) + (m_X(s))^2 + \lambda s \lambda(t - s) \\
&= \lambda s + \lambda^2 s^2 + \lambda s \lambda(t - s) = \lambda s + \lambda s \lambda t.
\end{aligned}
$$

Therefore

$$
K_X(s, t) = \lambda s + (\lambda s)(\lambda t) - (\lambda s)(\lambda t) = \lambda s.
$$

Hence if $s \leqslant t$, $K_X(s, t) = \lambda s$. By symmetry, $K_X(s, t) = \lambda t$ if $t \leqslant s$. Consequently

$$
K_X(s, t) = \lambda(s \wedge t) = \lambda \min(s, t).
$$

Let $X(t)$ and $Y(t)$ be two second-order processes. The *mutual-* or *cross-covariance function* $K_{XY}(s, t)$ of the processes $X(t)$ and $Y(t)$ is defined by

$$
K_{XY}(s, t) = \text{cov}(X(s), Y(t)) = E[X(s)Y(t)] - m_X(s)m_Y(t).
$$

Proposition 6.1.2. Let $X(t)$ and $Y(t)$ be two centered second-order processes [so that $m_X(t) = 0 = m_Y(t)$]. Then:

(i) $K_X(s, t) = K_X(t, s)$,

(ii) $K_{XY}(s, t) = K_{YX}(t, s)$,

(iii) $K_{X+Y}(s, t) = K_X(s, t) + K_{XY}(s, t) + K_{YX}(s, t) + K_Y(s, t)$,

(iv) $K_X(s, t)$ is positive definite, that is, for all $n \geqslant 1$, time points $t_1, \ldots, t_n \in T$ and real numbers $a_1, \ldots, a_n$,

$$
\sum_{i=1}^{n} \sum_{j=1}^{n} K_X(t_i, t_j) a_i a_j \geqslant 0.
$$

PROOF. Relations (i)–(iii) are clear. To see (iv), note that

$$
0 \leqslant E\left[\left(\sum_{i=1}^{n} X(t_i)a_i \right)^2 \right] = E\left[\sum_{i=1}^{n} \sum_{j=1}^{n} X(t_i)X(t_j)a_i a_j \right]
$$

$$
= \sum_{i=1}^{n} \sum_{j=1}^{n} K_X(t_i, t_j) a_i a_j,
$$

since $m_X(t) = 0$. $\square$

Definition 6.1.3. A sequence $\{X_n, n \geqslant 1\}$ of L_2-RVs is said to be *mean-square convergent* to a RV $X \in L_2$ if

$$E[|X_n - X|^2] \to 0, \qquad \text{as } n \to \infty.$$

Remark 6.1.4. Let $\{X_n, n \geqslant 1\}$ be a sequence of L_2-RVs. Then $E[|X_n - X_m|^2] \to 0$, as $m \to \infty$, $n \to \infty$ independently of each other, if and only if there is an RV $X \in L_2$ such that X_n is mean-square (MS) convergent to X. Moreover, $\|X_n\|^2 \to \|X\|^2$ and $E[X_n] \to E[X]$ as $n \to \infty$. Let $X_n \xrightarrow{\text{MS}} X$ and $Y_n \xrightarrow{\text{MS}} Y$. Using the Schwarz inequality, it is easy to see that

$$E[X_n Y_n] \to E[XY], \qquad \text{as } n \to \infty.$$

We can find the proof of each of these statements in a standard textbook on probability theory.

Theorem 6.1.5. (Loeve). *A sequence $\{X_n, n \geqslant 1\}$ of L_2 RVs converges in the mean-square sense if and only if*

$$E[X_n X_m] \to a \text{ finite limit } \alpha$$

as $m \to \infty$ and $n \to \infty$, independently of each other.

PROOF. The sufficiency follows from

$$E[(X_n - X_m)^2] = E[X_n^2] - 2E[X_n X_m] + E[X_m^2]$$

$$= \alpha - 2\alpha + \alpha = 0.$$

To see the necessity, let X_n converge in mean square to an $X \in L_2$. Then

$$E[X_n X_m] \to E[XX] = E[X^2] = \alpha, \qquad \text{say} . \qquad \square$$

6.2. Continuity

Definition 6.2.1. A stochastic process $X: T \times \Omega \to R$ is called *stochastically continuous* at a point $t_0 \in T$, if for any $\epsilon > 0$ we have $P\{|X(t) - X(t_0)| > \epsilon\} \to 0$ as $|t - t_0| \to 0$. If $X(t)$ is stochastically continuous at every point in T then it is called stochastically continuous on T. A process $X(t)$, $t \in T$, is called *stochastically uniformly continuous* on T if for arbitrary constants $\epsilon > 0$ and $\eta > 0$ there exists a $\delta > 0$ such that $P\{|X(t) - X(s)| > \epsilon\} < \eta$ as long as $|s - t| < \delta$.

Theorem 6.2.2. *If the process $X(t)$ is stochastically continuous on a compact T, then $X(t)$ is stochastically uniformly continuous.*

PROOF. Assume the contrary. Then there exists a pair $\epsilon > 0$ and $\eta > 0$ such that for any $\delta_n > 0$, and $s_n, t_n \in T$ with $|s_n - t_n| < \delta_n$, we have

$$P\{|X(s_n) - X(t_n)| > \epsilon\} > \eta.$$

Let us choose δ_n and s_n such that $\delta_n \to 0$ and $s_n \to s_0$. Then $t_n \to s_0$ and

$$\eta < P\{|X(s_n) - X(t_n)| > \epsilon\}$$
$$\leqslant P\{|X(s_n) - X(s_0)| > \tfrac{\epsilon}{2}\} + P\{|X(s_0) - X(t_n)| > \tfrac{\epsilon}{2}\}.$$

This contradicts the stochastic continuity of $X(t)$. This proves the theorem.

□

Definition 6.2.3. A process $X(t)$, $t \in T$, is called a *continuous process* if almost all of its sample functions are continuous on T.

As remarked earlier, we always assume that all our processes $X(t)$ are separable. Also, let $T = [0, 1] = I$.

Theorem 6.2.4. *Let $g(h)$ and $q(h)$ be two even functions nondecreasing in $h > 0$ such that*

$$\sum_{n \geqslant 1} g(2^{-n}) < \infty \qquad and \qquad \sum_{n \geqslant 1} 2^n q(2^{-n}) < \infty. \qquad (6.2.1)$$

Let $X(t)$, $t \in I$, be a process such that, for all t, $t + h \in I$ we have

$$P\{|X(t + h) - X(t)| \geqslant g(h)\} \leqslant q(h). \qquad (6.2.2)$$

Then $X(t)$ is a continuous process.

PROOF. The proof follows a standard polygonal approximation method. For every $n \geqslant 1$ consider the time points

$$t_{n,k} = k2^{-n}, \qquad k = 0, 1, \ldots, 2^n$$

6. Calculus with Stochastic Processes

and the polygonal approximation

$$X_n(t) = X(t_{n,k}) + 2^n(t - t_{n,k})\{X(t_{n,k+1}) - X(t_{n,k})\} \qquad (6.2.3)$$

for $t_{n,k} \leqslant t \leqslant t_{n,k+1}$ (see Figure 6.1). Then for $t_{n,k} \leqslant t \leqslant t_{n,k+1}$,

$$|X_{n+1}(t) - X_n(t)|$$
$$\leqslant 2^{-1}\{|X(t_{n+1,2k+1}) - X(t_{n+1,2k})| + |X(t_{n+1,2k+1}) - X(t_{n+1,2k+2})|\}. \qquad (6.2.4)$$

The RHS of this inequality is free of t in $I_{n,k} = [t_{n,k}, t_{n,k+1}]$. Consequently, the RHS of (6.2.4) dominates

$$\max_{t \in I_{n,k}} |X_{n+1}(t) - X_n(t)|.$$

Denote the RHS of (6.2.4) by $2^{-1}(A_1 + A_2)$. Now

$$P\left\{\max_{t \in I_{n,k}} |X_{n+1}(t) - X_n(t)| \geqslant g(2^{-n-1})\right\}$$
$$\leqslant \sum_{i=1}^{2} P\{A_i \geqslant g(2^{-n-1})\}$$
$$\leqslant 2q(2^{-n-1}),$$

and hence

$$P\left\{\max_{t \in I} |X_{n+1}(t) - X_n(t)| \geqslant g(2^{-n-1})\right\} \leqslant 2^{n+1}q(2^{-n-1}).$$

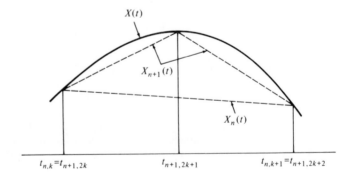

Figure 6.1

But by assumption (6.2.1), $\Sigma 2^n q(2^{-n}) < \infty$. Therefore, by the Borel–Cantelli lemma, $X_n(t)$ converges AS and the convergence is uniform in I. Since each $X_n(t)$ is continuous on I and the convergence is uniform, the limit $Y(t)$ is continuous on $[0, 1]$.

It remains to show that the processes $X(t)$ and $Y(t)$ are equivalent. By the separability of $X(t)$ we then conclude that $X(t)$ is a continuous process (see Section 6.1).

Since $X_{n+i}(t) = X(t)$ for all $t = t_{n,k}$, we have $P\{X(t) = Y(t)\} = 1$, for all $t = t_{n,k}$. Now let $t \neq t_{n,k}$ for every n and k. Then

$$t = \lim_{n \to \infty} t_{n,k_n} \qquad \text{when} \qquad 0 < t - t_{n,k_n} < 2^{-n},$$

and by hypothesis (6.2.2),

$$P\left\{ \left| X\left(t_{n,k_n}\right) - X(t) \right| \geqslant g\left(t - t_{n,k_n}\right) \right\} \leqslant q\left(t - t_{n,k_n}\right) \leqslant q(2^{-n}).$$

Again appealing to the Borel–Cantelli lemma we see that $X(t_{n,k_n}) \xrightarrow{\text{AS}} X(t)$. By the continuity of the process $Y(t)$, $X(t_{n,k_n}) = Y(t_{n,k_n}) \xrightarrow{\text{AS}} Y(t)$. Hence $X(t)$ and $Y(t)$ are equivalent processes, and the theorem follows. $\qquad \square$

Theorem 6.2.5. (Kolmogorov). *Let $X(t)$, $t \in I$, be a (separable) stochastic process. If*

$$E[|X(t + h) - X(t)|^2] \leqslant M|h| \, \|\log|h|\|^{-4},$$

where $t, t + h \in I$, $M > 0$, then $X(t)$ is a continuous process.

PROOF. This is a simple consequence of Theorem 6.2.4. Take $g(h) = \|\log|h|\|^{\alpha}$ with $1 < \alpha < \frac{3}{2}$ and apply Chebyshev's inequality. $\qquad \square$

Definition 6.2.6. A process $X(t)$, $t \in I$, is said to be *without discontinuities of the second kind* in I if almost all sample functions have the limits

$$X(s +) = \lim_{t \downarrow s} X(t), t \in [0, 1) \text{ and } X(s -) = \lim_{t \uparrow s} X(t), \quad t \in (0, 1].$$

The following theorem gives conditions under which a process $X(t)$ can have no discontinuities of the second kind. We omit the proof.

Theorem 6.2.7. (Cramer 1966). *Let $g(h)$ and $q(h)$ be two functions as prescribed in Theorem 6.2.4. If*

$$P\{\|[X(u) - X(t)][X(t) - X(s)]\| \geqslant g^2(h)\} \leqslant q(h)$$

for $0 \leqslant s < t < u \leqslant 1$, $u - s = h$, then $X(t)$ is equivalent to a process $Y(t)$ that has no discontinuities of the second kind.

Definition 6.2.8. A second-order stochastic process $X(t)$, $t \in T$, is said to be *mean-square continuous* at $s \in T$ if

$$\lim_{t \to s} E[|X(t) - X(s)|^2] = 0.$$

Theorem 6.2.9. *A second-order process $X(t)$, $t \in I$, is mean-square continuous at $t = \tau$ if and only if its covariance function $K(s,t)$ is continuous at $s = t = \tau$.*

PROOF. To see the sufficiency, note that

$$\lim_{h \to 0} E[|X(\tau + h) - X(\tau)|^2]$$

$$= \lim_{h \to 0} [K(\tau + h, \tau + h) - K(\tau + h, \tau) - K(\tau, \tau + h) + K(\tau, \tau)]$$

$$= 0, \qquad \text{by the continuity of } K \text{ at } (\tau, \tau).$$

We obtain the necessity as follows, by using the Schwarz inequality:

$$|K(\tau + h, \tau + h') - K(\tau, \tau)|$$

$$\leqslant E[|X(\tau + h) - X(\tau)| |X(\tau + h') - X(\tau)|]$$

$$+ E[|X(\tau + h) - X(\tau)| |X(\tau)|] + E[|X(\tau + h') - X(\tau)| |X(\tau)|]$$

$$\leqslant \{E[|X(\tau + h) - X(\tau)|^2] E[|X(\tau + h') - X(\tau)|^2]\}^{1/2}$$

$$+ \{E[|X(\tau + h) - X(\tau)|^2] E[|X(\tau)|^2]\}^{1/2}$$

$$+ \{E[|X(\tau + h') - X(\tau)|^2] E[|X(\tau)|^2]\}^{1/2}$$

$$\to 0, \text{ as } h \to 0, \qquad \text{by the mean-square continuity of } X(t).$$

This completes the proof. $\qquad\qquad\qquad\qquad\qquad\qquad\qquad\qquad$ $\square$

***Example* 6.2.10** *The mean-square continuity does not necessarily imply sample path continuity.* Poisson processes provide us with an example. Let $X(t)$ be a Poisson process with parameter λ. We know that every sample path of $X(t)$ is a step function with unit jumps (occurring at random times). A typical sample path is drawn in Figure 6.2. Therefore, $X(t)$ is a process with jump discontinuities only. For a Poisson process $K(s,t) = \lambda \min(s,t)$ [see Example 6.1.1 (2)]. Since $K(s,t)$ is continuous, the Poisson process $X(t)$ is mean-square continuous, by Theorem 6.2.9.

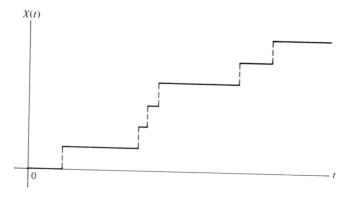

Figure 6.2

6.3. Differentiability

Definition 6.3.1. A stochastic process $X(t)$, $t \in I$, is said to be *sample-path differentiable* if almost all sample paths possess continuous derivatives in I.

Note that we require the sample derivatives to be continuous; that is, almost all sample paths are required to be continuously differentiable. Next we give a sufficient condition in order that a process be sample-path differentiable, but we omit the proof.

Theorem 6.3.2. *Let $g(h)$ and $q(h)$ be two functions as prescribed in Theorem 6.2.4, and let $X(t)$ be a real-valued process defined on $I = [0, 1]$. If*

$$P\{|X(t + h) + X(t - h) - 2X(t)| \leqslant g(h)\} \leqslant q(h),$$

for all $t - h$, t, $t + h \in I$, $h > 0$, then $X(t)$ is equivalent to a sample path differentiable process $Y(t)$, $t \in I$.

Definition 6.3.3. A second-order process $X(t)$, $t \in T$, is said to be *mean-square differentiable* at $t \in T$ if there exists a second-order process $Y(s)$, $s \in T$, such that

$$\lim_{h \to 0} E|h^{-1}[X(t + h) - X(t)] - Y(t)|^2 = 0.$$

We write $Y(t) = X'(t)$.

153

6. Calculus with Stochastic Processes

Theorem 6.3.4. *The process* $X(t)$, $t \in I$, *is mean-square differentiable at* t *if and only if the second generalized derivative* $\partial^2 K(t, u)/\partial t \, \partial u$ *exists at* (t, t).

The proof is similar to that of Theorem 6.2.9. For instance, the sufficiency follows from

$$E[h^{-1}\{X(t + h) - X(t)\}k^{-1}\{X(t + k) - X(t)\}]$$
$$= (hk)^{-1}\{K(t + h, t + k) - K(t + h, t) - K(t, t + k) + K(t, t)\}.$$

At this point we simply remark that the mean-square differentiability of a process does not necessarily imply the sample path differentiability [see Example 6.4.5 (2)].

6.4. Integration

Let $X(t)$, $t \in I = [0, 1]$, be a second-order process such that: (1) almost all sample functions of $\{X(t, \omega)\}$ are continuous and (2) the covariance function $K(s, t)$ of $X(t)$ is continuous on I^2. Also let $f(t)$, $t \in I$, be a deterministic piecewise continuous function. Then the integral

$$I_1(\omega) = \int_0^1 f(t)X(t, \omega) \, dt \tag{6.4.1}$$

exists in the usual sense for almost every $\omega \in \Omega$. Let us delete from Ω the set N of all ω for which the sample functions $X(\cdot, \omega)$ are not continuous. Then the integral (6.4.1) exists for every $\omega \in \Omega$. The reader who is familiar with measure theory can easily see that the stochastic integral in (6.4.1) is an RV. One can also define a stochastic process by

$$Y(t, \omega) = \int_0^t X(s, \omega) \, ds. \tag{6.4.2}$$

First we compute the expectations of $I_1(\omega)$ and $Y(t, \omega)$.

$$m_{I_1} = E[I_1(\omega)] = \int_0^1 f(t)E[X(t)] \, dt = \int_0^1 f(t)m_X(t) \, dt, \tag{6.4.3}$$

and similarly

$$m_Y(t) = \int_0^t m_X(s) \, ds. \tag{6.4.4}$$

Let $I_1 = \int_0^1 f_1(t)X(t) \, dt$ and $I_2 = \int_0^1 f_2(t)X(t) \, dt$, where $f_1(t)$ and $f_2(t)$ are any two continuous functions on I. Then

154

$$E\left[\int_0^1 f_1(t)X(t)\,dt \int_0^1 f_2(s)X(s)\,ds \right]$$

$$= E\left[\int_0^1 f_1(t) \int_0^1 f_2(s)X(t)X(s)\,ds\,dt \right]$$

$$= \int_0^1 f_1(t) \int_0^1 f_2(s)E[X(t)X(s)]\,ds\,dt,$$

and therefore

$$\mathrm{cov}\left\{ \int_0^1 f_1(t)X(t)\,dt, \int_0^1 f_2(s)X(s)\,ds \right\} = \int_0^1 f_1(t) \int_0^1 f_2(s)K(t,s)\,ds\,dt.$$

$$(6.4.5)$$

Consequently

$$K_Y(t,s) = \int_0^t \int_0^s K_X(\xi,\eta)\,d\eta\,d\xi, \qquad s,t \in I. \qquad (6.4.6)$$

Next we define stochastic integrals in the mean-square sense. Let $f(t)$ be a continuous deterministic function on $[0,1]$, and $X(t)$, $t \in [0,1]$, be a second-order process with zero mean function and covariance function $K(s,t)$. Now we proceed to define the integrals

$$J_1 = \int_0^1 f(t)X(t)\,dt \qquad \text{and} \qquad J_2 = \int_0^1 f(t)\,dX(t). \qquad (6.4.7)$$

Let $\mathcal{P}$: $0 = t_0 < t_1 < \cdots < t_n = 1$ be a partition of $I = [0,1]$. Define $|\mathcal{P}| = \max_{1 \leqslant i \leqslant n}(t_i - t_{i-1})$. Corresponding to such a partition, let

$$\mathcal{R}_1 = \sum_{i=0}^{n-1} f(t_i)X(t_i)[t_{i+1} - t_i]$$

$$\mathcal{R}_2 = \sum_{i=0}^{n-1} f(t_i)[X(t_{i+1}) - X(t_i)] \qquad (6.4.8)$$

be the Riemann/Riemann–Stieltjes sums. Clearly, these approximating sums are RVs. Let $\{\mathcal{P}_n\}$ be a sequence of partitions of I such that $|\mathcal{P}_n| \to 0$ as $n \to \infty$. If the sums $\mathcal{R}_1$ and $\mathcal{R}_2$ converge in the mean-square sense to the limits J_1 and J_2 respectively and the limits are the same for all sequences of partitions of the type described above, then the limits are defined as the stochastic integrals. Thus the stochastic integrals J_1 and J_2 are defined by

$$J_1 = \underset{|\mathcal{P}_n|\to 0}{\text{MS-lim}}\ \mathcal{R}_1 \qquad \text{and} \qquad J_2 = \underset{|\mathcal{P}_n|\to 0}{\text{MS-lim}}\ \mathcal{R}_2, \qquad (6.4.9)$$

where MS-lim denotes the limit in the mean-square sense.

155

Theorem 6.4.1. *Let $f(t)$ be a continuous function on $I = [0, 1]$ and $X(t)$ be a second-order process with zero mean function and the covariance function $K(s, t)$. Also let $K(s, t)$ be continuous on I^2 so that the Riemann integral*

$$R_1 = \int_0^1 \int_0^1 f(s)f(t)K(s, t)\, ds\, dt \qquad (6.4.10)$$

exists. Then, the mean-square integral J_1 exists, and

$$E[J_1] = 0 \quad and \quad \mathrm{var}(J_1) = R_1. \qquad (6.4.11)$$

PROOF. Let $\mathscr{P}_n: 0 = t_{n1} < \cdots < t_{nn} = 1$ and $\mathscr{P}'_n: 0 = t'_{n1} < \cdots < t'_{nn} = 1$, $n \geqslant 1$, be two sequences of partitions whose meshes $|\mathscr{P}_n|$ and $|\mathscr{P}'_n|$ approach 0 as $n \to \infty$. Let $\mathscr{R}_{1n}$ and $\mathscr{R}'_{1n}$ be the corresponding Riemann sums. Then

$$E[\mathscr{R}_{1n}\mathscr{R}'_{1n}] = \sum_{i=0}^{n-1}\sum_{j=0}^{n-1} f(t_{ni})f(t'_{nj})K(t_{ni}, t'_{nj})(t_{n(i+1)} - t_{ni})(t'_{n(j+1)} - t'_{nj}),$$

and letting $n \to \infty$, we obtain

$$R_1 = \int_0^1 \int_0^1 f(t)f(t')K(t, t')\, dt\, dt'$$

as the limit. This limit is independent of the sequence of partition of I. Now appealing to Loeve's Theorem 6.1.5, we see that the sum $\mathscr{R}_1$ converges in mean-square, and the limit is independent of the choice of the sequence of partitions. Hence the mean-square stochastic integral J_1 exists. The relations (6.4.11) are obvious [see also (6.4.5)]. $\qquad \square$

Theorem 6.4.2. *Let $X(t)$ be a second-order process with zero mean function and the covariance function $K(s, t)$ of bounded variation on I^2. If $f(t)$ is a continuous function on I so that the Riemann–Stieltjes integral*

$$R_2 = \int_0^1 \int_0^1 f(s)f(t)K(ds, dt) \qquad (6.4.12)$$

exists, then the mean-square integral $J_2 = \int_0^1 f(t)\, dX(t)$ exists with

$$E[J_2] = 0 \quad and \quad \mathrm{var}(J_2) = R_2. \qquad (6.4.13)$$

We omit the proof of this theorem because it is similar to that of Theorem 6.4.1.

Let the covariance function of $X(t)$, $t \in I$, be continuous on I^2. Then by Theorem 6.2.9 the process $X(t)$ is mean-square continuous. Let $X(t)$ be also (sample-) continuous on I. *If $f(t)$ is a continuous function or any suitable function such that the stochastic integral*

$$\int_0^1 f(t) X(t) \, dt$$

exists as a sample path integral denoted by I and as a mean-square integral denoted by J_1, then

$$P\{I_1 = J_1\} = 1. \tag{6.4.14}$$

The Riemann sum $\mathfrak{R}_1$ converges pathwise to I_1 and in mean-square to J_1. It is a well-known result in measure theory that when a sequence of RVs ξ_n converge to ξ in mean-square sense and to η with probability one, then ξ and η are equivalent RVs. This proves our claim.

Theorem 6.4.3. *Let $X(t)$ and $X'(t)$, $t \in T$, be two second-order processes with continuous covariance functions and such that $X'(t)$ is a process without discontinuities of the second kind and*

$$X(t) - X(a) = \int_a^t X'(s) \, ds, \quad t \in T, \tag{6.4.15}$$

where the integral exists sample-pathwise. Then $X'(t)$ is the mean-square derivative of $X(t)$.

PROOF. First note that

$$h^{-1}[X(t + h) - X(t)] - X'(t) = h^{-1} \int_t^{t+h} [X'(s) - X'(t)] \, ds,$$

and hence from (6.4.5),

$$\|h^{-1}[X(t + h) - X(t)] - X'(t)\|^2$$
$$= E[[h^{-1}\{X(t + h) - X(t)\} - X'(t)]^2]$$
$$= h^{-2} \int_t^{t+h} \int_t^{t+h} E[(X'(s) - X'(t))(X'(r) - X'(t))] \, dr \, ds.$$

$$\leqslant h^{-2} \int_t^{t+h} \int_t^{t+h} \{E[(X'(s) - X'(t))^2] E[(X'(r) - X'(t))^2]\}^{1/2} \, dr \, ds, \tag{6.4.16}$$

by the Schwarz inequality.

157

By the continuity of $K_{X'}(s,t)$, the process $X'(t)$ is mean-square continuous. Therefore, for an arbitrarily given $\epsilon > 0$ we can find a $\delta > 0$ such that

$$E[(X'(s) - X'(t))^2] < \epsilon \qquad \text{if } |t - s| < \delta. \qquad (6.4.17)$$

Choose an h such that $|h| < \delta$. Then using (6.4.17) in (6.4.16), we see that

$$\|h^{-1}[X(+h) - X(t)] - X'(t)\|^2 < \epsilon.$$

Since ϵ is arbitrary, this completes the proof. □

Theorem 6.4.4. *Let* $\{X(t), t \in I\}$ *be a mean-square differentiable process with derivative* $\{X'(t)\}$. *Then*

$$E\left[\sup_{t \in I} X^2(t)\right] \leqslant 2^{-1}\{\|X(0)\|^2 + \|X'(1)\|^2\} + \int_0^1 \|X(t)\| \|X'(t)\| \, dt.$$

PROOF. First observe that

$$\int_0^t X(s)X'(s) \, ds = 2^{-1}[X^2(t) - X^2(0)]$$

and

$$\int_t^1 X(s)X'(s) \, ds = 2^{-1}[X^2(1) - X^2(t)].$$

This implies, for any $t \in I$, that

$$2X^2(t) = X^2(0) + X^2(1) + 2\int_0^t X(s)X'(s) \, ds - 2\int_t^1 X(s)X'(s) \, ds$$

$$\leqslant X^2(0) + X^2(1) + 2\int_0^1 |X(s)X'(s)| \, ds,$$

and thus

$$E\left[\sup_{t \in I} X^2(t)\right] \leqslant 2^{-1}\{\|X(0)\|^2 + \|X(1)\|^2\} + \int_0^1 E|X(s)X'(s)| \, ds$$

$$\leqslant 2^{-1}\{\|X(0)\|^2 + \|X(1)\|^2\} + \int_0^1 \|X(t)\| \|X'(t)\| \, dt \quad \text{(why?)}.$$

This completes the proof. □

This theorem can be considered as an extension of the Chebyshev inequality from a RV to a stochastic process. So it can be used for the purposes similar to the applications of Chebyshev inequality.

Examples 6.4.5

EXAMPLE 1. Under the conditions of Theorem 6.4.4, compute $m_{X'}(t)$ and $K_{X'}(s, t)$.

From (6.4.15) it follows $m_X(t) - m_X(a) = \int_a^t m_{X'}(s)\,ds$, so that

$$\frac{d}{dt} m_X(t) = m_{X'}(t). \tag{6.4.18}$$

Now note that

$$E[X(s)(X(t) - X(a))] = \int_a^t E[X(s)X'(u)]\,du$$

and

$$m_X(s)[m_X(t) - m_X(a)] = \int_a^t m_X(s)m_{X'}(u)\,du.$$

Therefore,

$$K_X(s, t) - K_X(s, a) = \int_a^t K_{XX'}(s, u)\,du.$$

Consequently,

$$\frac{\partial}{\partial t} K_{XX}(s, t) = K_{XX'}(s, t),$$

and similarly

$$\frac{\partial}{\partial s} K_X(s, t) = K_{X'X}(s, t).$$

Combining these two relations, we have

$$K_{X'}(s, t) = K_{X'X'}(s, t) = \frac{\partial^2}{\partial s\,\partial t} K_{XX}(s, t) = \frac{\partial^2}{\partial s\,\partial t} K_X(s, t). \tag{6.4.19}$$

EXAMPLE 2. At the end of Section 6.3 we remarked that the mean-square differentiability of a process need not imply the sample-path differentiability. To see this, let us consider a Poisson process $N(t)$. From Example 6.2.10 it follows that $N(t)$ is a mean-square continuous process and is without discontinuities of the second kind. Define

$$X(t) = \int_0^t N(s)\,ds.$$

Then, clearly, $X(t)$ is a mean-square continuous process with mean-square derivative $N(t)$. But $X(t)$ is not sample-path differentiable since $N(t)$ is not sample continuous.

Let us now introduce two important classes of stochastic processes, namely, the Gaussian processes and Brownian motion processes. In Chapter 7 we study some properties of stationary Gaussian processes, whereas Chapter 9 is entirely devoted to Brownian motion processes. Here we use these processes to illustrate the notions introduced in this chapter.

Definition 6.4.6. A stochastic process $\{X(t), t \in T\}$ is called a *normal* or *Gaussian process* if for any integer $n \geqslant 1$ and any finite sequence $t_1 < t_2 < \cdots < t_n$ from T the RVs $X(t_1), \ldots, X(t_n)$ are jointly normally distributed.

The following result is a fundamental property of Gaussian RVs. As the proof of this result is easy, it is omitted.

Theorem 6.4.7. *Let $X_1, \ldots, X_n$ be n RVs jointly distributed as an n-dimensional normal random vector. If $Y_1, \ldots, Y_k$ are the linear combinations*

$$Y_1 = \sum_{i=1}^{n} a_{1i} X_i, \ldots, Y_k = \sum_{i=1}^{n} a_{ki} X_i$$

of $X_1, \ldots, X_n$, then $Y_1, \ldots, Y_k$ are jointly normally distributed.

On the basis of this theorem it is not difficult to see that Definition 6.4.6 is equivalent to Definition 6.4.8.

Definition 6.4.8. A stochastic process $\{X(t), t \in T\}$ is called a *Gaussian process* if every finite linear combination of the RVs $X(t), t \in T$, is normally distributed.

Example 6.4.9 In Example 6.1.1 (1) let us further assume that the RVs are normally distributed. Then the process

$$X(t) = \sum_{k=1}^{m} [\xi_k \cos \lambda_k t + \eta_k \sin \lambda_k t]$$

is a Gaussian process. To see this, consider a linear combination

$$\sum_{i=1}^{n} a_i X(t_i) = \sum_{i=1}^{n} \sum_{k=1}^{m} [\xi_k a_i \cos \lambda_k t_i + \eta_k a_i \sin \lambda_k t_i].$$

This is a linear combination of independent normal RVs and thus is normally distributed. Hence, by Definition 6.4.8, $X(t)$ is a Gaussian process.

Definition 6.4.10. A stochastic process $\{B(t), t \in R\}$ is called a *Brownian motion process* if: (1) $B(0) = 0$, (2) the increment $B(t) - B(s)$ over the interval $[s, t]$ is normally distributed with mean 0 and variance $\sigma^2(t - s)$, and (3) $\{B(t)\}$ is a process with independent increments.

***Example* 6.4.11** Show that a Brownian motion process $\{B(t)\}$ is a Gaussian process. Consider an arbitrary linear combination $\Sigma_{i=1}^n a_i B(t_i)$, where $a_i \in R$ and $0 \leqslant t_1 < \cdots < t_n$. Then

$$\sum_{i=1}^n a_i B(t_i) = \left(\sum_{k=1}^n a_k \right)[B(t_1) - B(0)] + \left(\sum_{k=2}^n a_k \right)[B(t_2) - B(t_1)] + \cdots$$

$$+ \cdots + \left(\sum_{k=n-1}^n a_k \right)[B(t_{n-1}) - B(t_{n-2})]$$

$$+ a_n[B(t_n) - B(t_{n-1})].$$

It follows from condition (3) of Definition 6.4.10 that $\Sigma_1^n a_i B(t_i)$ is expressed as a linear combination of independent normally distributed RVs, and hence is itself normal. Therefore, $B(t)$, $t \geqslant 0$, is a Gaussian process.

Theorem 6.4.12. *Let $\{X_n\}$ be a sequence of Gaussian RVs. If $X_n \to X$ in mean-square sense, then X is a Gaussian RV.*

The proof of this theorem is left as an exercise.

***Examples* 6.4.13**

EXAMPLE 1. Let $X(t)$ be a mean-square differentiable Gaussian process with derivative $X'(t)$, $t \in T$. Then $X'(t)$ is a Gaussian process.

We want to show that every linear combination $\Sigma_{i=1}^n a_i X'(t_i)$ is a Gaussian RV. First note that for each fixed h the RV

$$\sum_{i=1}^n a_i h^{-1}[X(t_i + h) - X(t_i)]$$

is Gaussian. But

$$\lim_{h \to 0} E\left[\left\{ \sum_{i=1}^n a_i h^{-1}[X(t_i + h) - X(t_i)] - \sum a_i X'(t_i) \right\}^2 \right] = 0.$$

Now, appealing to Theorem 6.4.12, we see that $\{X'(t)\}$ is a Gaussian process.

EXAMPLE 2. If $X(t)$ is a Gaussian process, then $\int_a^b X(t)\,dt$ is a Gaussian variable.

This follows from Theorem 6.4.12 and the fact that the Riemann sum $\sum_{i=0}^{n-1} X(t_i)[t_{i+1} - t_i]$, for any partition $a = t_0 < t_1 < \cdots < t_n = b$, is Gaussian.

EXAMPLE 3. In Example 2 we saw that the mean-square integral of a Gaussian process is a Gaussian variable. If the Gaussian process is mean-square differentiable, its derivative is a Gaussian process. But a Gaussian process need not be mean-square differentiable. In Example 6.4.11 we saw that a Brownian motion process is always a Gaussian process. We claim that *the Brownian motion $B(t)$ is not mean-square differentiable*.

If $B(t)$ is mean-square differentiable, there is a second-order process $B'(t)$ such that

$$\lim_{h \to 0} E[\{h^{-1} B(t + h) - B(t) - B'(t)\}^2] = 0$$

and consequently

$$\lim_{h \to 0} E[\{h^{-1}(B(t + h) - B(t))\}^2] = E[\{B'(t)\}^2] < \infty. \qquad (6.4.20)$$

But

$$E\{h^{-1}(B(t + h) - B(t))\}^2 = \frac{\sigma^2}{|h|} \to \infty, \qquad \text{as } h \to 0,$$

contradicting (6.4.20). Hence $B(t)$ is not mean-square differentiable.

Exercises

1. For each of the following stochastic processes compute the mean function $m(t)$ and the covariance function $K(s, t)$: (a) $X(t) = (At + B)$, where A and B are independent RVs with means a and b and variances σ_1^2 and σ_2^2, respectively, (b) $X(t) = (At^2 + Bt + C)$, where A, B, and C are independent RVs with mean 1 and variance 1, (c) $X(t) = \cos(at + b)$, where a is a positive constant and b is a RV uniformly distributed on $[0, 2\pi]$, and (d) $X(t) = n^{-1}\sum_{k=1}^n g(t, A_k)$, $0 \leqslant t \leqslant 1$, where $A_1, \ldots, A_n$ are independent RVs each uniformly distributed on $(0, 1)$ and $g(t, x)$, $(t, x) \in [0, 1] \times [0, 1]$, is defined by $g(t, x) = 1$ for $x \leqslant t$ and $g(t, x) = 0$ for $x > t$. In parts (e)–(h), $B(t)$ is a standard Brownian motion: (e) $X(t) = tB(t^{-1})$, $t > 0$, (f) $X(t) = a^{-1}B(a^2 t)$, $t \geqslant 0$, (g) $X(t) = [B(t)]^2$, $t \geqslant 0$, (h) $X(t) = B(t) - tB(1)$, $0 \leqslant t \leqslant 1$.

2. Let $\{A_n\}$, $n \geqslant 1$, be a sequence of IID RVs with mean μ and variance 1. Define the sequence $\{X_n\}$ by $X_n = n^{-1}\sum_{k=1}^n A_k$. Show that $\{X_n\}$ is a Cauchy sequence in $L_2(\Omega)$, that is, $E|X_n - X_m|^2 \to 0$ as $m, n \to \infty$, and that $E|X_n - \mu|^2 \to 0$ as $n \to \infty$.

3. Check whether each of the following sequences is mean-square convergent: (a) $\{X_n\}$, $n \geq 1$, is a sequence of independent RVs distributed according to $P\{X_n = 1\} = \frac{1}{n}$ and $P\{X_n = 0\} = 1 - n^{-1}$, and (b) $\{X_n\}$ is a sequence of independent RVs distributed according to $P\{X_n = n\} = n^{-2}$ and $P\{X_n = 0\} = 1 - n^{-2}$.

4. If a correlation function $C(s, t) = E[X(s)X(t)]$ is continuous on the diagonal of $T \times T$, show that it is continuous on $T \times T$.

5. Establish the following properties of mean-square differentiable processes: (a) the mean-square differentiability of $X(t)$ at a point t^* implies the mean-square continuity at t^*, (b) a linear combination of two mean-square differentiable processes is mean-square differentiable, and (c) if $g(t)$ is a differentiable deterministic function and $X(t)$ is a mean-square differentiable process, then

$$\frac{d}{dt}[g(t)X(t)] = X(t)\frac{dg}{dt} + g(t)\frac{dX}{dt}.$$

6. Investigate the properties of continuity, differentiability, and integrability (all in the mean-square sense), of the following processes: (a) process $X(t)$ in Exercise 1 (a), (b) process $X(t)$ in Exercise 1 (b), (c) process $X(t)$ in Exercise 1 (c), (d) process $X(t)$ in Exercise 1 (f), (e) $X(t)$ is a process with covariance function $K(s, t) = e^{-a(t-s)^2} \cos b(t - s)$, $a, b > 0$, (f) $X(t)$ is a process with covariance function $K(s, t) = [\sin a(t - s)]/(t - s)$, (g) $X(t)$ is a process with covariance function $K(s, t) = (a^2 + (t - s)^2)$, (h) $X(t)$ is a process with covariance $K(s, t) = e^{-a|t-s|}\{1 + a|t - s|\}$, and (i) $X(t)$ is a process with covariance function $K(s, t) = b^{-1}e^{-b|t-s|} - a^{-1}e^{-a|t-s|}$, where $a \geq b$.

In the following problems, $B(t)$ is a standard Brownian motion.

7. Find the mean and covariance functions of each of the following processes:

(a) $X(t) = \int_0^t B(s)\,ds, \quad t \geq 0$;

(b) $X(t) = \int_0^t s\,dB(s), \quad t \geq 0$;

(c) $X(t) = \int_{t-1}^t (t - s)\,dB(s), \quad t \in R$;

(d) $X(t) = \int_t^{t+1} [B(s) - B(t)]\,ds, \quad t \in R$.

8. Define, for $t \geq 0$ and a constant a,

$$X(t) = \int_0^t e^{a(t-s)}\,dB(s) \quad \text{and} \quad Y(t) = \int_0^t X(s)\,ds:$$

(a) find the means and variances of the processes $X(t)$ and $Y(t)$ and (b) show that

$$Y(t) = \int_0^t a^{-1}[e^{a(t-s)} - 1]\,dB(s), \quad t \geq 0.$$

7

Stationary Processes

7.1. Definition and Examples

Consider a Markov chain (MC) $X = \{X_n\}$, $n \geq 0$, with stationary transition probability matrix $[p(x,y)]$ and a stationary distribution $\pi(x)$, $x \in S$, where S is the state space. Assume that $\{\pi(x)\}$ is the initial distribution of the MC. Then $\{\pi(x)\}$ is a steady-state distribution, that is,

$$P\{X_n = x\} = \pi(x), \qquad x \in S, n \geq 0. \tag{7.1.1}$$

Because the RHS of (7.1.1) is independent of time, it follows that the one-dimensional distribution of the MC X is invariant under time shift. Next let us consider the m-dimensional distributions of X. It is well known to us that the m-dimensional distributions of an MC are determined by the initial distribution and the transition matrix [see Lemma 3.1.3 (iii)]. Thus we now have, for all positive integers k, that

$$P\left\{X_{n_i+k} = x_i, 1 \leq i \leq m\right\} = \pi(x_1) \prod_{i=1}^{m} p(n_i - n_{i-1}; x_{i-1}, x_i), \tag{7.1.2}$$

where $p(n; x, y)$ is the n-step transition probability, $n_1 < n_2 < \cdots < n_m$, and m is an integer ≥ 2. That is, the m-dimensional distributions are invariant under the time shift. Abstracting this invariance property, one can intuitively describe a stationary process as a stochastic process whose probabilistic laws are invariant under shifts in time.

165

Throughout this chapter $\{X_t, t \in T\}$ denotes a process whose parameter set T is defined with an additive semigroup structure. Typical examples of T are $T = \{0, 1, 2, \ldots\}$, $T = \{\cdots, -1, 0, 1, \ldots\}$, $T = R_+$, and $T = R$.

Definition 7.1.1. A stochastic process $X(t)$, $t \in T$, is called a *strictly stationary process* if the RVs

$$\{X(t_1), \ldots, X(t_n)\}$$

and the RVs

$$\{X(t_1 + h), \ldots, X(t_n + h)\}$$

have the same joint probability distributions for any positive integer n, any time points $t_1 < \cdots < t_n$ in T, and all $h \in T$.

The strict stationarity is defined in terms of the joint distribution. Therefore, it is not necessary that the process possesses even a mean function. Let now $X(t)$, $t \in T$, be a strictly stationary process that is also of second order. Since the one-dimensional distributions are invariant through time translations for all $h \in T$ we have

$$m(t) = E[X(t)] = E[X(t + h)] = m(t + h)$$

and hence the mean function $m(t)$ is constant. Similarly,

$$\sigma^2 = \mathrm{var}(X(0)) = \mathrm{var}(X(t)) = \mathrm{var}(X(t + h)) \qquad \text{for all } h \in T$$

and

$$E[X(0)X(t)] = E[X(s)X(t + s)],$$

so that the correlation and the covariance functions would depend on the time points s and t only through the difference $|t - s|$. In many practical applications the only use made of the stationarity is the constancy of the mean function and the dependence of covariance function $K(s, t)$ through the difference $|t - s|$. Besides, it is not always possible in practical situations to compute all the finite dimensional distributions in order to verify the stationarity. It is thus natural to generalize the notion of stationarity to the so-called wide sense stationarity. We remark here that this generalization by no means enlarges the class of strictly stationary processes (why?).

Definition 7.1.2. A stochastic process is called a *wide sense stationary process* if it is a second-order process with a constant mean function $m(t) \equiv m$ and a covariance function $K(s, t) = E[(X(t) - m)(X(s) - m)]$ that depends only on the difference $|t - s|$.

Wide sense stationary processes are also known as *covariance stationary*, *second-order stationary*, or *Khintchin stationary* processes. A wide sense stationary process is defined through its mean and covariance functions. Obviously, as these two functions do not in general determine all the other higher moments and all the finite dimensional distributions of the process, a wide sense stationarity need not imply strict stationarity. We mainly concentrate on wide sense stationary processes.

Examples 7.1.3

EXAMPLE 1. Let $\{X_n, n \geq 0\}$ be a sequence of IID RVs with mean 0 and variance 1. Then $\{X_n\}$ is a wide-sense-stationary process. Since the mean function $m(n) \equiv 0$ and

$$E[X_{n+k} X_n] = \begin{cases} 1 & \text{if } k = 0 \\ 0 & \text{if } k \neq 0 \end{cases}$$

(see also Example 7.2.6 (1)).

EXAMPLE 2. *Moving Average Process.* Let $\{X_n\}$ be a doubly infinite sequence $(n = 0, \pm 1, \pm 2, \ldots)$ of IID RVs with zero mean and unit variance. Define

$$Y_n = \sum_{k=0}^{K} a_k X_{n-k}, \qquad n = 0, \pm 1, \ldots, \tag{7.1.3}$$

where the coefficients a_k are given constants. Here K could be taken as infinite with an appropriate condition for the convergence of the resulting infinite series $\sum_{k=0}^{\infty} a_k X_{n-k}$. We consider only the finite K. The process Y_n defined by (7.1.3) (with finite or infinite K) is called a *moving average process*. The process $\{Y_n\}$ is a wide-sense-stationary process, since the mean function $m(n) \equiv 0$ and

$$E[Y_n Y_{n+m}] = E\left[\left(\sum_{j=0}^{K} a_j X_{n-j} \right) \left(\sum_{k=0}^{K} a_k X_{n+m-k} \right) \right]$$

$$= \begin{cases} a_K a_{K-m} + \cdots + a_{m+1} a_1 & \text{if } m \leq K - 1 \\ 0 & \text{if } m \geq K \end{cases}.$$

[developed further in Example 7.2.6 (2)].

EXAMPLE 3. Show that the process $X(t) = U \cos \lambda t + V \sin \lambda t$, $-\infty < t < \infty$, is wide sense stationary if and only if U and V are uncorrelated RVs with zero mean and equal variance [see also Examples 7.2.6 (3) and 7.3.7 (1)].

Let the RVs U and V satisfy the conditions stipulated above. Then

$$m(t) = E[U]\cos \lambda t + E[V]\sin \lambda t = 0$$

and

$$E[X(s)X(t + s)] = E[U^2 \cos \lambda s \cos \lambda(t + s) + V^2 \sin \lambda s \sin \lambda(t + s)]$$
$$+ E[UV](\cos \lambda s \sin \lambda(t + s) + \cos \lambda(t + s)\sin \lambda s)$$
$$= \sigma^2 \cos[\lambda(t + s) - \lambda s] = \sigma^2 \cos \lambda t,$$

where $\text{var}(U) = \sigma^2 = \text{var}(V)$. Hence $X(t)$ is wide sense stationary.

To see the necessity, let $X(t)$ be a wide-sense-stationary process. Set $u = E[U]$ and $v = E[V]$. Then

$$m = m(t) = u \cos \lambda t + v \sin \lambda t \qquad \text{for all } t \in R.$$

Let $m = 0$. Taking $t = 0$, we get $u = 0$ and taking, $t = (\pi/2\lambda)$, we get $v = 0$. If $m \neq 0$, then taking $t = 0$ and $t = (\pi/2\lambda)$, we see that $u = m = v$. Therefore

$$1 = \cos \lambda t + \sin \lambda t \qquad \text{for all } t.$$

This is not true for at least $t = (\pi/4\lambda)$. Hence m cannot be nonzero. If $m = 0$, we saw that $E[U] = 0 = E[V]$. Next

$$K(t) = E[U^2]\cos \lambda(t + s)\cos \lambda s + E[V^2]\sin \lambda(t + s)\sin \lambda s$$
$$+ E[UV]\sin \lambda(t + 2s) \qquad \text{for all } t.$$

The RHS should depend only on t. This is possible if and only if $E[UV] = 0$ and $E[U^2] = E[V^2]$. This gives the necessity of the stated conditions.

EXAMPLE 4. *Periodic Oscillation.* Define a complex-valued process $X(t, \omega)$ by $X(t, \omega) = X(\omega)f(t)$, $t \in R$, where X is a second-order centered real RV and f is a deterministic function. Show that $X(t)$ is wide sense stationary if and only if $f(t)$ is of the form $f(t) = ce^{i(\lambda t + \theta)}$, where $i = \sqrt{-1}$. [In the complex-valued process case the correlation function is defined by $E[X(s)\overline{X(t)}]$, where the bar denotes the complex conjugation.]

Since $E[X] = 0$, the mean function $m(t) \equiv 0$. Let $f(t) = ce^{i(\lambda t + \theta)}$. Then

$$E[X(s + t)\overline{X(s)}] = E[X^2]c^2 e^{i\{\lambda(t+s)+\theta\}} e^{-i(\lambda s + \theta)}$$
$$= c^2 \sigma^2 e^{i\lambda t},$$

where $\sigma^2 = E[X^2]$. Hence $\{X(t)\}$ is wide sense stationary. Note that the covariance function is proportional to the average energy of the oscillation per unit time with random amplitude.

To see the converse, let $X(t)$ be a wide-sense-stationary process. Then

$$E[X(s + t)\overline{X(s)}] = E[X^2]f(s + t)\overline{f(s)}$$

is necessarily independent of s. Taking $t = 0$, we see that

$$|f(s)|^2 = \text{constant} = c^2, \quad \text{say}.$$

Therefore, $f(s) = ce^{i\psi(s)}$, where $\psi(s)$ is a real function. Then

$$f(s + t)\overline{f(s)} = c^2 e^{i\psi(s+t)} e^{-i\psi(s)} = c^2 e^{i[\psi(s+t) - \psi(s)]}$$

is independent of s. Thus

$$\frac{d}{ds}[\psi(s + t) - \psi(s)] = 0 \quad \text{or} \quad \frac{d}{ds}\psi(s + t) = \frac{d}{ds}\psi(s).$$

Since this is true for all t,

$$\psi'(s) = \text{a constant} = \lambda, \quad \text{say}.$$

Then $\psi(s) = \lambda s + \theta$ and hence $f(t) = ce^{i(\lambda s + \theta)}$.

EXAMPLE 5. *Random Telegraph Signal*. The role of a random telegraph signal is basic in the construction of random-signal generators. A *random telegraph signal* is a stochastic process $X(t)$ defined by

$$X(t) = \xi(-1)^{N(t)}, \quad t \geqslant 0, \tag{7.1.4}$$

where the state space of $X(t)$ is the set $\{-1, 1\}$, the times at which the process changes the values between -1 and 1 are distributed according to a Poisson process $N(t)$ with intensity rate λ, and ξ is a random variable independent of $\{N(t)\}$ and such that $P\{\xi = 1\} = \frac{1}{2} = P\{\xi = -1\}$. A typical sample path of this process is given in Figure 7.1. First note that

$$E[X(t)] = E[\xi]E[(-1)^{N(t)}] \tag{7.1.5}$$

and

$$E[X(s)X(t)] = E[\xi^2]E[(-1)^{N(s)}(-1)^{N(t)}]. \tag{7.1.6}$$

Set $Y(t) = (-1)^{N(t)}$. Then we have to compute $E[Y(t)]$ and $E[Y(s)Y(t)]$. To compute these expectations, we need to find the probabilities $P\{Y(t) = 1\}$ and $P\{Y(t) = -1\}$. Now

169

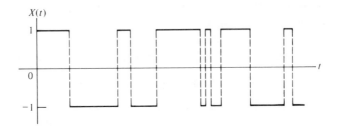

Figure 7.1

$$P\{Y(t) = 1\} = P\{\text{even number of changes in } [0, t]\}$$

$$= e^{-\lambda t} \sum_{k=0}^{\infty} \frac{(\lambda t)^{2k}}{(2k)!} = e^{-\lambda t} \cosh \lambda t \qquad (7.1.7)$$

$$P\{Y(t) = -1\} = P\{\text{odd number of changes in } [0, t]\}$$

$$= e^{-\lambda t} \sum_{k=0}^{\infty} \frac{(\lambda t)^{2k+1}}{(2k + 1)!} = e^{-\lambda t} \sinh \lambda t. \qquad (7.1.8)$$

Therefore

$$E[Y(t)] = 1P\{Y(t) = 1\} + (-1)P\{Y(t) = -1\}$$

$$= e^{-\lambda t}[\cosh \lambda t - \sinh \lambda t] = e^{-2\lambda t}. \qquad (7.1.9)$$

Using (7.1.9) in (7.1.5), $m(t) = E[X(t)] = e^{-2\lambda t} E[\xi] = 0$. Next

$$E[Y(s)Y(t)] = 1P\{Y(s) = 1 = Y(t) \quad \text{or} \quad Y(s) = -1 = Y(t)\}$$

$$+ (-1)P\{Y(s), Y(t) \text{ are of opposite signs}\}$$

$$= P\{Y(s) = 1 = Y(t)\} + P\{Y(s) = -1 = Y(t)\}$$

$$- P\{Y(s) = 1 = -Y(t)\}$$

$$- P\{Y(s) = -1 = -Y(t)\}. \qquad (7.1.10)$$

Now let $s < t$; the case $t < s$ follows similarly. Then

$$P\{Y(s) = 1 = Y(t)\} = P\{Y(s) = 1\} P\{Y(t) = 1 | Y(s) = 1\}$$

$$= e^{-\lambda s} \cosh \lambda s \ P\{\text{even number of changes in } (s, t)\}$$

$$= e^{-\lambda s} \cosh \lambda s \ e^{-\lambda(t-s)} \cosh \lambda(t - s). \qquad (7.1.11)$$

Similarly

$$P\{Y(s) = -1 = Y(t)\} = e^{-\lambda s}\sinh \lambda s \, e^{-\lambda(t-s)}\cosh \lambda(t-s), \qquad (7.1.12)$$

$$P\{Y(s) = -1 = -Y(t)\} = e^{-\lambda s}\sinh \lambda s \, e^{-\lambda(t-s)}\sinh \lambda(t-s), \qquad (7.1.13)$$

$$P\{Y(s) = 1 = -Y(t)\} = e^{-\lambda s}\cosh \lambda s \, e^{-\lambda(t-s)}\sinh \lambda(t-s). \qquad (7.1.14)$$

Using (7.1.11)–(7.1.14) in (7.1.10) and simplifying, we get

$$E[Y(s)Y(t)] = e^{-2\lambda(t-s)}, \qquad s < t.$$

Using this in (7.1.6) and noting that $E[\xi^2] = 1$, we obtain

$$K(s,t) = E[X(s)X(t)] = e^{-2\lambda|t-s|} = K(|t-s|).$$

This proves the wide sense stationarity of $X(t)$ [this topic is continued in Examples 7.2.6 (4) and 7.3.7 (2)].

EXAMPLE 6. *Random Binary Noise.* A stochastic process $X(t)$ is called a *random binary noise* if $X(t) = Y(t - U)$, where Y is a process taking values from $\{-1, 1\}$ on successive intervals of fixed length T and such that the RVs $Y(t)$ are independent for the values of t lying in nonoverlapping intervals and U is the random shift of time such that U is uniformly distributed on $(0, T)$ and is independent of $Y(t)$. We claim that $X(t)$ is wide sense stationary.

First note that $X(t) = Y(t - U)$ with

$$Y(t) = Y_n, \qquad \text{for} \qquad (n-1)T < t < nT.$$

where the RVs Y_n are IID with $P\{Y_k = 1\} = \frac{1}{2} = P\{Y_k = -1\}$. Since the RVs U and $Y(t)$ are independent we see that

$$E[Y(t-u)|U = u] = E[Y(t-u)] = 0$$

and consequently

$$m(t) = E[X(t)] = E[E[Y(t - U)|U]] = 0. \qquad (7.1.15)$$

Therefore, $K(s,t) = E[X(s)X(t)]$. It remains to compute $K(s,t)$.

The time axis is divided into successive intervals by the points 0, T, $2T$, $3T$, ... If s and t lie on two different intervals, it follows from the independence assumption on the RVs Y_n that $K(s,t) = 0$. Let $|t - s| > T$. Since U is uniform in $(0, T)$, it follows that $E[X(s)X(t)|U] = 0$, and consequently $E[X(s)X(t)] = E[E[X(s)X(t)|U]] = 0$. Now let $|t - s| < T$ and $s = nT$. Then t and s will lie in the same interval if

$$U + |t - s| < T \qquad \text{that is} \qquad U < T - |t - s|. \qquad (7.1.16)$$

171

The event (7.1.16) occurs with probability $(T - |t - s|)/T$. Therefore

$$E[X(s)X(t)] = \begin{cases} 1 - \dfrac{|t - s|}{T} & \text{if } |t - s| < T \\ 0 & \text{if } |t - s| > T \end{cases}. \qquad (7.1.17)$$

The relations (7.1.15) and (7.1.17) imply that $\{X(t)\}$ is wide sense stationary.

EXAMPLE 7. *Poisson Increment*. Let $N(t)$ be a Poisson process with intensity rate λ. Define

$$X(t) = N(t + 1) - N(t), \qquad t \geqslant 0.$$

Show that $X(t)$ is a wide-sense-stationary process.

$$m(t) = E[X(t)] = E[N(t + 1) - N(t)] = \lambda(t + 1) - \lambda t = \lambda \qquad (7.1.18)$$

for all t. Since a Poisson process has independent increments, it follows that $\text{cov}(X(s), X(t)) = 0$ if $|t - s| \geqslant 1$, and so let $s \leqslant t < s + 1$. Then

$$\text{cov}(X(s), X(t)) = \text{cov}(N(s + 1) - N(s), N(t + 1) - N(t))$$
$$= \text{cov}(N(s + 1) - N(t) + N(t) - N(s), N(t + 1) - N(s + 1) + N(s + 1)$$
$$- N(t)). \qquad (7.1.19)$$

But

$$\text{cov}(N(s + 1) - N(t), N(t + 1) - N(s + 1)) = 0$$
$$\text{cov}(N(t) - N(s), N(t + 1) - N(s + 1)) = 0$$
$$\text{cov}(N(t) - N(s), N(s + 1) - N(t)) = 0,$$

all following from the fact that $N(t)$ is a process with independent increments and $s \leqslant t < s + 1 \leqslant t + 1$. Using these in (7.1.19), we get

$$\text{cov}(X(s), X(t)) = \text{var}(N(s + 1) - N(t)) = \lambda(s + 1 - t)$$

Therefore

$$K(s, t) = \begin{cases} \lambda(1 - |t - s|) & \text{if } |t - s| < 1 \\ 0 & \text{if } |t - s| \geqslant 1 \end{cases}, \qquad (7.1.20)$$

a function depending only on the difference $|t - s|$. Relations (7.1.18) and (7.1.20) imply that $X(t)$ is wide sense stationary.

EXAMPLE 8. Let $\{X(t), -\infty < t < \infty\}$ be a Gaussian process. Then $\{X(t)\}$ is wide sense stationary if and only if it is strictly stationary.

The sufficiency is clear from the remarks we made after Definition 7.1.1. Let $X(t)$ be a wide-sense-stationary Gaussian process. Arbitrarily fix a real number h and define $Y(t) = X(t + h)$. Then $m_Y(t) = m_X(t)$ and $K_Y(s, t)$ $= K_X(s, t)$. Therefore, $\{Y(t)\}$ has the same joint distributions as $\{X(t)\}$. This proves that $X(t)$ is strictly stationary and, hence, also our claim.

Let $\{X(t), t \in R\}$ be a wide-sense-stationary process. From the criterion for the mean-square continuity of a stochastic process (see Theorem 6.2.9), it follows that $X(t)$ *is mean-square continuous for every t if and only if the covariance function $K(t)$ is continuous at $t = 0$.* We always assume that this continuity assumption is satisfied. Regarding the mean-square differentiability, *a wide-sense-stationary process $X(t)$, $-\infty < t < \infty$, is mean square differentiable if and only if the covariance function $K(t)$ is twice differentiable at $t = 0$.*

The periodic oscillation process is mean-square differentiable. But the random telegraph signal, binary noise, and Poisson increment processes are not mean-square differentiable.

Theorem 7.1.4. *Let $X(t)$, $t \in R$, be a separable wide-sense-stationary process with covariance function $K(\cdot)$ continuous at $t = 0$ (and hence everywhere). Then:*

(i)   *Almost all sample paths $X(\cdot, \omega)$ are continuous in any finite interval if*

$$2K(0) - K(t) - K(-t) = o(|t|/\|\log|t|\|^4)$$

for $|t| \to 0$.

(ii)   *Almost all sample paths are continuously differentiable in any finite interval if*

$$6K(0) - 4K(t) - 4K(-t) + K(2t) + K(-2t) = o(|t|^3/\|\log|t|\|^4).$$

7.2. Spectral Representation

Consider the stochastic process $X(t, \omega)$ of periodic oscillation [see Example 7.1.3 (4)]. The process is defined by separating the variables $X(t, \omega)$ $= X(\omega)e^{i\lambda t}$. (Note that we have absorbed $e^{i\theta}$ into $X(\omega)$ and hence X is a complex RV.) Separation of variables is a well-known method in the theory of partial-differential equations and where the general solution is represented as a superposition of particular solutions. The theory of stationary processes has a parallel to this situation. That is, the principle of superposition can be used to represent wide-sense-stationary processes. A similar representation is possible for the covariance function of the stationary process.

Consider the superposition

$$X(t) = \sum_{k=1}^{n} X_k e^{i\lambda_k t}$$

of n periodic oscillations with independent and centered random amplitudes X_k and different angular frequencies λ_k. The process $\{X(t)\}$ is wide sense stationary, and the covariance function $K(t)$ is given by

$$K(t) = \sum_{k=1}^{n} \sigma_k^2 e^{i\lambda_k t} = \sigma^2 \sum_{k=1}^{n} p_k e^{i\lambda_k t}, \qquad (7.2.1)$$

where σ_k^2 is the average power of the oscillation process, $\sigma^2 = \Sigma_{k=1}^{n} \sigma_k^2$ and $p_k = (\sigma_k^2/\sigma^2)$. Note that $\{p_k\}$, $1 \leqslant k \leqslant n$, is a probability distribution. Now expression (7.2.1) suggests the following representation

$$K(t) = \int e^{i\lambda t} \, dF(\lambda) \qquad (7.2.2)$$

where the integral is taken between suitable limits and F is a distribution function. Similarly, the form

$$X(t) = \sum_{k=-n}^{n} X_k e^{i\lambda_k t} \qquad (7.2.3)$$

suggests a representation

$$X(t) = \int_{-\infty}^{\infty} e^{i\lambda t} \, dY(\lambda), \qquad (7.2.4)$$

where $Y(\lambda)$ is a centered second-order process with orthogonal increments. We begin with Khintchin's theorem.

Theorem 7.2.1. (Khintchin). *A function $K(t)$ is the covariance function of a wide sense stationary and mean square continuous process with zero mean and unit variance if and only if it admits the following representation*

$$K(t) = \int_{-\infty}^{\infty} e^{i\lambda t} \, dF(\lambda),$$

where $F(\lambda)$ is some distribution function.

PROOF. (*Necessity*). Let $K(t)$ be the covariance function of a MS continuous wide-sense-stationary process. Then $K(t)$ is continuous for all t. Since $K(t) \leqslant K(0) = 1$, $K(t)$ is also bounded. We have seen in Chapter 6 that a covariance function is positive definite. That is,

$$\sum_{j=1}^{n} \sum_{i=1}^{n} K(t_i - t_j) z_i \bar{z}_j \geqslant 0$$

for any reals $t_1, \ldots, t_n$ and complex numbers $z_1, \ldots, z_n$. Now recall Bochner's theorem on the representation of characteristic functions (see Chapter 1). Since $K(t)$ is a continuous positive definite function with $K(0) = 1$, it admits, by Bochner theorem, the representation (7.2.2) for some distribution function $F(\lambda)$. $\qquad\square$

PROOF. (*Sufficiency*). Let $K(t)$ be the function defined by relation (7.2.2). We have to produce a wide-sense-stationary process $X(t)$ whose covariance function is $K(u)$. For an arbitrary integer $n > 0$ and real numbers $t_1, \ldots, t_n$, let $X(t_1), \ldots, X(t_n)$ be an n-dimensional random vector following the Gaussian distribution with

$$E[X(t_k)] = 0, \quad \text{var}(X(t_k)) = 1, \quad 1 \leqslant k \leqslant n$$

and the correlation coefficient of $X(t_i)$ and $X(t_j)$ given by

$$E[X(t_i)X(t_j)] = K(t_i - t_j). \tag{7.2.5}$$

The Gaussian process $X(t)$ obtained this way is wide sense stationary, by definition, and is the process that we are looking for. $\qquad\square$

In the case of real valued stationary process the representation (7.2.2) takes the form

$$K(t) = \int_{-\infty}^{\infty} \cos \lambda t \, dF(\lambda). \tag{7.2.6}$$

The spectral representation of a wide-sense-stationary process is due to Kolmogorov, who established (7.2.4) using Hilbert space methods. More direct methods are known now.

Theorem 7.2.2. *A centered wide-sense-stationary process $X(t)$ admits a spectral representation in the form*

$$X(t) = \int_{-\infty}^{\infty} e^{it\lambda} \, dY(\lambda), \tag{7.2.7}$$

where $Y(\lambda)$ is a centered second-order process with orthogonal increments. If $X(t)$ is real valued, the representation is given by

$$X(t) = \int \cos t\lambda \, dY(\lambda) + \int \sin t\lambda \, dZ(\lambda) \tag{7.2.8}$$

where Y and Z are uncorrelated to each other and are centered second-order processes with orthogonal increments.

175

PROOF. Let $F(\lambda)$ be the spectral distribution function of $X(t)$ obtained through Khintchin's formula (7.2.2). Let ξ and η be two continuity points of F. Recall that the continuity points of F are dense everywhere. So if we can define $Y(\lambda)$ at continuity points of F, then by continuity we can extend its definition at the discontinuity points (which forms an at most countable set).

Step 1. There is an obvious guess for the definition of $Y(\lambda)$. If (7.2.7) holds, then by analogy with the inversion formula for Fourier–Stieltjes integral we define, for $\xi < \eta$,

$$Y_T(\xi, \eta) = \frac{1}{2\pi} \int_{-T}^{T} X(t) \int_{\xi}^{\eta} e^{-it\lambda} \, d\lambda \, dt$$

$$= \frac{1}{2\pi} \int_{-T}^{T} X(t) \frac{e^{-it\eta} - e^{-it\xi}}{-it} \, dt. \tag{7.2.9}$$

We have to show that the integral in (7.2.9) exists and converges, in mean-square sense as $T \to \infty$, to a limit $Y(\xi, \eta)$. Now let $0 < S \leqslant T$. Then

$$E[|Y_T(\xi, \eta) - Y_S(\xi, \eta)|^2]$$

$$= E\left[\left|\int_{-T}^{T} X(t) \frac{e^{-it\eta} - e^{-it\xi}}{-2\pi it} \, dt - \int_{-S}^{S} X(t) \frac{e^{-it\eta} - e^{-it\xi}}{-2\pi it} \, dt\right|^2\right]$$

$$= E\left[\left|\int_{S<|t|<T} X(t) \frac{e^{-it\eta} - e^{-it\xi}}{-2\pi it} \, dt\right|^2\right]$$

$$= \int_{S<|t|<T} \int_{S<|s|<T} \frac{e^{-it\eta} - e^{-it\xi}}{-2\pi it} \frac{e^{is\eta} - e^{is\xi}}{2\pi is} E[X(t)\overline{X(s)}] \, dt \, ds$$

$$= \int_{S<|t|<T} \int_{S<|s|<T} \int_{-\infty}^{\infty} \frac{e^{-it\eta} - e^{-it\xi}}{-2\pi it} \frac{e^{is\eta} - e^{is\xi}}{2\pi is} e^{i\lambda(t-s)} \, dt \, ds \, dF(\lambda),$$

by Khintchin's formula (7.2.2),

$$= \int_{-\infty}^{\infty} \left|\int_{S<|t|<T} \frac{e^{-it\eta} - e^{-it\xi}}{-2\pi it} e^{i\lambda t} \, dt\right|^2 dF(\lambda). \tag{7.2.10}$$

We claim that the inner integral in (7.2.10) converges uniformly to zero as $S, T \to \infty$ if, for any fixed $\epsilon > 0$, $|\lambda - \eta| > \epsilon$ and $|\lambda - \xi| > \epsilon$. To see this, rewrite the inner integral as

$$\int\limits_{S<|t|<T} \frac{e^{-it\eta} - e^{-it\xi}}{-2\pi it} e^{i\lambda t} \, dt = \int\limits_{S<|t|<T} \int_{\xi}^{\eta} \frac{e^{-it(x-\lambda)}}{2\pi} \, dx \, dt. \qquad (7.2.11)$$

Then

$$\left| \int_S^T \int_\xi^\eta \cos t(x - \lambda) \, dx \, dt \right| = \left| \int_S^T t^{-1} [\sin t(\eta - \lambda) - \sin t(\xi - \lambda)] \, dt \right|$$

$$\leqslant \left| \frac{\cos t(\eta - \lambda)}{t(\eta - \lambda)} \right|_S^T \right| + \left| \int_S^T \frac{\cos t(\eta - \lambda)}{t^2(\eta - \lambda)} \, dt \right| + \left| \frac{\cos t(\xi - \lambda)}{t(\xi - \lambda)} \right|_S^T \right|$$

$$+ \left| \int_S^T \frac{\cos t(\xi - \lambda)}{t^2(\xi - \lambda)} \, dt \right|$$

$$\leqslant 2 \frac{S + T}{ST} [|\eta - \lambda|^{-1} + |\xi - \lambda|^{-1}].$$

Now let $S, T \to \infty$. Then

$$E[|Y_T(\xi, \eta) - Y_S(\xi, \eta)|^2] \to 0 \qquad \text{as } S, T \to \infty. \qquad (7.2.12)$$

Hence there is a second-order variable $Y(\xi, \eta)$ such that

$$E[|Y_T(\xi, \eta) - Y(\xi, \eta)|^2] \to 0 \qquad \text{as } T \to \infty. \qquad (7.2.13)$$

Step 2. If (ξ, η) and (α, β) are disjoint intervals, $Y(\xi, \eta)$ and $Y(\alpha, \beta)$ are orthogonal.

Let ξ, η, α, and β be any four continuity points of $F(\lambda)$. From (7.2.13) it follows that the correlation $E[Y(\xi, \eta)\overline{Y(\alpha, \beta)}]$ exists. Let us compute this correlation.

$$E[Y(\xi, \eta)\overline{Y(\alpha, \beta)}] = \lim_{T \to \infty} E[Y_T(\xi, \eta)\overline{Y_T(\alpha, \beta)}]$$

$$= \lim_{T \to \infty} \int_{-\infty}^{\infty} \int_0^T 2 \int_\xi^\eta \cos t(x - \lambda) \, dx \, dt \int_0^T 2 \int_\alpha^\beta \cos u(y - \lambda) \, dy \, du \, dF(\lambda).$$

$$\qquad (7.2.14)$$

Using the well-known fact that

$$\lim_{T \to \infty} \frac{1}{\pi} \int_0^T \frac{\sin at}{t} \, dt = \frac{1}{\pi} \int_0^\infty \frac{\sin at}{t} \, dt = \begin{cases} \frac{1}{2} & \text{for } a > 0 \\ 0 & \text{for } a = 0 \\ -\frac{1}{2} & \text{for } a < 0 \end{cases}$$

we obtain

$$\int_0^\infty \int_\xi^\eta \cos t(x - \lambda)\, dx\, dt = \int_0^\infty t^{-1}[\sin t(\eta - \lambda) - \sin t(\xi - \lambda)]\, dt$$

$$= \begin{cases} \pi & \text{if} & \xi < \lambda < \eta \\ 0 & \text{if} & \eta < \lambda \text{ or } \quad \lambda < \xi. \\ \frac{\pi}{2} & \text{if} & \lambda = \xi \text{ or } \quad \eta \end{cases} \qquad (7.2.15)$$

Using (7.2.15) in (7.2.14), we obtain

$$E[Y(\xi,\eta)\overline{Y(\alpha,\beta)}] = \begin{cases} \int\int_{\xi \vee \alpha}^{\eta \wedge \beta} dF(\lambda) & \text{if } \eta \wedge \beta > \xi \vee \alpha \\ 0 & \text{otherwise} \end{cases}. \qquad (7.2.16)$$

The expression (7.2.16) proves Step 2.

Step 3. There exists a second-order process $Y(\lambda)$ with orthogonal increments and

$$E[|Y(\eta) - Y(\xi)|^2] = F(\eta) - F(\xi). \qquad (7.2.17)$$

From (7.2.16) it follows that

$$E[|Y(\xi,\eta)|^2] = \int_\xi^\eta dF(\lambda) = F(\eta) - F(\xi). \qquad (7.2.18)$$

Moreover

$$E[|Y(\xi,\eta) - Y(\alpha,\eta)|^2] = |F(\xi) - F(\alpha)|$$

$$\to 0 \qquad \text{as } \alpha, \xi \to -\infty,$$

and consequently there exists a second-order process $Y(\cdot)$ such that

$$\lim_{\xi \to -\infty} E[|Y(\xi,\eta) - Y(\eta)|^2] = 0. \qquad (7.2.19)$$

Therefore

$$Y(\xi,\eta) = Y(\eta) - Y(\xi). \qquad (7.2.20)$$

Using expression (7.2.20) in (7.2.16), we see that $Y(\cdot)$ has orthogonal increments, and using (7.2.20) in (7.2.18), we obtain (7.2.17). This proves Step 3.

Step 4. We complete the proof of the theorem in this step. Let ξ and η be two continuity points of F with $\xi < \eta$. Then

$$E[X(t)\overline{(Y(\eta) - Y(\xi))}]$$

$$= \frac{1}{2\pi} \int_{-\infty}^\infty K(t - s) \int_\xi^\eta e^{isu}\, du\, ds \qquad \text{[by (7.2.9), (7.2.13), and (7.2.19)]}$$

$$= \frac{1}{\pi} \int_{-\infty}^{\infty} \int_0^{\infty} \int_{\xi}^{\eta} e^{it\lambda} \cos s(u - \lambda) \, du \, ds \, dF(\lambda) \qquad \text{[by (7.2.2)]}$$

$$= \frac{1}{\pi} \int_{-\infty}^{\infty} \int_0^{\infty} e^{it\lambda} s^{-1} [\sin s(\eta - \lambda) - \sin s(\xi - \lambda)] \, ds \, dF(\lambda)$$

$$= \int_{\xi}^{\eta} e^{it\lambda} dF(\lambda) \qquad \text{[by the value of } \int_0^{\infty} t^{-1} \sin at \, dt].$$

$$(7.2.21)$$

Since $Y(\lambda)$ is a process with orthogonal increment and satisfies (7.2.17), it is easy to see that the integral $\int_{-\infty}^{\infty} e^{it\lambda} dY(\lambda)$ exists as the mean-square limit of Riemann–Stieltjes sum. Using this fact along with (7.2.21) and Khintchin formula (7.2.2), we obtain

$$E\left[X(t) \overline{\int_{-\infty}^{\infty} e^{is\lambda} dY(\lambda)}\right] = \lim_{T \to \infty} \int_{-T}^{T} e^{i(t-s)\lambda} dF(\lambda)$$

$$= \int_{-\infty}^{\infty} e^{i(t-s)\lambda} dF(\lambda) \qquad (7.2.22)$$

$$= K(t - s),$$

and again for the same reasons we obtain

$$E\left[\int_{-\infty}^{\infty} e^{it\lambda} dY(\lambda) \overline{\int_{-\infty}^{\infty} e^{is\lambda} dY(\lambda)}\right] = \int_{-\infty}^{\infty} e^{i(t-s)\lambda} dF(\lambda) = K(t - s).$$

Therefore

$$E\left[\left|X(t) - \int_{-\infty}^{\infty} e^{it\lambda} dY(\lambda)\right|^2\right]$$

$$= E[|X(t)|^2] - 2\Re E\left[X(t) \overline{\int_{-\infty}^{\infty} e^{i\lambda t} dY(\lambda)}\right] + E\left[\left|\int_{-\infty}^{\infty} e^{it\lambda} dY(\lambda)\right|^2\right]$$

$$= 0 \qquad \text{[by (7.2.22)]}.$$

This gives our spectral representation (7.2.7) in the case when F is continuous.

If ξ is a discontinuity point of F, then $Y(\xi)$ can be defined by $Y(\xi) = \lim_{\eta \downarrow \xi} Y(\eta)$. All the essential formulas (7.2.17)–(7.2.22) hold so that the proof of (7.2.7) goes through. This completes the proof. $\qquad \square$

Next we make a few remarks that contain some relevant information.

Remark 7.2.3. Let $X(t)$ be a wide-sense-stationary process with covariance function $K(t)$. [Recall that we have been assuming that $X(t)$ is centered and mean-square continuous.] Then Khintchin's formula gives us a method to evaluate the spectral distribution function $F(\lambda)$ of $X(t)$. All we have to do is find the inverse Fourier–Stieltjes transform of $K(t)$. In several of the practical situations $|K(t)|$ approaches zero so rapidly as $|t| \to \infty$ that $\int_{-\infty}^{\infty} |K(t)|\, dt < \infty$. Consequently, it is possible to represent $K(t)$ as

$$K(t) = \int_{-\infty}^{\infty} e^{it\lambda} f(\lambda)\, d\lambda. \tag{7.2.23}$$

If (7.2.23) holds, then $F(\lambda) = \int_{-\infty}^{\lambda} f(x)\, dx$, and $f(\lambda) = F'(\lambda)$. The function $f(\lambda)$ is called the *spectral density function* or the *power spectrum* of $X(t)$ [or of $K(t)$]. Representation (7.2.23) avoids the Stieltjes integrals in (7.2.2). Inverting (7.2.23), we get

$$f(\lambda) = \frac{1}{2\pi} \int_{-\infty}^{\infty} e^{-i\lambda t} K(t)\, dt. \tag{7.2.24}$$

Remark 7.2.4. Consider the discrete time process $X_n = a_n X$, $n = 0, \pm 1, \pm 2, \ldots$, where X is a complex RV with zero mean and unit variance and $\{a_n\}$ is a doubly infinite sequence of complex numbers. It can be shown, as in Example 7.1.3 (4), that $\{X_n\}_{-\infty}^{\infty}$ is a wide-sense-stationary process if and only if $X_n = X e^{in\lambda}$. This represents a (discretized) harmonic oscillation with random amplitude and phase. However, $e^{in\lambda} = e^{in(\lambda + 2k\pi)}$ for integral values of k, so that the angular frequency is defined only to within an additive constant $2k\pi$. Consequently, we can assume that $\lambda \in [-\pi, \pi]$, thereby taking into account only the oscillations with frequencies lying in $[-\pi, \pi]$. Therefore, representation (7.2.7) reduces to

$$X_n = \int_{-\pi}^{\pi} e^{in\lambda}\, dY(\lambda), \tag{7.2.25}$$

and (7.2.23) takes the form

$$K(n) = \int_{-\pi}^{\pi} e^{in\lambda} f(\lambda)\, d\lambda, \tag{7.2.26}$$

provided that $\Sigma_{k=-\infty}^{\infty} |K(k)| < \infty$. Now (7.2.24) becomes

$$f(\lambda) = \frac{1}{2\pi} \sum_{k=-\infty}^{\infty} e^{-ik\lambda} K(k). \tag{7.2.27}$$

Remark 7.2.5. If $X(t)$ is a real-valued wide-sense-stationary process, we have [see (7.2.8)]

180

$$X(t) = \int_0^\infty \cos \lambda t \, dY(\lambda) + \int_0^\infty \sin \lambda t \, dZ(\lambda).$$

Here the processes Y and Z are given by

$$Y(\lambda) = \lim_{T \to \infty} \frac{1}{2\pi} \int_{-T}^T t^{-1} \sin \lambda t \, X(t) \, dt,$$

$$Z(\lambda) = \lim_{T \to \infty} \frac{1}{2\pi} \int_{-T}^T t^{-1} (1 - \cos \lambda t) X(t) \, dt. \tag{7.2.28}$$

Examples 7.2.6

EXAMPLE 1. Let $\{X_n\}_{-\infty}^\infty$ be a sequence of centered and uncorrelated RVs with unit variance [see Example 7.1.3 (1)]. Then $\{X_n\}$ is a stationary sequence with

$$K(0) = 1 \quad \text{and} \quad K(n) = 0 \quad \text{for } n \neq 0.$$

From (7.2.27) we have the constant power spectrum $f(\lambda) = 1/2\pi$. Actually, the wide-sense-stationary sequence of uncorrelated centered RVs is characterized by a constant power spectrum concentrated on $[-\pi, \pi]$.

EXAMPLE 2. *Moving Average Process.* In Example 7.1.3 (2) we considered a moving average process Y_n defined by a linear combination of uncorrelated RVs. Replacing the finite sum (7.1.3) bn infinite one, let us set

$$X_n = \sum_{k=0}^\infty a^k Y_{n-k},$$

where $a \in R$, $|a| < 1$, and all Y_n are centered uncorrelated RVs. As in Example 7.1.3 (2), we have

$$K(n) = \sum_{k \geqslant n} a^k a^{k-n} = \frac{a^n}{1 - a^2}, \quad n \geqslant 0,$$

so that the covariance function $K(n)$ is of the form

$$K(n) = A a^{|n|}, \qquad A > 0, \quad |a| < 1, \quad n = 0, \pm 1, \pm 2, \ldots . \tag{7.2.29}$$

Let us compute the power spectrum of this $K(\cdot)$.

$$f(\lambda) = \frac{1}{2\pi} \sum_{k=-\infty}^{\infty} K(k)e^{-ik\lambda}$$

$$= \frac{A}{2\pi} \left[\sum_{k=-\infty}^{-1} a^{-k}e^{-ik\lambda} + \sum_{k=0}^{\infty} a^k e^{-ik\lambda} \right]$$

$$= \frac{A}{2\pi} \left[\sum_{k\geqslant 1} a^k e^{ik\lambda} + \sum_{k\geqslant 0} a^k e^{-ik\lambda} \right] \tag{7.2.30}$$

$$= \frac{A}{2\pi} \frac{ae^{i\lambda}}{1 - ae^{i\lambda}} + \frac{1}{1 - ae^{-i\lambda}}$$

$$= \frac{A}{2\pi} \left[\frac{1 - a^2}{|e^{i\lambda} - a|^2} \right].$$

EXAMPLE 3. Consider the wide-sense-stationary process $X(t) = U \cos \lambda t + V \sin \lambda t$, where $\lambda > 0$ and U and V are uncorrelated RVs with zero mean and unit variance. Then, as seen in Example 7.1.3 (3), the covariance function $K(t)$ is given by $K(t) = \cos \lambda t$. Now the spectral distribution function of $X(t)$ is given by

$$F(u) = \begin{cases} 0 & \text{if } u \leqslant -\lambda \\ \frac{1}{2} & \text{if } -\lambda < u \leqslant \lambda. \\ 1 & \text{if } u > \lambda \end{cases} \tag{7.2.31}$$

This is obvious from (7.2.6).

EXAMPLE 4. *Random Telegraph Signal.* Consider the wide-sense-stationary process $X(t) = Y(-1)^{N(t)}$, where Y is an RV with $P\{Y = -1\} = P\{Y = 1\} = \frac{1}{2}$, $N(t)$ is a Poisson process with rate μ and, Y and $N(t)$ are independent. It is shown in Example 7.1.3 (5) that $K(t) = e^{-2\mu t}$.

Extend the domain of $K(\cdot)$ to negative t by defining $K(-t) = K(t)$. Then $K(t)$ takes the form $K(t) = e^{-2\mu|t|}$. Let us compute the power spectrum of the random telegraph signal process:

$$f(\lambda) = \frac{1}{2\pi} \int_{-\infty}^{\infty} e^{-i\lambda t} e^{-2\mu|t|} dt$$

$$= \frac{1}{2\pi} \int_{-\infty}^{0} e^{(2\mu - i\lambda)t} dt + \frac{1}{2\pi} \int_{0}^{\infty} e^{-(2\mu + i\lambda)t} dt$$

$$= \frac{1}{2\pi} \left[\frac{1}{2\mu - i\lambda} + \frac{1}{2\mu + i\lambda} \right] \tag{7.2.32}$$

$$= \frac{2\mu}{\pi(4\mu^2 + \lambda^2)}.$$

Figures 7.2 and 7.3 plot the covariance function $K(t) = e^{-2\mu|t|}$ and the corresponding power spectrum $f(\lambda)$.

The μ that appears in the functions K and f is the rate at which the changes in the telegraph signal occur. But the graph of $K(t)$ (see Figure 7.2) approaches the $2\mu t$ axis exponentially, and $K(t)$ is almost zero for values of t that are only a few multiples of $(2\mu)^{-1}$ or μ^{-1}. Thus the reciprocal μ^{-1} of the intensity rate of the Poisson type changes of the signal gives the length of time that is needed for the correlation between the signals $X(t)$ and $X(t + u)$ to taper off. This interpretation of the intensity rate μ of the Poisson process helps us to estimate (the parameter) μ. To do this, one has to find out for how long there is any accountable correlation between $X(0)$ and $X(t)$. Next consider the graph of $f(\lambda)$ on the $((\lambda/2\mu) \times f(\lambda))$ axes. At the origin $f(\cdot)$ takes the maximum value 1. For small values of $\lambda/2\mu$, the power spectrum stays close to this maximum value. As λ increases, the graph steadily approaches the $\lambda/2\mu$ axis. Processes with a flat power spectrum are of great importance (a white-noise process is an example of it—we study white-noise processes in Chapter 9). In the present case we can flatten the power spectrum by increasing the Poisson intensity rate and thereby increasing the length of the interval in which the power spectrum is approximately constant. In many practical problems we may not be interested for all λ in $-\infty < \lambda < \infty$, and so restrict our attention to a finite interval $[-T, T]$. By taking μ sufficiently large we can make $f(\lambda) \simeq f(0)$ in this interval $[-T, T]$. We remark further about the flat spectrum when we consider white-noise processes.

EXAMPLE 5. Find the power spectrum $f(\lambda)$ if the covariance $K(t)$ is given by

(a) $K(t) = Ae^{-a|t|}(1 + a|t|);$

(b) $K(t) = e^{-a|t|}(1 + a|t| + \frac{1}{3}a^2 t^2).$

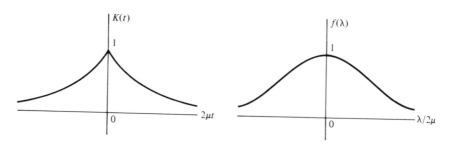

Figure 7.2   Figure 7.3

First set $\psi(a) = (A/2\pi) \int_{-\infty}^{\infty} e^{-a|t|} e^{-i\lambda t} \, dt$. We leave it to the reader to show that

$$\psi(a) = \frac{A}{\pi} \frac{a}{a^2 + \lambda^2}. \tag{7.2.33}$$

Now

$$\begin{aligned}
f(\lambda) &= \frac{A}{2\pi} \int_{-\infty}^{\infty} e^{-i\lambda t - a|t|} (1 + a|t|) \, dt \\
&= \frac{A}{2\pi} \int_{-\infty}^{\infty} e^{-i\lambda t - a|t|} \, dt + \frac{A}{2\pi} \int_{-\infty}^{\infty} a|t| e^{-i\lambda t - a|t|} \, dt \\
&= \psi(a) - a \frac{d\psi}{da} \\
&= \frac{A}{\pi} \frac{a}{a^2 + \lambda^2} - \frac{aA}{\pi} \frac{(a^2 + \lambda^2) - 2a^2}{(a^2 + \lambda^2)^2} \qquad \text{[using (7.2.33)]} \\
&= \frac{2A}{\pi} \frac{a^3}{(a^2 + \lambda^2)^2}.
\end{aligned}$$

Then set $\psi(\lambda) = \frac{1}{2\pi} \int_{-\infty}^{\infty} e^{-i\lambda t - a|t|} \, dt$. Note that

$$\psi(\lambda) = \frac{a}{\pi(a^2 + \lambda^2)}. \tag{7.2.34}$$

Now

$$\begin{aligned}
f(\lambda) &= \frac{1}{2\pi} \int_{-\infty}^{\infty} e^{-i\lambda t - a|t|} \left(1 + a|t| + \frac{a^2 t^2}{3} \right) dt \\
&= \psi(\lambda) - a \frac{d\psi}{da} + \frac{a^2}{3} \frac{d^2\psi}{da^2} \\
&= \frac{8a^5}{3\pi(\lambda^2 + a^2)^3} \qquad \text{[using (7.2.34)].}
\end{aligned}$$

EXAMPLE 6. *Random Binary Noise.* In Example 7.1.3. (6) we computed the covariance function of the random binary noise and obtained

$$K(t) = \begin{cases} (T - |t|)/T & \text{if } |t| < T \\ 0 & \text{otherwise} \end{cases}. \tag{7.2.35}$$

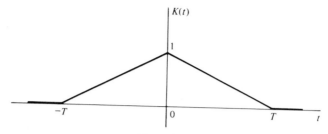

Figure 7.4

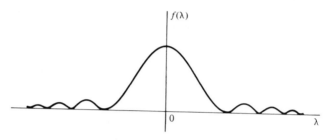

Figure 7.5

It can be shown that the power spectrum $f(\lambda)$ in this case is

$$f(\lambda) = \frac{2}{\pi T \lambda^2} \sin^2 \frac{\lambda T}{2}. \tag{7.2.36}$$

The graphs of (7.2.35) and (7.2.36) are given in Figures 7.4 and 7.5, respectively.

7.3. Ergodic Theory of Stationary Processes

Consider a physical process governed by a stationary process. Suppose we want to compute some statistical averages of the process and instead what is available is only a long-term observation of a single sample function. In such situations it is natural to ask whether it is possible to determine the statistical average from an appropriate time average corresponding to a single sample function. If the statistical or ensemble average of the process equals the time average of a sample function, the process will be called *ergodic*. This section presents some theorems that give conditions under which a stationary process becomes ergodic.

The ergodic theory has its origin in classical statistical mechanics. Consider a conservative dynamical system with n degrees of freedom and

185

whose equations of motion are given by

$$\frac{dq_i}{dt} = \frac{\partial H}{\partial p_i}, \qquad \frac{dp_i}{dt} = -\frac{\partial H}{\partial q_i}, \qquad i = 1, \ldots, n,$$

where $H = H(p_i(t), q_i(t); 1 \leqslant i \leqslant n)$ is the Hamiltonian of the system, $q_1, \ldots, q_n$ are the generalized coordinates, and $p_1, \ldots, p_n$ are the corresponding generalized momenta. The $2n$-dimensional space $S = \{(q_1, \ldots, q_n, p_1, \ldots, p_n)\}$ is called the *phase space* of the system. Let T_t be a transformation of S into itself such that $x_{s+t} = T_t x_s$ is the state of the system at time $s + t$ if x_s is the state of the system at time s. Liouville's theorem states that the volume is preserved under the transformations T_t. Because of the conservative nature of the dynamical system, a point in the region between two surfaces of constant energy will remain in the region forever. So a phase function

$$X(t) = X(q_1(t), \ldots, q_n(t), p_1(t), \ldots, p_n(t))$$

defines a strictly stationary process. Since conservative systems have constant energy (*ergos*) e, the possible trajectories lie on the surface $H = e$. In such a context the physicists wanted to establish the equality of space and time averages. These results came to be known as ergodic theorems.

Let $\{X_n\}$ be a stochastic process chosen to describe a physical process. One would like to measure from the observation of the process $\{X_n\}$ the statistical averages such as the mean function $m(n)$ and covariance function $K(m, n)$. An experimenter observes a single realization of the process for a long period $0 \leqslant k \leqslant N$ and wants to use this observation to estimate the statistical average, say, $E[X_n]$, $n \geqslant 0$. He forms the sample average $N^{-1} \Sigma_{k=0}^{N} X_k$ and wants to know if it converges to a limit as $N \to \infty$, where the convergence is in suitable probabilistic sense. Let $\{X_n\}$ be a strictly stationary process such that $E[|X_n|] < \infty$ for all n. Then $m(n)$ is a constant m. It is natural to ask if $N^{-1} \Sigma_0^N X_n \to m$, with probability one. Ergodic theorems give conditions for such a relation to hold.

Theorem 7.3.1. (Strong or Individual Ergodic Theorem). *Let $\{X_n\}_{-\infty}^{\infty}$ be a strictly stationary process with $E[|X_n|] < \infty$. Then the sample mean $n^{-1} \Sigma_1^n X_k$ converges to a limit with probability one. If, moreover, $\{X_n\}$ is a second-order process with bounded variance and such that the covariance function $K(n) \to 0$, as $n \to \infty$, then*

$$P\left\{ n^{-1} \sum_0^{n-1} X_k \to m = E[X_n] \right\} = 1. \qquad (7.3.1)$$

PROOF. Set $S_{ab} = (b - a)^{-1}(X_a + X_{a+1} + \cdots + X_{b-1})$. The first part of the theorem claims that S_{0n} converges with probability one to a limit as $n \to \infty$. Let C denote the contrary event, that is, $C = \{\omega: S_{0n}(\omega)$ does not converge$\}$. Then we have to show that $P\{C\} = 0$. Let us assume the contrary, that $P\{C\} > 0$, and arrive at a contradiction (which will establish the first part of the theorem).

Set $S^* = \limsup\limits_{n \to \infty} S_{0n}$ and $S_* = \liminf\limits_{n \to \infty} S_{0n}$. Consider a particular realization $X_\omega(n) = x_n$, $n \geq 0$, from C. Set in this case, $S^* = s^*$ and $S_* = s_*$. Then there exist rationals α and β such that $s_* < \alpha < \beta < s^*$. Consider all such intervals $C_n = (\alpha_n, \beta_n)$ with rational end points. By assumption, $P\{C\} > 0$; therefore, there is a $k > 0$ such that $S_* < \alpha_k < \eta_k < S^*$, and $P\{C_k\} > 0$. Corresponding to such a realization, consider the integral interval (a, b) for which $S_{ab} > \beta$ and $S_{ac} \leq \beta$ for all $c \in (a, b)$. Call this interval a "special interval WRT β."

We claim that two special intervals do not overlap. Let (a, b) and (c, d) be two special intervals such that $a < c < b < d$. Since (a, b) is special,

$$S_{ab} > \beta \qquad \text{and} \qquad S_{ai} \leq \beta \qquad \text{for all } i \in (a, b). \qquad (7.3.2)$$

Noting $S_{ab} = (b - a)^{-1}[(c - a)S_{ac} + (b - c)S_{cb}]$, it follows from $S_{ab} > \beta$ that either $S_{ac} > \beta$ or $S_{cb} > \beta$. But $S_{ac} \not> \beta$ by (7.3.2). Similarly, $S_{cb} \not> \beta$, since (c, d) is special. This shows that (a, b) and (c, d) are nonoverlapping.

Let us now define the so-called ρ-special interval. A special interval (a, b) is called ρ-special if the rank $(b - a) > \rho$ and it is not contained in any other special interval whose rank is greater than ρ. From what is shown above, it is easy to see that two ρ-special interval lie outside of each other.

Let C_ρ denote the event

$$\{S_* < \alpha < \beta < S^*, \text{ and there is a } \tau \leq \rho \text{ with } S_{0\tau} > \beta\}.$$

Then C is the limit of such events and $P\{C\} = \lim\limits_{\rho \to \infty} P\{C_\rho\}$. Since we assumed that $P\{C\} > 0$, there is a ρ such that $P\{C_\rho\} > 0$.

Our next step is to further decompose C_ρ into mutually exclusive events. If the event C_ρ occurs, one can find a least τ_0 among all $\tau \leq \rho$ such that $S_{0\tau} > \beta$. Now the interval $(0, \tau_0)$ is either ρ-special or is contained in a ρ-special interval $a \leq 0 < b$, and vice versa. Set $-a = k$ and $b - a = m$. Since the ρ-special intervals do not overlap, the event C_ρ is decomposed into C_{km}, corresponding to the intervals $(-k, -k + m)$. That is,

$$C = \bigcup_k \bigcup_m C_{km}, \qquad 1 \leq m \leq \rho, \qquad 0 \leq k \leq m - 1.$$

Since $\{X_n\}$ is strictly stationary, the event C_{0m} transforms into C_{km} under $p' = (p + k)$ such that

$$P\{C_{km}\} = P\{C_{0m}\} \quad \text{and} \quad E[X_0|C_{km}] = E[X_k|C_{0m}]. \quad (7.3.3)$$

Also

$$
\begin{aligned}
P\{C_\rho\} E[X_0|C_\rho] &= \sum_{m=1}^{\rho} \sum_{k=0}^{m-1} P\{C_{km}\} E[X_0|C_{km}] \\
&= \sum_{m=1}^{\rho} P\{C_{0m}\} \sum_{k=0}^{m-1} E[X_k|C_{0m}] \quad \text{[by (7.3.3)]} \\
&= \sum_{m=1}^{\rho} P\{C_{0m}\} E[mS_{0m}|C_{0m}] \\
&> \sum_{m=1}^{\rho} P\{C_{0m}\} m\beta \quad \text{(since } S_{0m} > \beta \text{ on } C_{0m}) \\
&= \beta \sum_{m=1}^{\rho} \sum_{k=0}^{m-1} P\{C_{0m}\} \\
&= \beta P\{C_\rho\}.
\end{aligned}
$$

Hence $E[X_0|C_\rho] > \beta$. Therefore, $E[X_0|C] \geqslant \beta$, since the events C_ρ converge to C.

Working similarly with α (instead of β), one can show that $E[X_0|C] \leqslant \alpha$. That is, we have shown that

$$E[X_0|C] \leqslant \alpha < \beta \leqslant E[X_0|C],$$

if $P\{C\} > 0$. This is the contradiction we have been looking for. This proves the first part of the theorem.

To prove the second half of the theorem, let the hypothesis $K(n) \to 0$, as $n \to \infty$, hold. The claim is that

$$P\{S_{0n} \to m\} = 1, \quad m = E[X_n].$$

For the moment let us assume that the variance of S_{0n} goes to zero as $n \to \infty$, that is, $\text{var}(S_{0n}) \to 0$. Then by Chebyshev's inequality, $S_{0n} \to m$ in probability. But in the first half of the theorem we have shown that S_{0n} converges with probability one. Hence the theorem will be proved once we establish that $\text{var}(S_{0n}) \to 0$, as $n \to \infty$. Now setting $\sigma^2 = \text{var}(X_0) = K(0)$,

$$
\begin{aligned}
\text{var}(S_{0n}) &= E\left[n^{-1} \sum_{k=0}^{n-1} (X_k - m)\right]^2 \\
&= n^{-2}[n \, \text{var}(X_0) + 2 \sum_{0 \leqslant i < j \leqslant n-1} K(j-i)] \\
&= n^{-2}\left[n\sigma^2 + 2 \sum_{i=1}^{n-1} (n-i)K(i)\right].
\end{aligned}
$$

Since $K(n) \to 0$, as $n \to \infty$, for any $\epsilon > 0$, we can find an $N > 0$ such that $|K(m)| \leqslant \epsilon$, for all $m \geqslant N$. Therefore

$$\text{var}(S_{0n}) \leqslant n^{-2} \left[n\sigma^2 + 2 \sum_{i=1}^{N-1} (n-i) + 2\epsilon \sum_{i=N}^{n-1} (n-i) \right]$$

$$\leqslant n^{-2} \left[n\sigma^2 + 2(N-1)(n-1) + 2\epsilon(n-N-1)(n-N) \right].$$

Choosing a sufficiently large n and small $\epsilon > 0$, we can now make $\text{var}(S_{0n})$ arbitrarily small. This is what we sought to establish and hence this proves the theorem. □

It should be remarked here that under the conditions of the individual ergodic theorem it is also true that the sample mean converges in mean; that is, there is an RV ξ such that $n^{-1}\Sigma_{k=1}^n X_k \to \xi$ with probability one and $\lim_{n\to\infty} E[|(n^{-1}\Sigma_{k=1}^n X_k) - \xi|] = 0$. Here, of course, we do not assume the existence of finite variance of X_n.

Definition 7.3.2. Let $\mathfrak{S}$ denote the space of all sequences $\{x_n, n \geqslant 0\}$ of real numbers. An operation $T: \mathfrak{S} \to \mathfrak{S}$ is called a *shift operator* if $T(x_0, x_1, \ldots) = (x_1, x_2, \ldots)$. A subset $A \subset \mathfrak{S}$ is said to be *shift invariant* if $TA \subset A$.

Definition 7.3.3. A strictly stationary process $\{X_n\}$ is said to be *ergodic* if $P\{(X_0, X_1, \ldots) \in A\} = 0$ or 1 for every shift-invariant set A.

Theorem 7.3.4. *Let $\{X_n\}$ be a strictly stationary ergodic process with finite mean m. Then*

$$P\{ \lim_{n\to\infty} n^{-1} \sum_{k=1}^n X_k = m \} = 1.$$

PROOF. For each real number r let us define the set

$$A_r = \left\{ \mathbf{x} \in \mathfrak{S}: \lim_{n\to\infty} n^{-1} \sum_{k=1}^n x_k \leqslant r \right\}.$$

It can easily be verified that A_r is shift invariant and consequently

$$P\{(X_0, X_1, \ldots) \in A_r\} = P\left\{ \lim_{n\to\infty} n^{-1} \sum_{k=1}^n X_k \leqslant r \right\}.$$

By the individual ergodic theorem $P\{\lim_{n\to\infty} n^{-1}\Sigma_{k=1}^n X_k = \xi\} = 1$, for

189

some RV ξ. Because of this and the fact that A_r is shift invariant, we have $P\{\xi \leqslant r\} = 0$ or 1 for every real r. Therefore, ξ is a constant RV. But, as remarked earlier, it is also true that

$$\lim_{n \to \infty} E\left[\left|\left(n^{-1} \sum_{k=1}^{n} X_k\right) - \xi\right|\right] = 0.$$

Consequently, $E[\xi] = E[X_n] = m$, and hence $\xi = m$. This completes the proof. $\qquad\qquad\qquad\qquad\qquad\qquad\qquad\qquad\qquad\qquad\qquad\qquad\qquad\square$

Theorem 7.3.5. (Mean-Square Ergodic Theorem). *Let* $\{X_n\}$ *be a wide-sense-stationary process with mean* $m = E[X_n]$. *Then*

$$\lim_{n \to \infty} E\left[\left|n^{-1} \sum_{k=}^{n} X_k - m - Y^*(0)\right|^2\right] = 0,$$

where $Y^*(0) = Y(0) - Y(0 -)$ *and* Y *is the process that appears in the spectral representation* (7.2.25) *of the process* $\{X_n\}$. *Consequently, the limiting time average* $\lim_{n \to \infty} n^{-1} \sum_{k=1}^{n} X_k$ *equals the space average* m *if and only if* $Y^*(0) = 0$.

PROOF. Let F be the spectral distribution function of $\{X_n\}$. Define Z and G as follows:

$$Z(\lambda) = \begin{cases} Y(\lambda) & \text{if } \lambda < 0 \\ Y(\lambda) - Y^*(0) & \text{if } \lambda \geqslant 0 \end{cases},$$

$$G(\lambda) = \begin{cases} F(\lambda) & \text{if } \lambda < 0 \\ F(\lambda) - F(0) + F(0 -) & \text{if } \lambda \geqslant 0 \end{cases}.$$

Then $Z(\lambda)$ is a process with independent increments and

$$E[|Z(\eta) - Z(\xi)|^2] = G(\eta) - G(\xi).$$

and G is continuous at the origin. Note that

$$X_n - m = \int_{-\pi}^{\pi} e^{in\lambda} \, dY(\lambda) = \int_{-\pi}^{\pi} e^{in\lambda} \, dZ(\lambda) + Y^*(0).$$

(Recall that in Theorem 7.2.2 we assumed that $m = 0$.) Now

$$E\left[\left|n^{-1} \sum_{k=1}^{n} X_k - m - Y^*(0)\right|^2\right]$$

$$= E\left[\left|n^{-1} \sum_{k=1}^{n} \int_{-\pi}^{\pi} e^{ik\lambda} \, dZ(\lambda)\right|^2\right]$$

190

$$= n^{-2} \int_{-\pi}^{\pi} \left\{ \left[\sin^2 \frac{n\lambda}{2} \right] \Big/ \left[\sin^2 \frac{\lambda}{2} \right] \right\} dG(\lambda)$$

$$\leqslant G(\epsilon) - G(-\epsilon) + n^{-2} \left[\sin^2 \frac{\epsilon}{2} \right]^{-1} (G(\pi) - G(-\pi)),$$

for small $\epsilon > 0$. Since ϵ is arbitrarily small and G is continuous at 0, as noted above, we can make the last expression arbitrarily small, by taking sufficiently large n. This completes the proof. □

Thus far we have restricted our attention to the stationary sequences. Theorem 7.3.6 deals with a wide-sense-stationary process $\{X(t)\}$. It gives a necessary and sufficient condition in order that the space average $m = E[X(t)]$ is equal to the limiting time average

$$\overline{X} = \lim_{T \to \infty} (2T)^{-1} \int_{-T}^{T} X(t) \, dt,$$

with probability one. We omit the proof.

Theorem 7.3.6. *Let $\{X(t)\}$ be a real-valued wide-sense-stationary process with mean m and covariance function $K(t)$. Then*

$$\overline{X} = \lim_{T \to \infty} (2T)^{-1} \int_{-T}^{T} X(t) \, dt = m$$

with probability one if and only if

$$\lim_{T \to \infty} T^{-1} \int_{0}^{2T} \left(1 - \frac{t}{2T} \right) K(t) \, dt = 0. \tag{7.3.4}$$

Examples 7.3.7

EXAMPLE 1. Consider the wide-sense-stationary process $X(t) = U \cos \lambda t + V \sin \lambda t$ that we treated in Examples 7.1.3 (3) and 7.2.6 (3). The covariance function of $\{X(t)\}$ is $K(t) = \sigma^2 \cos \lambda t$. We claim here that $K(t)$ satisfies the condition (7.3.4), and hence $\{X(t)\}$ satisfies the conclusion of Theorem 7.3.6. But this is clear from

$$\left| T^{-1} \int_{0}^{2T} \left(1 - \frac{t}{2T} \right) \sigma^2 \cos \lambda t \, dt \right| = \left| \frac{\sigma^2}{2\lambda^2 T^2} (1 - \cos 2\lambda T) \right|.$$

191

EXAMPLE 2. *Random Telegraph Signal.* Consider the random telegraph signal process $\{X(t)\}$ studied in Examples 7.1.3 (5) and 7.2.6 (4). We have shown that the covariance function of this process is given by $K(t) = e^{-2\lambda|t|}$. We claim that this process satisfies the condition (7.3.4) and hence the conclusion of the ergodic Theorem 7.3.6. For

$$T^{-1} \int_0^{2T} \left(1 - \frac{t}{2T}\right) e^{-2\lambda t} \, dt = \frac{1}{2\lambda T} - \frac{1 - e^{-4\lambda T}}{8\lambda^2 T^2}$$

$$\to 0 \quad \text{as } T \to \infty.$$

Exercises

1. Let $X(t) = \sin Ut$, where U is uniformly distributed on $[0, 2\pi]$. (a) If $t = 1, 2, \ldots$, show that $\{X(t)\}$, $t = 1, 2, \ldots$, is a wide-sense-stationary process but not strictly stationary. (b) Let $t \in [0, \infty)$. Show that $\{X(t), t \geqslant 0\}$ is neither strictly stationary nor wide sense stationary.

2. Let $\{X(t)\}$, $t \geqslant 0$, be a second-order stationary process with covariance function $K(\cdot)$. Set $\sigma^2(T) = T^{-1} \int_0^T X(s) \, ds$. Show that

$$\sigma^2(T) = 2T^{-2} \int_0^T (T - s) K(s) \, ds, \quad K(s) = \frac{1}{2} \frac{d^2}{ds^2} [s^2 \sigma^2(s)].$$

3. Let $X(t)$ be a wide-sense-stationary process with covariance function $K_X(\cdot)$. Set $Y(t) = T^{-1} \int_t^{t+T} X(s) \, ds$, $t \geqslant 0$, and T is fixed. Compute the covariance function of $Y(\cdot)$ if

$$\text{(a) } K_X(s) = e^{-\alpha|s|},$$

$$\text{(b) } K_X(s) = \begin{cases} 1 - |s| & \text{if } |s| \leqslant 1 \\ 0 & \text{otherwise} \end{cases}.$$

4. Let $\{X_n, n \geqslant 0\}$ be a stationary, Gaussian, and Markov process with zero mean function. Show that the covariance function $K(\cdot)$ is of the form $K(m) = \sigma^2 a^{|m|}$ for some fixed a with $|a| \leqslant 1$.

5. Let $X_n = \sum_{k=1}^N \sigma_k \sqrt{2} \cos(a_k n - U_k)$, where σ_k and a_k are positive constants, $k = 1, \ldots, N$, and $U_1, \ldots, U_N$ are independent RVs uniformly distributed on $(0, 2\pi)$. Show that $\{X_n\}$ is a wide-sense-stationary process.

6. Let $\{B(t), t \in R\}$ be a standard Brownian motion and set $X(t) = \int_{-\infty}^t e^{-\alpha(t-s)} \, dB(s)$, $t \in R$, $\alpha > 0$. Show that $X(t)$, $t \in R$, is a wide-sense-stationary process.

7. Let $\{B(t), t \geqslant 0\}$ be a standard Brownian motion. Show that $X(t) = B(t + 1) - B(t)$, $t \geqslant 0$, is wide sense stationary.

8. Let $X_n = (X + na \mod 1)$, $n = 0, \pm 1, \pm 2, \ldots$, where X is an RV uniformly distributed on $[0, 1]$ and a is a fixed irrational number. Show that $\{X_n\}$ is strictly stationary.

9. Let $X_n, n = 0, \pm 1, \ldots$, be a Gaussian process satisfying the equation $\Sigma_{k=0}^{m} a_k X_{n-k} = \Sigma_{k=0}^{m} b_k Y_{n-k}$, $a_0 \neq 0$, $b_0 \neq 0$, where all Y_n are independent unit normal RVs. Find conditions under which $\{X_n\}$ will be a stationary solution of this equation.

10. Let $\{X_n\}$ be a Gaussian stationary process with covariance function $K(m)$. Show that $\{X_n\}$ satisfies the stochastic difference equation $\Sigma_{k=0}^{N} a_k X_{n-k} = 0$, $a_0 = 1$, if and only if $\Sigma_{k=0}^{N} a_k K(n - k) = 0$.

11. Find the general Gaussian stationary process $\{X_n\}$ solving: (a) $(X_{n+2} + X_n) = 0$ and (b) $(X_{n+3} - X_{n+2} + X_{n+1} - X_n) = 0$. [Answer: (a) $X_{n+2} = U \cos(\pi n/2) + V \sin(\pi n/2)$; (b) $X_n = U \cos(n\pi/2) + V \sin(n\pi/2) + W$.]

12. Let $\{X(t)\}$ be a Markov and Gaussian stationary process. Show that its covariance function is of the form $K(s) = e^{-a|s|}$. Show also that the power spectrum of K is proportional to a Cauchy density. Discuss the discrete parameter case given in Exercise 4.

13. Let $\{X(t)\}$, $t \in R$, be a continuous wide-sense-stationary process with unknown mean m and covariance function $K(s) = ae^{-b|s|}$, $t \in R$, where $a > 0$ and $b > 0$. For fixed $T > 0$ set $\overline{X} = T^{-1} \int_0^T X(s) \, ds$. Show that $E\overline{X} = m$, that is, $\overline{X}$ is an unbiased estimator of m, and that

$$\text{var}(\overline{X}) = 2a[(bT)^{-1} - (bT)^{-2}(1 - e^{-bT})].$$

14. Let $\{X(t)\}$, $t \in R$, be a wide-sense-stationary process and $X^{(n)}(t)$ the nth derivative of $X(t)$ that is assumed to exist. Show that $X^{(n)}(t)$ is wide sense stationary with

$$K_n(s) = (-1)^n K^{(2n)}(s),$$

where K_n is the covariance function of $X^{(n)}$.

15. Show that the spectral density of the process $X(t)$, $t \in R$, defined in Exercise 7 is given by $\sigma^2(1 - \cos \lambda)/\pi\lambda^2$.

16. Let $\{X_n\}$, $n = 0, \pm 1, \ldots$, be a strictly stationary MC with finite number of states, transition probability matrix M, and zero mean function, where M has only simple eigenvalues. Find the spectral distribution of X_n, in terms of the eigenvalues of M.

193

17. Show that the spectral density of:

(a) $K(t) = \sigma^2 e^{-a|t|} \cos bt$ is $f(\lambda) = \dfrac{a\sigma^2(\lambda^2 + a^2 + b^2)}{\pi[(\lambda^2 + a^2 - b^2)^2 + 4a^2 b^2]}$,

(b) $K(t) = \sigma^2 e^{-a|t|}[\cos bt - ab^{-1} \sin b|t|]$ is

$$f(\lambda) = \dfrac{2\sigma^2 a\lambda^2}{\pi[(\lambda^2 + a^2 + b^2) - 4b^2\lambda^2]},$$

(c) $K(t) = \sigma^2 e^{-a|t|}[\cos bt + ab^{-1} \sin b|t|]$ is

$$f(\lambda) = \dfrac{2\sigma^2 a(a^2 + b^2)}{\pi[(\lambda^2 + a^2 - b^2)^2 + 4a^2 b^2]},$$

(d) $K(t) = \sigma^2 e^{-a|t|}(1 + a|t| - 2a^2 t^2 + 3^{-1} a^3 |t|^3)$ is

$$f(\lambda) = \dfrac{16\sigma^2 a^3 \lambda^4}{\pi(\lambda^2 + a^2)^4}.$$

18. Show that the covariance function of:

(a) $f(\lambda) = \displaystyle\sum_{k=1}^{n} \dfrac{a_k}{\lambda^2 + b_k^2}$ is $K(t) = \pi \displaystyle\sum_{k=1}^{n} a_k b_k^{-1} e^{-b_k |t|}$,

(b) $f(\lambda) = \begin{cases} a & \text{for } |\lambda| \leqslant b \\ 0 & \text{for } |\lambda| > b \end{cases}$ is $K(t) = 2at^{-1} \sin bt$,

(c) $f(\lambda) = \begin{cases} 0 & \text{for } |\lambda| < a \text{ or } |\lambda| > 2a \\ b^2 & \text{for } a \leqslant |\lambda| \leqslant 2a \end{cases}$ is $K(t) = 2b^2 t^{-1}(2 \cos at - 1)\sin at$.

19. Is the process $X_n = X$, $n = 0, \pm 1, \ldots$ ergodic?

20. Is the process X_n defined in Exercise 8 ergodic?

21. Show that a process $\{X_n\}$ of IID RVs is ergodic.

22. Show that the moving average process defined in Example 7.1.3 (2) is ergodic.

23. Let X_0 be an RV with probability density function given by $g(x) = 2x$ for $0 \leqslant x \leqslant 1$ and 0 elsewhere, and X_{n+1} is an RV uniformly distributed on $(1 - X_n, 1]$, given $X_0, \ldots, X_n$. Show that $\{X_n\}$ is an ergodic process.

8

Martingales

8.1. Definitions and Examples

Every gambler is naturally interested in finding some betting strategy that would give him a net expected gain after making a series of bets. But it can be shown mathematically that no betting system could convert a sequence of "fair" games into an advantageous one unless the gambler has infinite amounts of time and money. A possible interpretation of a "fair" game is that the gambler's expected fortune in game $n + 1$, given that the history up to game n inclusive, is the same as his fortune in game n. The notion of martingale corresponds to this interpretation. The term *martingale* is actually a French acronym for the betting system of doubling the bets until the gambler wins a game. But this should not mislead the reader into thinking that the martingale theory is something restricted to gambling strategies. The martingale theory is a powerful tool in probability theory and this chapter presents only an introduction to this topic. The concept of martingale is due to P. Levy, who introduced it in terms of consecutive sums of RVs. It was J. L. Doob who explored it systematically and brought to light its unexpected potentialities (Doob 1953, Meyer 1966).

Consider a gambler who is playing a sequence of games in each of which he wins with probability $\frac{1}{2}$ or loses with probability $\frac{1}{2}$. Let $\{Y_n\}$, $n \geq 1$, be a sequence of IID RVs denoting the outcome of each game such that

$$P\{Y_n = 1\} = \frac{1}{2} = P\{Y_n = -1\}. \qquad (8.1.1)$$

Here $\{Y_n = 1\}$ (resp. $\{Y_n = -1\}$) denotes the event that the gambler wins (resp. loses) the nth game, $n \geqslant 1$. If the gambler employs a betting strategy based on the past history of the game, his successive bets can be described by a sequence of RVs.

$$b_n = b_n(Y_1, \ldots, Y_{n-1}), \qquad n \geqslant 2. \qquad (8.1.2)$$

Let X_0 be the initial fortune of the gambler. Then

$$X_n = X_0 + \sum_{i=1}^{n} b_i Y_i \qquad (8.1.3)$$

gives his fortune at the end of nth game. We claim that

$$E[X_{n+1} | Y_1, \ldots, Y_n] = X_n. \qquad (8.1.4)$$

To prove this claim, first observe from (8.1.3) that

$$X_{n+1} = X_n + b_{n+1} Y_{n+1}$$

so that

$$E[X_{n+1} | Y_1, \ldots, Y_n] = E[X_n | Y_1, \ldots, Y_n] + E[b_{n+1} Y_{n+1} | Y_1, \ldots, Y_n]$$

$$= X_n + b_{n+1} E[Y_{n+1} | Y_1, \ldots, Y_n],$$

since X_n and b_{n+1} are determined by $Y_1, \ldots, Y_n$;

$$= X_n + b_{n+1} E[Y_{n+1}],$$

since $\{Y_n\}$ is an independent sequence,

$$= X_n, \qquad \text{since } E[Y_{n+1}] = 0 \text{ for all } n \geqslant 0.$$

This proves that if the gambler has an equal chance of winning or losing a game and his betting strategy depends on the past history of the game, the game is "fair" as interpreted earlier. Note also that the expected winnings in any game is zero. Hence these betting strategies do not help to change fair games into favorable games. Having learned this, a gambler should no longer be fascinated by some of these strategies that are only seemingly favorable.

Definition 8.1.1. A stochastic process $\{X_n, n \geqslant 0\}$ is said to be a *martingale* with respect to a process $\{Y_n, n \geqslant 0\}$ if, for all $n \geqslant 0$,

$$E[|X_n|] < \infty \qquad \text{and} \qquad E[X_{n+1} | Y_0, \ldots, Y_n] = X_n. \qquad (8.1.5)$$

We call $\{X_n\}$ a *submartingale* WRT $\{Y_n\}$ if, for all $n \geqslant 0$, X_n is a function of $(Y_0, \ldots, Y_n)$,

$$E[X_n^+] < \infty \quad \text{and} \quad E[X_{n+1} | Y_0, \ldots, Y_n] \geq X_n, \quad (8.1.6)$$

where $X_n^+ = \max\{0, X_n\}$.

We call $\{X_n\}$ a *supermartingale* WRT $\{Y_n\}$ if, for all $n \geq 0$, X_n is a function of $(Y_0, \ldots, Y_n)$,

$$E[X_n^-] < \infty \quad \text{and} \quad E[X_{n+1} | Y_0, \ldots, Y_n] \leq X_n, \quad (8.1.7)$$

where $X_n^- = \min\{0, X_n\}$.

Whereas a martingale describes a fair game, the submartingales and supermartingales describe favorable and unfavorable games, respectively.

Next we define the same notions WRT σ-algebras. (The reader is advised to briefly review the relevant material from Chapter 1.) Let $(\Omega, \mathcal{C}, P)$ be a complete probability space on which all our random variables are defined, and let $\{\mathcal{C}_n, n \geq 0\}$ be a sequence of σ-subalgebras of $\mathcal{C}$ such that $\mathcal{C}_n \subset \mathcal{C}_{n+1}$, $n \geq 0$. A stochastic process X_n, $n \geq 0$, is said to be *adapted* to an increasing sequence $\{\mathcal{C}_n\}$ of σ-algebras if, for every $n \geq 0$, X_n is measurable WRT $\mathcal{C}_n$, that is, $\{X_n \leq x\} \in \mathcal{C}_n$ for every $x \in R$. In Definition 8.1.1 we assumed, while defining a submartingale, say, that X_n is a function of $(Y_0, \ldots, Y_n)$. Let $\mathcal{B}_n$ be the σ-algebra generated by the RVs $Y_0, \ldots, Y_n$, $n \geq 0$. Then $\mathcal{B}_n$, $n \geq 0$, is an increasing sequence of σ-algebras. By saying that X_n is a function of $Y_0, \ldots, Y_n$ we actually mean that $\{X_n\}$ is adapted to $\{\mathcal{B}_n\}$.

Definition 8.1.2. A stochastic process $\{X_n, n \geq 0\}$ that is adapted to an increasing family $\{\mathcal{C}_n, n \geq 0\}$ of σ-algebras is called a *martingale* if, for all $n \geq 0$,

$$E[|X_n|] < \infty \quad \text{and} \quad E[X_{n+1} | \mathcal{C}_n] = X_n. \quad (8.1.8)$$

The adapted sequence $\{X_n, \mathcal{C}_n, n \geq 0\}$ is called a *submartingale* if, for all $n \geq 0$,

$$E[X_n^+] < \infty \quad \text{and} \quad E[X_{n+1} | \mathcal{C}_n] \geq X_n. \quad (8.1.9)$$

A supermartingale is defined analogously.

Before presenting any example we consider some immediate consequences of these definitions.

Proposition 8.1.3. (i) An adapted sequence $\{X_n, \mathcal{C}_n, n \geq 0\}$ is a submartingale if and only if $\{-X_n, \mathcal{C}_n, n \geq 0\}$ is a supermartingale.

(ii) If $\{X_n, \mathcal{C}_n\}$ and $\{Y_n, \mathcal{C}_n\}$ are two submartingales and a and b are two positive constants, then $\{aX_n + bY_n, \mathcal{C}_n\}$ is a submartingale.

(iii) If $\{X_n, \mathcal{C}_n\}$ and $\{Y_n, \mathcal{C}_n\}$ are two submartingales (resp. supermartingales), then $\{\max\{X_n, Y_n\}, \mathcal{C}_n\}$ (resp. $\{\min\{X_n, Y_n\}, \mathcal{C}_n\}$) is a submartingale (resp. supermartingale).

The proof is easy and is left as an exercise. In this and similar propositions the σ-algebras $\{\mathcal{C}_n\}$ can be replaced by a process $\{Y_n\}$, (i.e., $\{X_n\}$ is a submartingale WRT $\{Y_n\}$ instead of WRT $\{\mathcal{C}_n\}$).

Proposition 8.1.4. Let $\{X_n\}$ be a martingale WRT $\{\mathcal{C}_n\}$ (or WRT $\{Y_n\}$). Then:
(i) $E[X_{n+k}|\mathcal{C}_n] = X_n$, for every $k \geqslant 0$, and (ii) $E[X_n] = E[X_0]$, for all $n \geqslant 0$.

PROOF. The proof of (i) goes by induction. The relation in (i) is certainly true for $k = 0$, since X_n is $\mathcal{C}_n$-measurable, and also for $k = 1$, by the definition of a martingale. Let us assume that it holds for some i, that is,

$$E[X_{n+i}|\mathcal{C}_n] = X_n. \tag{8.1.10}$$

Then

$$E[X_{n+i+1}|\mathcal{C}_n] = E[E[X_{n+i+1}|\mathcal{C}_{n+i}]|\mathcal{C}_n]$$
$$= E[X_{n+i}|\mathcal{C}_n], \quad \text{since } \{X_n\} \text{ is a martingale,}$$
$$= X_n, \quad \text{by (8.1.10).}$$

By induction, (8.1.10) holds for all $i \geqslant 0$. The proof of (ii) is:

$$E[X_n] = E[E[X_n|\mathcal{C}_0]] = E[X_0], \quad \text{by (i)}. \qquad \square$$

Proposition 8.1.5. The adapted family $\{X_n, \mathcal{C}_n\}$ is a (sub-, super-) martingale if and only if

$$E[X_{n+1} I_A](\geqslant, \leqslant) = E[X_n I_A], \tag{8.1.11}$$

for all $A \in \mathcal{C}_n$, $n \geqslant 1, 2, \ldots$.

PROOF. Let $A \in \mathcal{C}_n$. Then

$$\int_A X_{n+1} \, dP = \int_A E[X_{n+1}|\mathcal{C}_n] \, dP,$$

by the definition of conditional expectation,

$$(\geqslant, \leqslant) = \int_A X_n \, dP,$$

by the definition of (sub-, super-) martingale. $\qquad \square$

Consequently, $E[X_n]$ is (increasing, decreasing) constant for a (sub-, super-) martingale $\{X_n\}$.

We need Jensen's inequality to prove Theorem 8.1.6, which gives us a method of constructing submartingales from (sub-) martingales. Let us first recall Jensen's inequality. Let $\psi: [a,b] \to R$ be a convex function, that is, $\psi(\alpha x + (1 - \alpha)y) \leqslant \alpha\psi(x) + (1 - \alpha)\psi(y)$ for all $x, y \in [a,b]$ and all $\alpha \in [0, 1]$. If X is an RV with finite expectation and $\mathscr{B}$ is a σ-subalgebra of $\mathcal{Q}$, then

$$\psi E[X|\mathscr{B}] \leqslant E[\psi(X)|\mathscr{B}]. \tag{8.1.12}$$

Theorem 8.1.6. (i) *Let $\{X_n, \mathcal{Q}_n\}$ be a martingale and $\psi: R \to R$ be a convex function such that $E[|\psi(X_n)|] < \infty$ for all n. Then $\{\psi(X_n), \mathcal{Q}_n\}$ is a submartingale and consequently, $\{|X_n|^p\}$ is a submartingale if $E[|X_n|^p] < \infty$, where $p \geqslant 1$.*

(ii) *Let $\{X_n, \mathcal{Q}_n\}$ be a submartingale and $\psi: R \to R$ be an increasing, convex function such that $E[|\psi(X_n)|] < \infty$, for all $n \geqslant 0$. Then $\{\psi(X_n), \mathcal{Q}_n\}$ is a submartingale and consequently $\{X_n^+\}$ is a submartingale.*

PROOF.   By Jensen's inequality and the assumptions in (i),

$$E[\psi(X_{n+1})|\mathcal{Q}_n] \geqslant \psi E[X_{n+1}|\mathcal{Q}_n] = \psi(X_n),$$

since $\{X_n\}$ is a martingale. Hence $\{\psi(X_{n+1})\}$ is a submartingale; (ii) follows similarly. $\square$

Theorem 8.1.7. (Halmos's Optional Skipping Theorem). *Let $\{Y_n, \mathcal{Q}_n\}$ be a submartingale and $\{Z_n\}$ a sequence of RVs defined by*

$$Z_n = \begin{cases} 1 & if \ (Y_1, \ldots, Y_n) \in B_n \\ 0 & if \ (Y_1, \ldots, Y_n) \notin B_n \end{cases} \tag{8.1.13}$$

for arbitrarily chosen $B_n \in \mathscr{B}(R^n)$. Define

$$X_1 = Y_1, \qquad X_2 = X_1 + Z_1(Y_2 - Y_1), \ldots,$$
$$X_n = X_{n-1} + Z_{n-1}(Y_n - Y_{n-1}), \ldots. \tag{8.1.14}$$

Then $\{X_n, \mathcal{Q}_n\}$ is a submartingale and $E[X_n] \leqslant E[Y_n]$, for all $n \geqslant 1$. If $\{Y_n, \mathcal{Q}_n\}$ is a martingale, then $\{X_n, \mathcal{Q}_n\}$ is a martingale and $E[X_n] = E[Y_n]$ for all n.

199

PROOF. We prove the theorem only for the submartingale case; the martingale case is analogous. Now

$$E[X_{n+1}|\mathcal{Q}_n] = E[X_n + Z_n(Y_{n+1} - Y_n)|\mathcal{Q}_n]$$

$$= X_n + Z_n E[Y_{n+1} - Y_n|\mathcal{Q}_n],$$

since X_n and Z_n are functions of $Y_1, \ldots, Y_n$

and thus are $\mathcal{Q}_n$-measurable,

$$\geqslant X_n + Z_n(Y_n - Y_n), \qquad \text{since } \{Y_n\} \text{ is a submartingale .}$$

We show by induction that $E[X_n] \leqslant E[Y_n]$. From (8.1.14), $E[X_1] = E[Y_1]$. Let us assume that $E[Y_k - X_k] \geqslant 0$. Then

$$E[Y_{k+1} - X_{k+1}] = E[E[Y_{k+1} - X_{k+1}|\mathcal{Q}_k]]$$

$$= E[E[Y_{k+1} - X_k - Z_k(Y_{k+1} - Y_k)|\mathcal{Q}_k]]$$

$$= E[E[(1 - Z_k)(Y_{k+1} - Y_k) + (Y_k - X_k)|\mathcal{Q}_k]]$$

$$= E[(1 - Z_k)E[Y_{k+1} - Y_k|\mathcal{Q}_k] + E[Y_k - X_k|\mathcal{Q}_k]]$$

$$\geqslant E[E[Y_k - X_k|\mathcal{Q}_k]], \qquad \text{since } \{Y_n\} \text{ is a submartingale,}$$

$$= E[Y_k - X_k] \geqslant 0. \qquad \square$$

The intuitive meaning of the theorem is as follows. Let $\{Y_n\}$ denote the fortune, after the nth game, of a gambler when he uses no skipping strategy and $\{X_n\}$ the fortune when he uses a skipping strategy. The RV $Z_n = 1$ if he bets in game n and $Z_n = 0$ if he passes game n. What the theorem says is that if the game is initially favorable (submartingale) or fair (martingale), it remains favorable or fair, and no skipping strategy can increase the expected winning.

Examples 8.1.8

EXAMPLE 1. *Doob–Levy Martingale*. Let X be an RV with $E[|X|] < \infty$ and $\{\mathcal{Q}_n\}$ an increasing sequence of σ-subalgebras of $\mathcal{Q}$. Define $X_n = E[X|\mathcal{Q}_n]$. Then $\{X_n, \mathcal{Q}_n\}$ is a martingale. First

$$E[|X_n|] = E[|E[X|\mathcal{Q}_n]|] \leqslant E[E[|X||\mathcal{Q}_n]] = E[|X|] < \infty.$$

Next

$$E[X_{n+1}|\mathcal{Q}_n] = E[E[X|\mathcal{Q}_{n+1}]|\mathcal{Q}_n]$$

$$= E[X|\mathcal{Q}_n], \qquad \text{since } \mathcal{Q}_n \subset \mathcal{Q}_{n+1},$$

$$= X_n.$$

Now let $\{\mathcal{Q}_n\}$ be a decreasing sequence of σ-algebras, that is, $\mathcal{Q}_n \supset \mathcal{Q}_{n+1}$, $n \geqslant 0$, and $X_n = E[X|\mathcal{Q}_n]$. Then $\{X_n, \mathcal{Q}_n\}$ is a *reverse martingale*, that is,

$$E[X_n|\mathcal{Q}_{n+1}] = X_{n+1}. \tag{8.1.15}$$

Using the suitable inequality signs in (8.1.15) one can define reverse sub- and super-martingales. To see (8.1.15):

$$E[X_n|\mathcal{Q}_{n+1}] = E[E[X|\mathcal{Q}_n]|\mathcal{Q}_{n+1}]$$

$$= E[X|\mathcal{Q}_{n+1}], \quad \text{since } \mathcal{Q}_{n+1} \subset \mathcal{Q}_n,$$

$$= X_{n+1}.$$

EXAMPLE 2. *Successive Sums of Independent RVs.*

(i) Let $Y_0 = 0$ and $\{Y_n, n \geqslant 1\}$ be a sequence of independent centered RVs, that is, $E[|Y_n|] < \infty$ and $E[Y_n] = 0$. Define $X_0 = 0$ and $X_n = \sum_{k=1}^n Y_k$. Then $\{X_n\}$ is a martingale WRT $\{Y_n\}$. First $E[|X_n|] \leqslant \sum_{k=1}^n E[|Y_k|] < \infty$. Now, denoting $Y_0, \ldots, Y_n$ by $\mathbf{Y}_n$, we have

$$E[X_{n+1}|\mathbf{Y}_n] = E[X_n + Y_{n+1}|\mathbf{Y}_n]$$

$$= X_n + E[Y_{n+1}|\mathbf{Y}_n], \quad \text{since } X_n \text{ is a function of } \mathbf{Y}_n,$$

$$= X_n + E[Y_{n+1}], \quad \text{since all } Y_n \text{ are independent,}$$

$$= X_n, \quad \text{since } Y_{n+1} \text{ is centered}.$$

(ii) Next let $\{Y_n\}$, $n \geqslant 1$, be independent RVs with $E[|Y_n|] < \infty$ and $E[Y_n] = m_n \neq 0$, for all $n \geqslant 1$. Define $X_n = \Pi_{k=1}^n (Y_k/m_k)$. Then $\{X_n\}$ is a martingale WRT $\{Y_n\}$. Clearly, $E[|X_n|] < \infty$. Now

$$E[X_{n+1}|\mathbf{Y}_n] = E\left[\left(\frac{X_n Y_{n+1}}{m_{n+1}}\right)|\mathbf{Y}_n\right]$$

$$= X_n E\left[\frac{Y_{n+1}}{m_{n+1}}\right] = X_n.$$

EXAMPLE 3. *Pólya's Urn.* Consider an urn initially containing r red and b black balls. Repeated drawings are made from this urn as follows: after each drawing the ball drawn is replaced along with a balls of the same color. Here r, b and a are positive integers. Let $\{Y_n\}$ be a sequence of RVs such that $Y_n = 1$ if the nth ball drawn is red and $Y_n = 0$ if the nth ball drawn is black. Let r_n and b_n be the number of red and black balls, respectively, in the urn after the nth draw has been completed. Define X_n as the proportion of red balls in the urn at the completion of the nth draw, that is, $X_n = r_n/(r_n + b_n)$. Then $\{X_n\}$ is a martingale WRT $\{Y_n\}$.

Noting that $0 \leqslant X_n \leqslant 1$, we have $E[|X_n|] < \infty$. Next observe that a red ball can be drawn in the $(n + 1)$th draw with probability $r_n/(r_n + b_n)$ and in that case $r_{n+1} = (r_n + a)$ and $b_{n+1} = b_n$. Similarly, $r_{n+1} = r_n$ and $b_{n+1} = (b_n + a)$ with probability $b_n/(r_n + b_n)$. Therefore,

$$E[X_{n+1}|\mathbf{Y}_n] = \frac{r_n + a}{r_n + b_n + a} \frac{r_n}{r_n + b_n} + \frac{r_n}{r_n + b_n + a} \frac{b_n}{r_n + b_n}$$

$$= \frac{r_n}{r_n + b_n} = X_n,$$

where $\mathbf{Y}_n$ denotes $\{Y_0, Y_1, \ldots, Y_n\}$ as in Example 2. Since $\{X_n\}$ is a martingale, we have, by Proposition 8.1.4 (ii), that $E[X_n] = E[X_1] = r/(r + b)$, for all $n \geqslant 1$ [see also Example 8.2.7 (1)].

EXAMPLE 4. Let X_1 be uniformly distributed over $[0, 1]$. We successively define a sequence as follows. If it is given that $X_1 = x_1, \ldots, X_{n-1} = x_{n-1}$, then define X_n as an RV uniformly distributed on $[0, x_{n-1}]$. The sequence $\{X_n\}$ is a supermartingale.

First note that $E[X_n^-] = 0$ and then that

$$E[X_{n+1}|X_1, \ldots, X_n] = E[X_{n+1}|X_n].$$

Since X_{n+1} is uniformly distributed over $[0, X_n]$,

$$E[X_{n+1}|\mathbf{X}_n] = \tfrac{1}{2}X_n \leqslant X_n \qquad [\text{recall } \mathbf{X}_n = \{X_1, \ldots, X_n\}]$$

and hence $\{X_n\}$ is a supermartingale. Therefore, $E[X_n]$ is decreasing, as $n \uparrow$, (as already observed in Proposition 8.1.5). Now

$$E[X_{n+1}] = E[E[X_{n+1}|X_1, \ldots, X_n]]$$

$$= \tfrac{1}{2}E[X_n] = \cdots$$

$$= 2^{-n}E[X_1] = 2^{-(n+1)},$$

which is decreasing as n increases [further discussion is continued in Example 8.2.7 (2)].

EXAMPLE 5. Let $\{X_n\}$ be a Markov chain with the set of rationals in $(0, 1)$ as the state space S and the transition probabilities described as follows. For fixed rationals $0 < a \leqslant b < 1$, if $x \in S$ and $X_n = x$, then $X_{n+1} = ax$ with probability $1 - x$ and $X_{n+1} = ax + 1 - b$ with probability x. Then $\{X_n\}$ is a martingale or supermartingale if $a = b$ or $a < b$, respectively.

Let $a = b$. Then

$$E[X_{n+1}|X_0, \ldots, X_n] = E[X_{n+1}|X_n]$$

$$= aX_n(1 - X_n) + (aX_n + 1 - b)X_n$$

$$= X_n - (b - a)X_n$$

$$= X_n, \quad \text{since} \quad a = b.$$

Hence $\{X_n\}$ is a martingale if $a = b$. The case $a < b$, which gives us a supermartingale, follows similarly [see also Example 8.2.7 (3)].

EXAMPLE 6. We consider here an MC that is homogeneous in time and space. Let $\{X_n\}$, $n \geqslant 0$, be an MC with state space $S = \{\cdots, -1, 0, 1, \ldots\}$ and the transition probabilities given by $p(x, y) = p(y - x)$, where $p(u) \geqslant 0$ and $\Sigma_{u \in S} p(u) = 1$. First we show that $X_n = (X_0 + Y_1 + \cdots + Y_n)$, where $X_0, Y_1, \ldots, Y_n$ are IID RVs with the common distribution $P\{Y_k = u\} = p(u)$, $u \in S$. Next we claim that $\{\xi^{X_n}\}$ is a martingale if $\Sigma_{u \in S} p(u)\xi^u = 1$.

Let $\{X_n\}$ be the MC given above. Define $Y_n = (X_n - X_{n-1})$, $n \geqslant 1$. Then X_n has the desired representation provided that we show $X_0, Y_1, \ldots, Y_n$ to be independent RVs with the common distribution $P\{Y_k = u\} = p(u)$. This follows from

$$P\{Y_{n+1} = x_{n+1}|X_0 = x_0, Y_1 = x_1, \ldots, Y_n = x_n\}$$

$$= P\left\{X_{n+1} = \sum_{k=0}^{n+1} x_k | X_0 = x_0, X_1 = x_0 + x_1, \ldots, X_n = \sum_{k=0}^{n} x_k\right\}$$

$$= P\left\{X_{n+1} = \sum_{k=0}^{n+1} x_k | X_n = \sum_{k=0}^{n} x_k\right\}$$

$$= p(x_{n+1}).$$

This proves our first claim. It then follows from

$$E[\xi^{X_{n+1}}|X_0 = x_0, \ldots, X_n = x_n] = E[\xi^{X_{n+1}}|X_n = x_n]$$

$$= \sum_u p(u)\xi^{x_n + u}$$

$$= \xi^{x_n}, \quad \text{since} \quad \sum p(u)\xi^u = 1,$$

that $\{\xi^{X_n}\}$ is a martingale if $\Sigma|p(u)\xi^u| < \infty$ and $\Sigma p(u)\xi^u = 1$.

EXAMPLE 7. *Branching Chain.* Let $\{X_n\}$ be a branching chain with state space $S = \{0, 1, \ldots\}$, $X_0 \equiv 1$, and $X_{n+1} = (Y_1 + \cdots + Y_{X_n})$, a sum of X_n IID RVs with $P\{Y_k = u\} = p(u)$ for all $k \geqslant 1$, and $u \geqslant 0$. Let $\mu = E[Y_k]$.

8. Martingales

Show that: (i) $\{\mu^{-n}X_n\}$ is a martingale if $0 < \mu < \infty$ and (ii) $\{\xi^{X_n}\}$ is a martingale if ξ is a fixed point of the probability-generating function $g(\xi) = \Sigma_u p(u)\xi^u, \xi \geq 0$.

(i)
$$E[\mu^{-n-1}X_{n+1}|X_0 = x_0, \ldots, X_n = x_n]$$
$$= E[\mu^{-n-1}X_{n+1}|X_n = x_n]$$
$$= \sum_u p(x_n, u)u\mu^{-n-1}$$
$$= \mu^{-n-1}\sum_{u \geq 0} uP\left\{Y_1 + \cdots + Y_{x_n} = u\right\}$$
$$= \mu^{-n-1}E\left[Y_1 + \cdots + Y_{x_n}\right]$$
$$= \mu^{-n-1}x_n\mu$$
$$= \mu^{-n}x_n$$

This shows that $\{\mu^{-n}X_n\}$ is a martingale.

(ii)
$$E[\xi^{X_{n+1}}|X_0 = x_0, \ldots, X_n = x_n]$$
$$= E[\xi^{X_{n+1}}|X_n = x_n]$$
$$= \sum_{u \geq 0} p(x_n, u)\xi^u$$
$$= \sum_{u \geq 0} \xi^u P\left\{Y_1 + \cdots + Y_{x_n} = u\right\}$$
$$= E[\xi^{Y_1 + \cdots + Y_{x_n}}]$$
$$= \{E[\xi^{Y_1}]\}^{x_n}$$
$$= \{g(\xi)\}^{x_n} = \xi^{x_n},$$

where we used the fact that ξ is a fixed point of $g(\xi)$, that is, $g(\xi) = \xi$ [discussion continued in Example 8.2.7 (4)].

EXAMPLE 8. In the last three examples we generated martingales from MCs. Some general methods for inducing martingales are known. These are done by using the so-called concordant function or right regular functions and the eigenfunctions of the transition probability matrix. Let $\{Y_n\}$ be an MC with transition matrix $\mathbf{M} = [p(x,y)]$, $x, y \in S$, the state space. A function f on S is called a *concordant function* or a *right regular sequence* if

$$f(y) = \sum_{x \in S} p(y, x)f(x). \tag{8.1.16}$$

Recall that $f(\cdot)$ is a right eigenfunction corresponding to an eigenvalue λ if

$$\lambda f(y) = \sum_{x \in S} p(y, x) f(x) \qquad (8.1.17)$$

for all y. Show that: (i) $\{X_n = f(Y_n)\}$ is a martingale if f is a bounded concordant function and (ii) $\{X_n = \lambda^{-n} f(Y_n)\}$ is a martingale if f is a right eigenfunction WRT the eigenvalue λ and $E[|f(Y_n)|] < \infty$ for all n. [By replacing the equality in (8.1.16) with $\leqslant$ or $\geqslant$, one can define a sub- or super-regular sequence. Then accordingly $f(Y_n)$ becomes a sub- or super-martingale.] Now let us now prove (i).

By assumption f is a bounded function. Therefore, $E[|f(Y_n)|] < \infty$. It follows from

$$E[f(Y_{n+1})|Y_0, \ldots, Y_n] = E[f(Y_{n+1})|Y_n]$$

$$= \sum_{x \in S} p(Y_n, x) f(x)$$

$$= f(Y_n), \qquad \text{by (8.1.16)},$$

that $\{f(Y_n)\}$ is a martingale WRT the MC $\{Y_n\}$. Next we prove (ii). By assumption, $E[|f(Y_n)|] < \infty$. The martingale property follows as above:

$$E[\lambda^{-n-1} f(Y_{n+1})|Y_0, \ldots, Y_n] = E[\lambda^{-n-1} f(Y_{n+1})|Y_n]$$

$$= \lambda^{-n} \lambda^{-1} \sum_x p(Y_n, x) f(x)$$

$$= \lambda^{-n} f(Y_n), \qquad \text{by (8.1.17)}.$$

EXAMPLE 9. Let $\{Y_n\}, n \geqslant 1$, be a sequence of IID RVs following the standard normal distribution. Fix an $\alpha \in R$. Define $S_n = \Sigma_{k=1}^n Y_k$ and $X_n^\alpha = \exp(\alpha S_n - n\alpha^2/2)$. Then $\{X_n^\alpha\}$ is a martingale WRT $\{Y_n\}$. If $F(\alpha)$ is a distribution function on R, then $\{\xi_n = \int X_n^\alpha \, dF(\alpha)\}$ is a martingale.

The second claim easily follows from the first. So we only prove that $\{X_n^\alpha\}$ is a martingale. Since S_n is normally distributed with mean 0 and variance n, it is easy to show that $E[|X_n^\alpha|] < \infty$. Now

$$E[X_{n+1}^\alpha|Y_0, \ldots, Y_n] = E\left[\exp\left\{\alpha S_{n+1} - \frac{(n+1)\alpha^2}{2}\right\} | Y_0, \ldots, Y_n\right]$$

$$= E\left[\exp\left\{\alpha S_n - \frac{n\alpha^2}{2}\right\} \exp\left\{\alpha Y_{n+1} - \frac{\alpha^2}{2}\right\} | Y_0, \ldots, Y_n\right]$$

$$= \exp\left\{\alpha S_n - \frac{n\alpha^2}{2}\right\} E\left[\exp\left\{\alpha Y_{n+1} - \frac{\alpha^2}{2}\right\}\right],$$

since $\exp\{\alpha S_n - n\alpha^2/2\}$ is a function of $Y_0, \ldots, Y_n$ and $\exp\{\alpha Y_{n+1} - \alpha^2/2\}$ is independent of $Y_0, \ldots, Y_n$. But

$$E\left[\exp\left(\alpha Y_{n+1} - \frac{\alpha^2}{2}\right)\right] = \frac{1}{\sqrt{2\pi}} \int e^{\alpha x - \alpha^2/2} e^{-x^2/2} \, dx$$

$$= \frac{1}{\sqrt{2\pi}} \int e^{-\frac{1}{2}(x-\alpha)^2} \, dx = 1.$$

Using this in the last equality, we see that $\{X_n^\alpha\}$ is a martingale.

EXAMPLE 10. The exponential martingale considered in the last example is actually a special case of *Wald's martingale*. Let $Y_0 = 0$ and $\{Y_n\}$ be a sequence of IID RVs with a finite-moment generating function $\mu(\alpha) = E[e^{\alpha Y}]$ for some $\alpha \neq 0$. Set $X_0 = 1$ and $X_n = [\mu(\alpha)]^{-n} \exp\{\alpha \Sigma_{k=1}^n Y_k\}$. Then $\{X_n\}$ is a martingale WRT $\{Y_n\}$.

To establish that $\{X_n\}$ is a martingale, we appeal to Example 8. First observe that $\{S_n = \Sigma_{k=1}^n Y_k\}$ is an MC with state space $S = R$. We claim that $f(x) = e^{\alpha x}$ is an eigenfunction corresponding to the eigenvalue $\mu(\alpha)$. To see this, let F be the common distribution of the RVs $Y_n, n \geq 1$. Then the transition distribution is given by

$$P\{S_{n+1} \leq y | S_n = x\} = F(y - x),$$

and the analog of (8.1.17) is obtained as follows:

$$\int e^{\alpha y} \, d_y F(y - x) = e^{\alpha x} \int e^{\alpha u} \, dF(u) = e^{\alpha x} \mu(\alpha) = \mu(\alpha) f(x).$$

Example 8 now shows that $\{X_n\}$ is a martingale. To see that Example 9 is a special case, let F be the unit normal distribution. Then $\mu(\alpha) = \exp[\alpha^2/2]$, and

$$X_n = \exp\left[\alpha S_n - \frac{n\alpha^2}{2}\right].$$

EXAMPLE 11. *Likelihood Ratios*. The study of likelihood ratios arises in the theory of testing statistical hypotheses. Let $\{Y_n\}$ be a random sample, a sequence of IID RVs. The common density of the RVs Y_k is unknown. So we make the (null) hypothesis that f_0 is the common density and test this assumption against an alternative hypothesis that the common density is f_1. One of the test procedures uses the so-called likelihood ratios defined by

$$X_n = \prod_{k=0}^n \frac{f_1(Y_k)}{f_0(Y_k)}, \qquad n = 0, 1, 2, \ldots. \qquad (8.1.18)$$

Here we assume that $f_0(x) > 0$ for all x. If f_0 is the true density of the random sample (or the population), then $\{X_n\}$ is a martingale WRT the random sample $\{Y_n\}$. To prove this, let us assume that the null hypothesis is true. Then

$$E[X_{n+1} | Y_0, \ldots, Y_n] = E\left[X_n \frac{f_1(Y_{n+1})}{f_0(Y_{n+1})} | Y_0, \ldots, Y_n\right]$$

$$= X_n E\left[\frac{f_1(Y_{n+1})}{f_0(Y_{n+1})}\right] \quad \text{(why?)}$$

$$= X_n \int [f_1(y)/f_0(y)] f_0(y)\, dy$$

$$= X_n \int f_1(y)\, dy = X_n.$$

EXAMPLE 12. Let P and Q be two probability measures on $(\Omega, \mathcal{Q})$ and $\{Y_n\}$ a sequence of RVs that are Borel measurable, that is, $\{\omega: Y_n(\omega) \leqslant y\} \in \mathcal{Q}$ for every $y \in R$. By $p_n(y_1, \ldots, y_n)$ and $q_n(y_1, \ldots, y_n)$, $n \geqslant 1$, denote the joint density of $Y_1, \ldots, Y_n$ WRT P and Q. Define

$$X_n = \begin{cases} q_n(Y_1, \ldots, Y_n)/p_n(Y_1, \ldots, Y_n) & \text{if } p_n(Y_1, \ldots, Y_n) > 0 \\ 0 & \text{if } p_n(Y_1, \ldots, Y_n) = 0 \end{cases}.$$

Then $\{X_n\}$ is a supermartingale WRT $\{Y_n\}$.

First observe that

$$E[|X_n|] = \int\limits_{\{p_n > 0\}} \left[\frac{q_n(\mathbf{x})}{p_n(\mathbf{x})}\right] p_n(\mathbf{x})\, d\mathbf{x} \leqslant \int q_n(\mathbf{x})\, d\mathbf{x} = 1.$$

Let $A = \{(Y_1, \ldots, Y_n) \in B\}$, $B \in \mathcal{B}(R^n)$. Then

$$\int_A X_{n+1}\, dP = \int\limits_{\{\mathbf{x} \in B, p_{n+1}(\mathbf{y}) > 0\}} \left[\frac{q_{n+1}(\mathbf{y})}{p_{n+1}(\mathbf{y})}\right] p_{n+1}(\mathbf{y})\, d\mathbf{y},$$

where $\mathbf{x} = (x_1, \ldots, x_n)$, $\mathbf{y} = (x_1, \ldots, x_n, x_{n+1}) = (\mathbf{x}, x_{n+1})$. Now, if $\mathbf{x} \in B$ and $p_n(\mathbf{x}) = 0$, then $p_{n+1}(\mathbf{y}) = 0$ for almost all x_{n+1}. Therefore

$$\int\limits_{\{\mathbf{x} \in B, p_n(\mathbf{x}) > 0, p_{n+1}(\mathbf{y}) > 0\}} q_{n+1}(\mathbf{y})\, d\mathbf{y} \leqslant \int\limits_{\{\mathbf{x} \in B, p_n(\mathbf{x}) > 0\}} q_n(\mathbf{x})\, d\mathbf{x}$$

$$= \int\limits_{\{\mathbf{x} \in B, p_n(\mathbf{x}) > 0\}} [q_n(\mathbf{x})/p_n(\mathbf{x})] p_n(\mathbf{x})\, d\mathbf{x}$$

$$= \int_A X_n\, dP,$$

and hence $\int_A X_{n+1} dP \leqslant \int_A X_n dP$ for all events A determined by $Y_1, \ldots,$ Y_n. Therefore, $\{X_n\}$ is a supermartingale WRT $\{Y_n\}$.

EXAMPLE 13. Let $\{X_n\}$, $n \geqslant 0$, be a martingale and also an MC with a finite state space $S = \{0, 1, \ldots, N\}$ and transition probability matrix $[p(i,j)]$, $i, j \in S$. Show that the states 0 and N are absorbing.

Using the martingale and Markov property of $\{X_n\}$, we see that

$$X_n = E[X_{n+1}|X_0, \ldots, X_n] = E[X_{n+1}|X_n]$$

$$= \sum_{k=0}^{N} kp(X_n, k).$$

This shows that the identity function $f(x) = x$ on S is a concordant function and thus

$$x = \sum_{y=0}^{N} p(x,y)y, \qquad x \in S.$$

In particular, $0 = \Sigma_{y=0}^{N} p(0,y)y$. This implies that $0 = p(0,y)$ for $y = 1, 2,$ $\ldots, N$. Since the row sums of a transition matrix is one, we get $p(0,0) = 1$, and the state 0 is absorbing. It follows similarly that N is absorbing.

EXAMPLE 14. If $\{X_n\}$ is a martingale and $E[X_n^2] < \infty$ for all n, show that the increments $X_0, X_1 - X_0, X_2 - X_1, \ldots$ are orthogonal, that is,

$$E[(X_i - X_{i-1})(X_j - X_{j-1})] = 0, \qquad i \neq j.$$

Let $i < j$. Then

$$E[(X_i - X_{i-1})(X_j - X_{j-1})]$$

$$= E[E[(X_i - X_{i-1})(X_j - X_{j-1})|X_0, \ldots, X_i]]$$

$$= E[(X_i - X_{i-1})E[(X_j - X_{j-1})|X_0, \ldots, X_i]]$$

$$= E[(X_i - X_{i-1})(X_i - X_i)], \qquad \text{by martingale property,}$$

$$= 0.$$

8.2. Martingale Convergence Theorems

The convergence theorems presented in this section are of fundamental importance in the theory of stochastic processes and find many applications in probability theory. As the proofs of these theorems may not be simple enough for several senior-level students, they can omit the proofs of some or all of the theorems presented in this section and concentrate on the examples.

Let $\{Y_n\}$, $n \geq 0$, be a sequence of centered second-order RVs. Such a sequence is called a *partingale* if the successive partial sums $X_n = Y_0 + \cdots + Y_n$, form a martingale. Partingales are also known as *conditionally independent sequences*. Consider the following three statements:

1. $\{Y_n\}$ is a sequence of independent RVs.
2. $\{Y_n\}$ is a partingale.
3. $\{Y_n\}$ is a sequence of orthogonal RVs.

From Examples 8.1.8 (2) and 8.1.8 (14) we see that

$$(1) \Rightarrow (2) \Rightarrow (3). \tag{8.2.1}$$

Thus the notion of partingale is intermediate between the independence and orthogonality of RVs, and one can expect to extend the theorems such as laws of large numbers to the partingale case. Such extensions exist. [As an application of the martingale convergence theorem, we later prove the strong law of large numbers.] We simply remark here that the implications in (8.3.1) cannot be reversed in general.

Theorem 8.2.1. (Doob's Submartingale Convergence Theorem). *If* $\{X_n, \mathcal{Q}_n\}$, $n \geq 1$, *is a submartingale such that* $\sup_{n \geq 1} E[X_n^+] < \infty$, *there exists a RV* X_∞ *such that* $E[\|X_\infty\|] < \infty$ *and* $X_n \to X_\infty$ *with probability one.*

To establish this theorem, we need Doob's inequality for the expected number of upcrossings of a level. Let $\{X_n\}$, $n = 1, \ldots, N$, be a submartingale and $a < b$ be two levels (real numbers). Let $t_1(\omega)$ be the first integer in $\{1, \ldots, N\}$ such that $X_{t_1(\omega)}(\omega) \leq a$, let $t_2(\omega)$ be the first integer $\geq t_1(\omega)$ and such that $X_{t_2(\omega)}(\omega) \geq b$, let $t_3(\omega)$ be the first integer $\geq t_2(\omega)$ and such that $X_{t_3(\omega)}(\omega) \leq a$, and so on. If M is the number of finite t_k, define the number $U(a, b)$ of upcrossings of the interval (a, b) by

$$U(a,b) = \begin{cases} \dfrac{M}{2} & \text{if } M \text{ is even} \\[2mm] \dfrac{M-1}{2} & \text{if } M \text{ is odd} \end{cases}.$$

Doob's Upcrossing Inequality 8.2.2. *Let* $\{X_n\}$, $1 \leq n \leq N$, *be a submartingale and* $U(a, b)$ *be the number of upcrossings of the interval* (a, b). *Then*

$$E[U(a,b)] \leq (b - a)^{-1} E[(X_N - a)^-]. \tag{8.2.2}$$

209

PROOF. For simplicity let $a = 0$ and $\{X_n\}$, $1 \leqslant k \leqslant N$, be a nonnegative submartingale. Let all $\mathbf{t}_k$ be as in the definition of the number of upcrossings. If $k < \mathbf{t}_1$, define $Z_k = 0$; if $\mathbf{t}_1 \leqslant k < \mathbf{t}_2$, define $Z_k = 1$; if $\mathbf{t}_2 \leqslant k < \mathbf{t}_3$, define $Z_k = 0$; and so on. Define

$$\xi_1 = X_1, \qquad \xi_n = \xi_{n-1} + Z_{n-1}(X_n - X_{n-1}), \qquad n \geqslant 2.$$

Then ξ_n denotes the total increase during upcrossings. Thus $\xi_n \geqslant bU(0, b)$. From the definition of RVs Z_k it is clear that Z_k is a function of $X_1, \ldots, X_k$. Appealing now to Halmos's optional skipping theorem, we see that $\{\xi_n, 1 \leqslant n \leqslant N\}$ is a submartingale WRT $\{X_n\}$ and $E[\xi_n] \leqslant E[X_n]$ so that

$$bE[U(0, b)] \leqslant E[\xi_n] \leqslant E[X_n].$$

This proves Doob's inequality. $\qquad\qquad\qquad\qquad\qquad\qquad\qquad\qquad\square$

Proof of Theorem 8.2.1. Define sets A and $A(a, b)$ by

$$A = \{\omega: X_n(\omega) \text{ does not converge to a finite or infinite limit}\}$$

$$A(a, b) = \{\omega: \liminf_{n \to \infty} X_n(\omega) < a < b < \limsup_{n \to \infty} X_n(\omega)\}.$$

Then

$$P(A) = P\left\{ \bigcup_{\substack{a < b \\ a,b \in Q}} A(a, b) \right\},$$

where Q is the set of rationals. We claim that $P(A) = 0$. On the contrary, if $P\{A(a, b)\} > 0$ for some rationals a and b, then at least on that set of positive probability the submartingale is making an infinite number of upcrossings of (a, b) and hence $E[U(a, b)] = \infty$. Let $U_n(a, b)$ denote the number of upcrossings of (a, b) by $X_1, \ldots, X_n$. Then $U_n(a, b)$ monotonically increases to $U(a, b)$. Therefore, $E[U_n(a, b)] \uparrow E[U(a, b)] = \infty$. However, by the hypothesis that $\sup_n E[X_n^+] < \infty$ and Doob's upcrossing inequality (8.2.2), we have

$$E[U_n(a, b)] \leqslant (b - a)^{-1} E[(X_n - a)^+]$$

$$\leqslant (b - a)^{-1}\left\{ \sup_n E[X_n^+] + |a| \right\}$$

$$< \infty,$$

which is a contradiction. Therefore, $P(A) = 0$ and $X_n \to X_\infty$ with probability one.

It remains to show that $E[|X_\infty|] < \infty$. First note that $E[X_n] \geqslant E[X_1]$ by submartingale property (Proposition 8.1.5). From the hypothesis of the theorem it now follows that

$$E[|X_n|] \leqslant 2 \sup_n E[X_n^+] - E[X_1] < \infty,$$

and thus Fatou's lemma (see Chapter 1) gives us

$$E[|X_\infty|] \leqslant \liminf_{n \to \infty} E[|X_n|] < \infty.$$

This proves the theorem. □

Definition 8.2.3. A stochastic process $\{X_n\}$ is said to be *uniformly integrable* if

$$\lim_{a \uparrow \infty} \sup_{n \geqslant 1} E\left[|X_n| I_{\{|X_n| > a\}}\right] = 0, \qquad (8.2.3)$$

which in terms of integral takes the form

$$\int_{\{|X_n| > a\}} |X_n| \, dP \xrightarrow{a \uparrow \infty} 0 \qquad \text{uniformly in } n. \qquad (8.2.4)$$

The convergence in probability or with probability one of a sequence of RVs does not in general imply the convergence in mean or in mean square. With the uniform integrability of the sequence of RVs, we can overcome this problem. We state the precise result without proof.

Theorem 8.2.4. (i) *Let $\{X_n\}$, $n \geqslant 1$, be a sequence of RVs such that $X_n \to X$ with probability one (or in probability) and $\{|X_n|^p, n \geqslant 1\}$, $p \geqslant 1$, is an uniformly integrable sequence. Then $E[|X_n - X|^p] \to 0$, as $n \to \infty$.*

(ii) *If $\{X_n\}$ is an uniformly integrable sequence of RVs such that $X_n \to X$ with probability one, then $E[|X_n - X|] \to 0$, as $n \to \infty$.*

Theorem 8.2.5. (Levy's Martingale Convergence Theorem.) *Let $\mathcal{C}_n$, $n \geqslant 0$, be an increasing sequence of σ-subalgebras of $\mathcal{C}$ and $\mathcal{C}_\infty$ be the σ-algebra generated by $\cup_{n \geqslant 0} \mathcal{C}_n$. Let X be an RV with $E[|X|] < \infty$, and set $X_n = E[X|\mathcal{C}_n]$, $n \geqslant 0$. Then:*

(i) *$\{X_n, n \geqslant 0\}$ is an uniformly integrable sequence.*

(ii) *$X_n \to E[X|\mathcal{C}_\infty]$ with probability one and in mean.*

PROOF. (i) The uniform integrability of $\{X_n\}$ follows from the Markov inequality and absolute continuity of an integral (see Chapter 1), as follows:

8. Martingales

$$0 \leqslant P\{|X_n| \geqslant a\} \leqslant a^{-1}E[|X_n|], \qquad \text{the Markov inequality,}$$

$$= a^{-1}E[|E[X|\mathcal{Q}_n]|] \leqslant a^{-1}E[E[|X||\mathcal{Q}_n]]$$

$$= a^{-1}E[|X|] \to 0 \qquad \text{uniformly in } n, \text{ as } a \uparrow \infty,$$

and

$$E\left[|X_n|I_{\{|X_n|\geqslant a\}}\right] \leqslant E\left[I_{\{|X_n|\geqslant a\}}E[|X||\mathcal{Q}_n]\right]$$

$$= E\left[E\left[|X|I_{\{|X_n|\geqslant a\}}|\mathcal{Q}_n\right]\right], \qquad \text{since } X_n \text{ is } \mathcal{Q}_n\text{-measurable,}$$

$$= E\left[|X|I_{\{|X_n|\geqslant a\}}\right] = \int_{\{|X_n|\geqslant a\}} |X|\,dP.$$

This proves (i).

(ii) By Example 8.1.8 (1), $\{X_n, \mathcal{Q}_n\}$ is a martingale, the so-called Doob–Levy martingale, and by part (i) it is uniformly integrable. Since $E[|X_n|] \leqslant E[E[|X||\mathcal{Q}_n]] = E[|X|] < \infty$ (the hypothesis in Doob's submartingale convergence theorem), it follows that X_n converges with probability one to an RV X_∞ with $E[|X_\infty|] < \infty$. By the uniform integrability of $\{X_n\}$ and Theorem 8.2.4 it follows that $E[|X_n - X_\infty|] \to 0$, as $n \to \infty$.

It remains to show that $X_\infty = E[X|\mathcal{Q}_\infty]$. Since $E[|X_n - X_\infty|] \to 0$, it follows that $E[X_n] \to E[X_\infty]$ as $n \to \infty$. Therefore, if A is an arbitrary set in $\mathcal{Q}_n$, then

$$\int_A X\,dP = \int_A E[X|\mathcal{Q}_n]\,dP, \qquad \text{by the definition of conditional expectation,}$$

$$= \int_A X_n\,dP, \qquad \text{by the definition of } X_n,$$

$$\to \int_A X_\infty\,dP.$$

Consequently, $E[XI_A] = E[X_\infty I_A]$ for all $A \in \cup_{n \geqslant 1}\mathcal{Q}_n$ and hence for all $A \in \mathcal{Q}_\infty$. Noting that X_∞ is $\mathcal{Q}_\infty$-measurable and hence by the definition of conditional expectation, $X_\infty = E[X|\mathcal{Q}_\infty]$. □

Theorem 8.2.6. *Let $\{X_n, \mathcal{Q}_n, n \geqslant 0\}$ be a martingale or a nonnegative submartingale such that $E[|X_n|^2] \leqslant K < \infty$, for all $n \geqslant 0$. Then $\{X_n\}$ converges as $n \to \infty$ to a limit RV X_∞ both with probability one and in mean-square:*

$$P\left\{\lim_{n\to\infty} X_n = X_\infty\right\} = 1 \qquad \text{and}$$

$$\lim_{n\to\infty} E[|X_n - X_\infty|^2] = 0. \tag{8.2.5}$$

PROOF. Let the hypotheses of the theorem hold. First we claim that $\{X_n, n \geq 0\}$ is a uniformly integrable sequence. Toward this end it suffices to show that $E[|X_n| I_{\{|X_n| \geq a\}}]$ can be made arbitrarily small for large $a > 0$. So let us arbitrarily fix an $\epsilon > 0$ and set $a = (K/\epsilon)$, where $K = \sup_n E[|X_n|^2]$, which is finite by hypothesis. Then

$$E\left[|X_n| I_{\{|X_n| \geq a\}}\right] \leq \frac{1}{a} E\left[|X_n|^2 I_{\{|X_n| \geq a\}}\right]$$

$$\leq \frac{1}{a} E[|X_n|^2]$$

$$\leq \frac{K}{a} = \epsilon.$$

This proves the uniform integrability of $\{X_n\}$.

Next we claim that $\sup_n E[|X_n|] < \infty$. For an arbitrary $\epsilon > 0$, it follows from the uniform integrability that

$$E[|X_n|] = E\left[|X_n| I_{\{|X_n| \geq a\}}\right] + E\left[|X_n| I_{\{|X_n| < a\}}\right]$$

$$\leq \epsilon + a P\{|X_n| < a\} \leq \epsilon + a.$$

Thus the hypothesis of Doob's submartingale convergence theorem is satisfied. Hence X_n converges, as $n \to \infty$, to an RV X_∞ with probability one and $E[|X_\infty|] < \infty$. Also, by Theorem 8.2.4, $E[|X_n - X_\infty|] \to 0$ as $n \to \infty$.

We now claim that the completed sequence $\{X_n, n = 0, 1, \ldots, \infty\}$ is a martingale or nonnegative submartingale according as the process $\{X_n, n \geq 0\}$ is a martingale or a nonnegative submartingale. To see this, consider an arbitrary set A in $\mathcal{C}_n$. It follows from Proposition 8.1.5 that, for all $m \geq n$,

$$E[X_n I_A] = (\leq) E[X_m I_A].$$

Because X_n converges to X_∞ in mean, by letting $m \to \infty$ we get

$$E[X_n I_A](\leq) = E[X_\infty I_A].$$

Appealing again to Proposition 8.1.5, we obtain that

$$E[X_\infty | \mathcal{C}_n](\geq) = X_n.$$

This proves that the completed sequence $\{X_n, n = 0, 1, \ldots, \infty\}$ is a martingale (or submartingale).

Now it follows from Theorem 8.1.6 that the sequence $\{|X_n|^2, 0 \leq n \leq \infty\}$ is a nonnegative submartingale in both the cases where $\{X_n, 0 \leq n \leq \infty\}$ is a martingale or nonnegative submartingale.

8. Martingales

Next we claim that $\{X_n^2, 0 \leqslant n \leqslant \infty\}$ is uniformly integrable. This follows from

$$P\{X_n^2 \geqslant a\} \leqslant a^{-1} E[X_n^2], \qquad \text{by Markov inequality,}$$

$$\leqslant a^{-1} E[X_\infty^2], \qquad \text{by Proposition 8.1.5,}$$

$$\xrightarrow[a\uparrow\infty]{} 0, \qquad \text{uniformly in } n,$$

and

$$E\left[X_n^2 I_{\{X_n^2 \geqslant a\}} \right] \leqslant E[X_n^2] \leqslant K < \infty.$$

Finally appealing to Theorem 8.2.4, we obtain that $E[|X_n - X_\infty|^2] \to 0$ as $n \to \infty$, and this completes the proof. $\qquad\square$

Examples 8.2.7

EXAMPLE 1. *Pólya's Urn.* In Example 8.1.8 (3) we saw that the proportions $\{X_n\}$ of red balls in the urn under Pólya's urn scheme form a martingale. The proportions X_n are uniformly bounded $0 \leqslant X_n \leqslant 1$ for all n. Therefore, the sequence is uniformly integrable and hence there exists an RV X_∞ such that $X_n \to X_\infty$ with probability one and $E[X_\infty] = \lim_n E[X_n] = E[X_1]$ $= r/(r + b)$ by the martingale property.

EXAMPLE 2. Example 8.1.8 (4) gives us a supermartingale. Without invoking any convergence theorem we can show that $X_n \to 0$ with probability one. In Example 8.1.8 (4) we saw that $E[X_n] = 2^{-n}$. Thus

$$E\left[\sum_{n\geqslant 1} X_n \right] = \sum_{n\geqslant 1} E[X_n] = \sum_{n\geqslant 1} 2^{-n} < \infty,$$

and hence $\sum_{n\geqslant 1} X_n < \infty$ with probability one (recall that a nonnegative RV with finite mean is finite-valued with probability one). Hence $X_n \to 0$ with probability one.

EXAMPLE 3. Let $\{X_n\}$, $n \geqslant 0$, be the MC considered in Example 8.1.8 (5). The state space S of this chain is the set of rationals in $(0,1)$. Fix two rationals $0 < a \leqslant b < 1$ and set for $x \in S$,

$$P\{X_{n+1} = ax | X_n = x\} = 1 - x = 1 - P\{X_{n+1} = ax + 1 - b | X_n = x\}.$$

We saw that $\{X_n\}$ becomes a martingale if $a = b$. In this case, show that $X_n \to X_\infty$ with probability one where $X_\infty = 0$ or 1 and that $P\{X_\infty = 1\}$ $= E[X_0]$. If $a < b$, then $\{X_n\}$ becomes a supermartingale. Show now that $X_\infty \to 0$ with probability one.

Let $a = b$. Then $\{X_n\}$ is a uniformly bounded martingale and hence is uniformly integrable. Therefore, there exists an RV X_∞ such that $P\{\lim_{n\to\infty} X_n = X_\infty\} = 1$ and $E[|X_n - X_\infty|] \to 0$, as $n \to \infty$. Now, since

$$|X_{n+1} - X_n| = (1 - a)X_n \quad \text{or} \ (1 - a)(1 - X_n),$$

we have that, if $X_n(\omega) \to \xi(\omega)$,

$$|X_{n+1} - X_n| \to (1 - a)\xi \quad \text{or} \ (1 - a)(1 - \xi).$$

Since $\{X_n\}$ converges with probability one, $X_{n+1} - X_n \to 0$ with probability one. Consequently, $\xi = 0$ or 1 and hence $X_\infty = 0$ or 1 with probability one. Next, from the martingale property and the convergence in mean,

$$E[X_\infty] = \lim_{n\to\infty} E[X_n] = E[X_0].$$

But $E[X_\infty] = 1 P\{X_\infty = 1\} + 0 P\{X_\infty = 0\} = P\{X_\infty = 1\}$.

Now let $a < b$. First note that $E[X_n]$ is a decreasing function of n by Proposition 8.1.5 (because X_n is a supermartingale). It is clear also from

$$E[X_{n+1}] = (1 + a - b)E[X_n] = cE[X_n] = \cdots = c^{n+1}E[X_0],$$

where $0 < c = (1 + a - b) < 1$. Hence $E[X_n] \to 0$ as $n \to \infty$ and

$$E[X_\infty] = \lim_{n\to\infty} E[X_n] = 0.$$

Since $X_\infty \geqslant 0$, this gives $X_\infty = 0$ with probability one.

EXAMPLE 4. *Branching Chain.* Here we consider the application of martingale convergence theorem to the branching chain [see Example 8.1.8 (7); also Section 3.7]. As in Example 8.1.8. (7), let $X_0 \equiv 1$ and $X_{n+1} = Y_1 + \cdots + Y_{X_n}$ with IID RVs $\{Y_k\}$ and $P\{Y_k = u\} = p(u)$. Let $\mu = E[Y_k] > 0$. Let us also assume that $p(0) + p(1) < 1$. We now establish Watson–Steffensen Theorem 3.7.5 using the martingale convergence theorem.

Let us first consider the case when $\mu < 1$. In this case we claim that the population eventually becomes extinct with probability one. First observe that

$$E[X_{n+1} | X_n = k] = kE[Y_1] = k\mu.$$

Therefore, $E[X_{n+1} | X_n] = \mu X_n$ and $E[X_{n+1}] = \mu E[X_n]$. If $\mu < 1$, then $P\{\lim_{n\to\infty} X_n = 0\} = 1$. Since the process X_n assume only nonnegative integral values, X_n becomes 0 eventually.

Next let $\mu > 1$. We claim that

$$P\{X_n \text{ is 0 eventually}\} = \xi = 1 - P\{X_n \to \infty\}, \tag{8.2.6}$$

215

where ξ is the unique fixed point in $[0, 1)$ of the probability-generating function of the RVs Y_k. (See the proof of Theorem 3.7.5 for the uniqueness of ξ.)

First we take the case $\xi \in (0, 1)$. Recall, from Example 8.1.8 (7), that $\{\xi^{X_n}\}$ is a nonnegative martingale. Therefore, $P\{\lim_n X_n = X_\infty\} = 1$ for some X_∞. To establish (8.2.6), we first show that the probability of X_∞ assuming any finite positive integral value m is zero. If $P\{X_\infty = m\} > 0$, there is an integer $N > 0$ such that $P\{X_n = m \text{ for all } n \geqslant N\} > 0$. Set $p = P\{X_{n+1} = m | X_n = m\}$. Because $P\{X_{n+1} = 0 | X_n = m\} = [p(0)]^m > 0$, it follows that $p < 1$. Now

$$P\{X_n = m \quad \text{for all} \quad n \geqslant N\}$$

$$= P\{X_N = m\} P\{X_{N+1} = m | X_N = m\}$$

$$\times P\{X_{N+2} = m | X_k = m, k = N, N+1\} \cdots$$

$$= P\{X_N = m\} p P\{X_{N+2} = m | X_{N+1} = m\} P\{X_{N+3} = m | X_{N+2} = m\} \cdots,$$

by the Markov property,

$$= \lim_{k \to \infty} p^k P\{X_N = m\}$$

$$= 0, \quad \text{since } p < 1.$$

This leaves 0 and ∞ as the only possible values of X_∞. Recalling that $\xi \in (0, 1)$, we note that $\{\xi^{X_n}\}$ is uniformly bounded and hence uniformly integrable. Therefore, from the mean convergence of the martingale $\{\xi^{X_n}\}$, $n \geqslant 0$, we get

$$E[\xi^{X_\infty}] = \lim_{n \to \infty} E[\xi^{X_n}] = E[\xi^{X_0}] = E[\xi] = \xi.$$

But

$$\xi = E[\xi^{X_\infty}] = \xi^0 P\{X_\infty = 0\} + \lim_{k \to \infty} \xi^k P\{X_\infty = \infty\}$$

$$= P\{X_\infty = 0\}, \quad \text{since } 0 < \xi < 1,$$

and consequently $P\{X_\infty = \infty\} = 1 - \xi$. This proves (8.2.6).

Now we take $\xi = 0$. Since ξ satisfies $\xi = \Sigma_{u \geqslant 0} p(u) \xi^u$, it follows that $p(0) = 0$; that is, each individual in the population gives birth to at least one. Therefore,

$$1 = X_0 \leqslant X_1 \leqslant \cdots \leqslant X_n \leqslant X_{n+1} \leqslant \cdots.$$

and consequently $X_n \uparrow X$, for some limit X. Here

$$P\{X < \infty\} = P\{X_n \text{ eventually is a finite RV}\}$$
$$= \sum_{m \geqslant 1} P\{X_n = m \text{ for all large } n\}$$
$$= 0, \qquad \text{as before,}$$

and hence $P\{\lim_{n \to \infty} X_n = \infty\} = 1$.

As the final case, let $\mu = 1$. We claim then that the population becomes extinct with probability one. Since $\{X_n\}$ is a nonnegative martingale (note $\mu = 1$ in Example 8.1.8 (1) (i),

$$P\{X_n \to \text{ to a finite limit}\} = 1.$$

As in the last case, this finite value is zero almost surely. Therefore, $X_n \to 0$ with probability one.

EXAMPLE 5. Let $\{Y_n\}$ be an irreducible recurrent MC with transition matrix $[p(x,y)]$. Show that every bounded concordant function f is a constant function.

Let f be a bounded concordant function on the state space of the MC $\{Y_n\}$, that is, $f(x) = \sum_{y \in S} p(x,y) f(y)$. By Example 8.1.8 (8) (i), $\{X_n = f(Y_n)\}$ is a martingale that is bounded because f is so (by assumption). Therefore, X_n converges with probability one. Because of the recurrence, the MC visits all the states infinitely often. Consequently $X_n = f(x)$ and $X_n = f(y)$ for infinitely many n and for any two states x and y. Since X_n converges to a limit with probability one, $f(x) = f(y)$ for all pairs x and y and consequently f is a constant function.

EXAMPLE 6. *A Counterexample.* Let $\{X_n\}$ be a submartingale. Doob's convergence theorem states that X_n converges with probability one if $\sup_n E[X_n^+] < \infty$. This is not a necessary condition. In other words, there are martingale that converge with probability one, but $\sup_n E[|X_n|] = \infty$.

Let $\{a_n\}$, $n \geqslant 1$, be a sequence of positive numbers and $\{p_n\}$, $n \geqslant 1$, be another sequence with $0 < p_n < \frac{1}{2}$ for all n. Suppose that $\sum_{n \geqslant 1} p_n < \infty$ and $\sum_{n \geqslant 1} a_n p_n = \infty$. Define an MC as follows: $X_1 = 0$. If $X_n \neq 0$ for some $n > 1$, then set $X_{n+1} = X_n$. That is, once the chain leaves the 0 state, it is absorbed in the state it jumps to. Let $\{0, \pm a_n, n \geqslant 1\}$ be the state space. If $X_n = 0$, then let

$$X_{n+1} = \begin{cases} a_{n+1} & \text{with probability } p_{n+1} \\ -a_{n+1} & \text{with probability } p_{n+1} \\ 0 & \text{with probability } 1 - 2p_{n+1} \end{cases}$$

217

First we claim that $\{X_n\}$ is a martingale. This follows from the computations

$$E[X_{n+1}|X_n = a_n] = a_n, \qquad E[X_{n+1}|X_n = -a_n] = -a_n,$$

$$E[X_{n+1}|X_n = 0] = a_{n+1}p_{n+1} + (-a_{n+1})p_{n+1} + 0(1 - 2p_{n+1}) = 0.$$

Next we show that $E[|X_n|] \to \infty$, as $n \to \infty$. To see this let us compute $E[|X_n|]$ for the first few n values.

$$E[|X_1|] = 0, \qquad E[|X_2|] = a_2 p_2 + |-a_2|p_2 + 0 = 2a_2 p_2,$$

$$E[|X_3|] = 2a_2 p_2 + (1 - 2p_2)2a_3 p_3, \qquad \text{(since } X_2 = 0 \text{ or } \pm a_2),$$

$$E[|X_4|] = 2a_2 p_2 + (1 - 2p_2)2a_3 p_3 + (1 - 2p_2)(1 - 2p_3)2a_4 p_4$$

$$\cdots.$$

Therefore,

$$\lim_{n \to \infty} E[|X_n|] \geqslant \left\{ \prod_{n \geqslant 1} (1 - 2p_n) \right\} 2 \sum_{n \geqslant 2} a_n p_n.$$

Since $\Sigma p_n < \infty$ and $0 < p_n < \frac{1}{2}$, we have $\Pi_{n \geqslant 1}(1 - 2p_n) > 0$. Hence it follows from the assumption $\Sigma_{n \geqslant 1} a_n p_n = \infty$ that $\lim_{n \to \infty} E[|X_n|] = \infty$.

Finally we claim that X_n converges everywhere despite the fact that $E[|X_n|] \to \infty$. From the defined transitions of $\{X_n\}$, it follows that either

$$X_n(\omega) = 0 \qquad \text{for all } n \text{ and all } \omega \text{ or}$$

$$X_n(\omega) = a_k \qquad \text{for some } k, \quad \text{all } n \geqslant k, \quad \text{and all } \omega.$$

Hence X_n converges everywhere.

A long line of theorems in probability theory can be established using the martingale convergence theorems. As a sample we apply it only to prove a strong law of large numbers.

Theorem 8.2.8. (Strong Law of Large Numbers). *If* $\{X_n\}$, $n \geqslant 1$, *is a sequence of IID RVs such that* $E[|X_n|] < \infty$ *and* $E[X_n] = \mu$ *for all* n *and if* $S_n = \Sigma_{k=1}^n X_k$, *then* $n^{-1}S_n$ *converges to* μ *with probability one and in mean.*

PROOF. Under the conditions of the theorem, both X_1 and $(X_1, \ldots, X_n)$ are independent of $(X_{n+1}, X_{n+2}, \ldots)$. Therefore, (X_1, S_n) is independent of $(X_{n+1}, X_{n+2}, \ldots)$. Since $X_{n+k} = S_{n+k} - S_{n+k-1}$, $k \geqslant 1$, we have

$$E[X_1|S_n] = E[X_1|S_n, S_{n+1}, \ldots]. \tag{8.2.7}$$

Now, if B is any Borel set on the real line,

$$E\left[X_1 I_{\{S_n \in B\}}\right] = \int_R \cdots \int_R x_1 I_B\left(\sum_{k=1}^{n} x_k\right) dF(x_1) \cdots dF(x_n),$$

where F is the common DF of all X_n,

$$= \frac{1}{n} E\left[\left(\sum_{k=1}^{n} X_k\right) I_{\{S_n \in B\}}\right], \qquad \text{by Fubini theorem,}$$

$$= E\left[n^{-1} S_n I_{\{S_n \in B\}}\right].$$

Hence $E[X_1 | S_n] = n^{-1} S_n$. Using this in (8.2.7), we obtain

$$E[X_1 | S_n, S_{n+1}, \ldots] = n^{-1} S_n. \tag{8.2.8}$$

Let $\mathcal{C}^n$ be the σ-algebra generated by $\{X_n, X_{n+1}, \ldots\}$ and $\mathcal{C}^\infty$ be the tail σ-algebra $\cap_{n \geq 1} \mathcal{C}^n$. Having seen the proof of Levy's martingale convergence, it should not be difficult to prove that

$$n^{-1} S_n = E[X_1 | S_n, S_{n+1}, \ldots] \to E[X_1 | \mathcal{C}^\infty]$$

with probability one and in mean. Since $\lim_{n \to \infty} n^{-1} S_n$ is a function measurable WRT the tail σ-algebra $\mathcal{C}^\infty$, it must be a constant by Kolmogorov's zero–one law. From $E[|n^{-1} S_n - E[X_1 | \mathcal{C}^\infty]|] \to 0$ as $n \to \infty$, it follows that $E[n^{-1} S_n] \equiv E[X_1] = \mu$, and consequently the said constant is μ. $\qquad\square$

8.3. Optional Sampling Theorem

Consider a discrete-time stochastic process $\{X_n\}$, $n \geq 0$, and let $\mathcal{C}_n$ be the σ-algebra generated by $\{X_0, \ldots, X_n\}$. Recall that a mapping $t: \Omega \to \{0, 1, \ldots, \infty\}$ is called a *stopping time* WRT $\{X_n\}$ (or WRT $\{\mathcal{C}_n\}$) if the event $\{t = n\}$ is completely determined by $\{X_0, \ldots, X_n\}$ (or is a set in $\mathcal{C}_n$). An event A is said to be *prior to* t if $A \cap \{t \leq n\} \in \mathcal{C}_n$. The collection $\mathcal{C}_t$ of all the events prior to t is a σ-algebra. A stopping time t relative to $\{X_n\}$ is said to be *finite* (resp. *bounded*) if $P\{t < \infty\} = 1$ (resp. there is a finite $N > 0$ such that $P\{t \leq N\} = 1$).

On several occasions in this chapter we have seen that betting strategies will not convert a fair game (martingale) into a favorable game (submartingale). Let $\{X_n\}$, $n \geq 0$, be a martingale and t a bounded stopping time. Let us assume that the gambler plans to quit at a random time t. If $E[X_t] = E[X_0]$, then the gambler's strategy does not help to improve his expected

219

fortune. The same could be said about the expected fortunes of the gambler at any two stopping times. In general, if $\{t_n\}$ is an increasing sequence of stopping times relative to $\{X_n\}$, then by optional stopping strategy one can consider the sequence $\{X_{t_n}\}$ and inquire as to whether the martingale property preserved in passing from the martingale sequence $\{X_n\}$ to $\{X_{t_n}\}$. Under some mild conditions the answer is in the affirmative.

Definition 8.3.1. Let t be a stopping time WRT a process $\{X_n\}$, $n \geqslant 0$. By the *process X^t stopped at* t we mean the stochastic process defined by

$$X^t = \{X_n^t\} = \{X_{t \wedge n}\} \qquad (8.3.1)$$

where $t \wedge n = \min(t, n)$.

Theorem 8.3.2. (Stopped Martingale). *Let* $\{X_n, \mathcal{Q}_n\}$, $n \geqslant 0$, *be a (sub-) martingale and* t *be a stopping time* WRT $\{X_n\}$. *Then the stopped process* X^t *is a (sub-) martingale* WRT $\{\mathcal{Q}_n\}$.

PROOF. We prove only the martingale case and the submartingale case is handled similarly.

It follows from $|X_n^t| \leqslant \Sigma_{k=0}^n |X_k|$ that $E[|X_n^t|] < \infty$. Noting that $X_n^t = \Sigma_{k=0}^{n-1} X_k I_{\{t=k\}} + X_n I_{\{t \geqslant n\}}$, we obtain

$$E[X_{n+1}^t - X_n^t | \mathcal{Q}_n] = E[(X_{n+1} I_{\{t \geqslant n+1\}} - X_n I_{\{t \geqslant n\}} + X_n I_{\{t=n\}}) | \mathcal{Q}_n]$$

$$= E[(X_{n+1} - X_n) I_{\{t \geqslant n+1\}} | \mathcal{Q}_n]$$

$$= I_{\{t \geqslant n+1\}} E[X_{n+1} - X_n | \mathcal{Q}_n]$$

$$= 0.$$

Hence $\{X_n^t\}$ is a martingale. $\qquad \square$

Theorem 8.3.3. (Optional Stopping). *Let* $\{X_n, \mathcal{Q}_n, n \geqslant 0\}$ *be a submartingale (martingale) and* s *and* t *be two bounded stopping times* WRT $\{X_n\}$ *such that* $P\{s \leqslant t\} = 1$. *Then* $E[|X_s|] < \infty$, $E[|X_t|] < \infty$ *and*

$$E[X_t | \mathcal{Q}_s] \geqslant (=) X_s; \qquad (8.3.2)$$

consequently, $E[X_t] \geqslant (=) E[X_s]$.

PROOF. We give the proof only in the submartingale case. Define a process $\{Y_n\}$ by

$$Y_n = X_n^t - X_n^s = X_{t \wedge n} - X_{s \wedge n}.$$

Since s and t are bounded stopping times, there is an $N > 0$ such that $P\{s \leqslant t \leqslant N\} = 1$. Now it is clear that

$$Y_0 = 0 \quad \text{and} \quad Y_k = X_t - X_s \quad \text{for all } k \geqslant N.$$

We claim that $\{Y_n\}$ is a submartingale. As in Theorem 8.3.2, let us write

$$Y_n = \sum_{k=1}^{n} X_{k-1}(I_{\{s < k-1 = t\}} - I_{\{s = k-1 < t\}}) + X_n I_{\{s < n \leqslant t\}}.$$

Then,

$$E[Y_{n+1} - Y_n | \mathcal{Q}_n] = E[(X_{n+1} - X_n)I_{\{s < n+1 \leqslant t\}} | \mathcal{Q}_n]$$

$$= I_{\{s < n+1 \leqslant t\}} E[X_{n+1} - X_n | \mathcal{Q}_n]$$

$$\geqslant 0, \quad \text{by submartingale property.}$$

This proves the claim that $\{Y_n\}$ is a submartingale. So, by Proposition 8.1.5, $E[X_t - X_s] = E[Y_k] \geqslant E[Y_0] = 0, k \geqslant N$.

Now let A be an arbitrary element in $\mathcal{Q}_s$ and set

$$t^A(\omega) = \begin{cases} t(\omega) & \text{if } \omega \in A \\ \infty & \text{if } \omega \notin A; \end{cases} \quad s^* = s^A \wedge N \text{ and } t^* = t^A \wedge N.$$

Then

$$0 \leqslant E[X_{t^*} - X_{s^*}] = E[(X_t - X_s)I_A].$$

From this and Proposition 8.1.5 we obtain (8.3.2), and this completes the proof. $\qquad \square$

The following optional sampling theorem (Theorem 8.3.4) is an immediate consequence of this theorem.

Theorem 8.3.4. (Optional Sampling Theorem). *Let $\{X_n\}$ be a martingale WRT the increasing sequence $\{\mathcal{Q}_n\}$, $n \geqslant 0$, and $t_1 \leqslant t_2 \leqslant \cdots \leqslant$ be an increasing sequence of bounded stopping times. Then the optionally sampled process $\{X_{t_n}\}$ is a martingale WRT $\{\mathcal{Q}_{t_n}\}$.*

In these results we assumed that the stopping times are bounded. In Theorem 8.3.5 we relax this condition. However, we need to impose some restrictions on the martingale.

Theorem 8.3.5. (Optional Stopping Theorem). *Let $\{X_n\}$ be a martingale and* $\mathbf{t}$ *a finite stopping time. If:*

$$\text{(a)} \ E[|X_{\mathbf{t}}|] < \infty \quad and \quad \text{(b)} \ \lim_{n\to\infty} E[X_n I_{\{\mathbf{t}>n\}}] = 0, \qquad (8.3.3)$$

then $E[X_{\mathbf{t}}] = E[X_0]$.

PROOF.   First we note that

$$E[X_{\mathbf{t}}] = E[X_{\mathbf{t}} I_{\{\mathbf{t}\leqslant n\}}] + E[X_{\mathbf{t}} I_{\{\mathbf{t}>n\}}]$$
$$= E[X_{\mathbf{t}\wedge n}] - E[X_n I_{\{\mathbf{t}>n\}}] + E[X_{\mathbf{t}} I_{\{\mathbf{t}>n\}}]. \qquad (8.3.4)$$

It follows from Theorem 8.3.3 or directly from

$$E[X_{\mathbf{t}\wedge n}] = \sum_{i=0}^{n-1} E[X_{\mathbf{t}} I_{\{\mathbf{t}=i\}}] + E[X_n I_{\{\mathbf{t}\geqslant n\}}]$$

$$= \sum_{i=0}^{n-1} E[X_i I_{\{\mathbf{t}=i\}}] + E[X_n I_{\{\mathbf{t}\geqslant n\}}]$$

$$= \sum_{i=0}^{n-1} E[I_{\{\mathbf{t}=i\}} E[X_n | X_0, \ldots, X_i]] + E[X_n I_{\{\mathbf{t}\geqslant n\}}],$$

by martingale property,

$$= \sum_{i=0}^{n-1} E[X_n I_{\{\mathbf{t}=i\}}] + E[X_n I_{\{\mathbf{t}\geqslant n\}}]$$

$$= E[X_n] = E[X_0], \qquad \text{by Proposition 8.1.4,}$$

that $E[X_{\mathbf{t}\wedge n}] = E[X_0]$.

It remains to show that the last two expectations on the RHS of (8.3.4) vanish (as $n \to \infty$). By the hypothesis (8.3.3 (b)), $\lim_{n\to\infty} E[X_n I_{\{\mathbf{t}>n\}}] = 0$. The proof will be completed once we show that $\lim_{n\to\infty} E[X_{\mathbf{t}} I_{\{\mathbf{t}>n\}}] = 0$.

By condition (8.3.3 (a)), $E[|X_{\mathbf{t}}|] < \infty$. Note that

$$E[|X_{\mathbf{t}}|] \geqslant E[|X_{\mathbf{t}}| I_{\{\mathbf{t}\leqslant n\}}]. \qquad (8.3.5)$$

Also

$$E[|X_{\mathbf{t}}| I_{\{\mathbf{t}\leqslant n\}}] = \sum_{i=0}^{n} E[|X_{\mathbf{t}}| | \mathbf{t} = i] P\{\mathbf{t} = i\}. \qquad (8.3.6)$$

Letting $n \to \infty$ in (8.3.6),

$$\lim_{n\to\infty} E[|X_{\mathbf{t}}| I_{\{\mathbf{t}\leqslant n\}}] = \sum_{i\geqslant 0} E[|X_{\mathbf{t}}| | \mathbf{t} = i] P\{\mathbf{t} = i\} = E[|X_{\mathbf{t}}|]. \quad (8.3.7)$$

Therefore, it follows from (8.3.5) and (8.3.7) that

$$\lim_{n \to \infty} E[X_t I_{\{t > n\}}] = 0.$$

This proves the theorem. $\qquad \Box$

Examples 8.3.6

EXAMPLE 1. *Random Walk Revisited.* Let $\{J_n\}$ be a sequence of IID RVs such that

$$P\{J = 1\} = p = 1 - P\{J = -1\}.$$

Let $X_n = \Sigma_{k=1}^n J_k$ be the associated simple random walk (RW). Let a and b be two positive integers that denote the initial capitals of two gamblers; call them Tom and Dick, respectively. The game ends as soon as one of them wins all the fortune of the other. The duration t of the game is a stopping time, and we have shown in Chapter 2 that $P\{t < \infty\} = 1$. Set $c = (a + b)$. If ρ denotes the probability of ruin for Dick, then show that

$$\rho = \begin{cases} a/(a + b) & \text{if } p = \frac{1}{2} \\ (1 - s^a)/(1 - s^{a+b}) & \text{if } p \neq \frac{1}{2} \end{cases},$$

where $s = (q/p)$.

Let $p = \frac{1}{2}$. Then $E[J_k] = 0$ for all k. Therefore, $\{X_n\}$ is a martingale by Example 8.1.8 (2). Noting that $|X_n| \leq c$, we obtain $E[X_t] = E[X_1] = 0$ by the optional stopping theorem. But

$$E[X_t] = b\rho - a(1 - \rho)$$

(X_n is the resulting gain at the end of game n). Thus

$$\rho = \frac{a}{a + b} \qquad \text{if } p = \frac{1}{2}.$$

Now let $p \neq \frac{1}{2}$. Set $s = (q/p)$. First note that $E[s^{J_n}] = (sp + s^{-1}q) = (q + p) = 1$, for all $n \geq 1$. Then it follows, as in Example 8.1.8 (6), that $\{s^{J_n}\}$ is a martingale. Set $Y_n = s^{J_n}$, $n \geq 1$, and observe that $|Y_t| \leq s^c \vee s^{-c}(= \max(s^c, s^{-c}))$ and for $k < t$, $|Y_k| \leq s^b \vee s^{-a}$. Appealing now to the optional stopping theorem, we obtain $E[Y_t] = E[Y_1] = 1$. But $E[Y_t] = \rho s^b + (1 - \rho)s^{-a}$. Thus

$$\rho = \frac{1 - s^a}{1 - s^{a+b}}, \qquad \text{if } p \neq \frac{1}{2}.$$

223

EXAMPLE 2. *Return to the Proof of Theorem* 2.4.5 (i). Let $p \geqslant \frac{1}{2}$ and $\mathbf{t}_\alpha$ be the time of first passage through α, where α is a positive integer. Let $g(s)$ denote the probability generating function of $\mathbf{t}_\alpha$, that is, $g(s) = E[s^{\mathbf{t}_\alpha}]$. First note that

$$E[r^{J_n}] = pr + qr^{-1} = u, \qquad \text{say .} \tag{8.3.8}$$

If $r > 1$, then $u > 1$, $(p \geqslant \frac{1}{2})$. Now define $Y_n = u^{-n} r^{X_n}$, $n \geqslant 1$. Then $\{Y_n\}$ is a martingale by Example 8.1.8 (2) (ii). Since $X_0 = 0$ and $\alpha > 0$, it follows that $X_n < \alpha$ for all $n < \mathbf{t}_\alpha$ and hence $|Y_{\mathbf{t}_\alpha}| \leqslant r^\alpha u^{-\mathbf{t}_\alpha} \leqslant r^\alpha$ for $r > 1$. By appealing to the optional stopping theorem, we get

$$E\left[Y_{\mathbf{t}_\alpha}\right] = E[Y_1] = E[u^{-1} r^{X_1}] = 1, \qquad \text{by (8.3.8),}$$

$$E[u^{-\mathbf{t}_\alpha}] = r^{-\alpha} \tag{8.3.9}$$

for $r > 1$, since $P\{X_{\mathbf{t}_\alpha} = \alpha\} = 1$. For $0 < s < 1$ we set

$$r = \frac{2qs}{1 - (1 - 4pqs^2)^{\frac{1}{2}}}, \tag{8.3.10}$$

which solves $(pr + qr^{-1}) = s$. Note that $r > 1$. Now it follows from (8.3.9) and (8.3.10) that

$$g(s) = E[s^{\mathbf{t}_\alpha}] = (2qs)^{-\alpha}[1 - (1 - 4pqs^2)^{\frac{1}{2}}]^\alpha.$$

This proves Theorem 2.4.5 (i).

Exercises

1. Tom plays a card game using a deck of five red and five black cards. Five cards are drawn in succession without replacement. In the kth draw, $1 \leqslant k \leqslant 5$, Tom wins if the card drawn is red and loses otherwise. He uses the following betting strategy. He bets $5 when the deck contains more red cards than black cards and bets $1 in all other draws. Find Tom's expected winnings.

2. If Tom always bets $5 in this game, find his expected winnings. Which of these two strategies is better?

3. Let $\{Y_n\}$, $n \geqslant 1$, be a sequence of centered independent RVs. Which of the following sequences $\{X_n\}$ are martingales WRT $\{Y_n\}$: (a) $X_n = n^{-1}(Y_1 + \cdots + Y_n)$, (b) $X_n = Y_1 Y_2 \cdots Y_n$, (c) $X_n = (Y_1 Y_2 \cdots Y_n)^{1/n}$, (d) $X_n = e^{Y_1 + \cdots + Y_n}$?

4. Let $\{Y_n\}$ be a sequence of independent unit normal RVs and $S_n = (Y_1 + \cdots + Y_n)$. Set

$$X_n = \frac{1}{\sqrt{n+1}} \exp\left\{ \frac{S_n^2}{2(n+1)} \right\}.$$

Show that $\{X_n\}$ is a martingale WRT $\{Y_n\}$.

5. Let $\{Y_n\}$ be a sequence of independent RVs identically distributed according to $P\{Y = 1\} = p$, $P\{y = -1\} = q = (1-p)$, and $S_n = (Y_1 + \cdots + Y_n)$. Show that $X_n = [q/p]^{S_n}$, $n \geq 1$, is a martingale WRT $\{Y_n\}$.

6. Let $\{Y_n\}$, $n \geq 0$, be an MC with transition probabilities $p(0,0) = 1$, $p(x, x+1) = p$, and $p(x, 0) = q$, where $p > 0$, $(p + q) = 1$, and $x = 1, 2, \ldots$. For arbitrarily fixed constants, a and b define

$$X_n = \begin{cases} ap^{1-Y_n} + b[1 - p^{1-Y_n}] & \text{if } Y_n > 0 \\ b & \text{if } Y_n = 0 \end{cases}.$$

Show that $\{X_n\}$ is a martingale WRT $\{Y_n\}$.

7. Let $Y \in N(\mu, \sigma^2)$. Given that $Y = y$, let $\{Y_n\}$ be a sequence of independent Gaussian RVs with mean y and variance 1. Set $S_n = (Y_1 + Y_2 + \cdots + Y_n)$. Show that $X_n = (\mu\sigma^{-2} + S_n)/(n + \sigma^{-2})$, $n \geq 1$, is a martingale WRT $\{Y_n\}$.

8. Consider a branching process Y_n defined by $Y_{n+1} = (I_n + Z_{n,1} + \cdots + Z_{n,Y_n})$, where $Z_{n,k}$ denotes the number of offspring of the kth individual in the nth generation and I_n is the number of immigrations in the nth generation. Let $E[I_n] = a$ and $E[Z_{n,k}] = m \neq 1$. Set $X_n = m^{-n}[Y_n - a(1 - m^n)(1 - m)^{-1}]$. Show that $\{X_n\}$ is a martingale.

9. Let $\{Y_n\}$, $n \geq 0$, be an MC on $S = \{0, 1, \ldots, N\}$ with transition probabilities

$$p(x, y) = \binom{2x}{y}\binom{2N - 2x}{N - y} / \binom{2x}{N}.$$

For what value of a is the sequence $X_n = a^{-n} Y_n(N - Y_n)$, $n \geq 0$, a martingale WRT $\{Y_n\}$?

10. Consider a sequence $\{A_{nk}\}$, $n \geq 1$, of successively finer partitions of Ω by measurable sets A_{nk} in a σ-algebra $\mathcal{C}$. Let $\mathcal{C}_n$ be the σ-algebra generated by $A_{n1}, A_{n2}, \ldots$. Note that $\mathcal{C}_n \subset \mathcal{C}_{n+1}$, $n \geq 1$. Let $P(A_{nk}) > 0$ for all k and n. Let $m(\cdot)$ be a countably additive set function on $\mathcal{C}$. Define $X_n(\omega) = m(A_{nk})/P(A_{nk})$ if $\omega \in A_{nk}$ ($k \geq 1$ and $n \geq 1$). Show that $\{X_n\}$ is a martingale WRT $\{\mathcal{C}_n\}$.

11. Let $\{X_n\}$ be a submartingale. Then show that, for any $x > 0$,

(a) $P\left\{ \max_{1 \leq k \leq n} X_k > x \right\} \leq x^{-1} E|X_n|$

(b) $P\left\{ \min_{1 \leq k \leq n} X_k < -x \right\} \leq x^{-1}[E|X_n| - E[X_1]]$.

Using this and Proposition 8.1.6, establish Kolmogorov's extension of Chebyshev's inequality; namely, if $\{X_n\}$ is a second-order martingale, then

$$P\left\{ \max_{1 \leqslant k \leqslant n} |X_k| > \epsilon \right\} \leqslant \epsilon^{-2} E[X_n^2].$$

12. Let $\{X_n\}$ be a martingale such that $E[X_n^2] \leqslant M < \infty$, for all n, and $\lim_{n \to \infty} \sup_{k \geqslant 1} |E[X_n X_{n+k}] - E[X_n]E[X_{n+k}]| = 0$. Show that X_n converges in mean square to a constant.

13. Let $\{X_n\}$ be a sequence of independent centered RVs. Use martingale convergence to establish that the sequence of partial sums $S_n = (X_1 + \cdots + X_n)$ converges with probability one if $\Sigma_{n \geqslant 1} E[X_n^2] < \infty$.

14. Give an example of a martingale $\{X_n\}$ such that $X_n \to -\infty$ with probability one.

15. Give an example of a positive martingale $\{X_n\}$ that is not uniformly integrable.

16. Let $\{X_n\}$ be an MC and R a set of recurrent states of the chain. Let $T_1 < T_2 < \cdots$ be the times of successive visits to R. Show that $\{X_{T_n}\}$ is an MC.

17. Let $\{X_n\}$ be a sequence of IID RVs following the distribution $P\{X = 1\} = p = 1 - P\{X = -1\}$. Which of the following T are stopping times?

    (a) $T = \min\{n \geqslant 1: X_n = 1\}$ or $T = \infty$ if no such n exists .

    (b) $T = \min\left\{ n \geqslant 1: \sum_1^{n+1} X_k = 0 \right\}$ or $T = \infty$ if no such n exists .

    (c) $T = \min\left\{ n \geqslant 1: \sum_1^n X_k > 0 \right\}$ or $T = \infty$ if no such n exists .

18. Let $\{X_n\}$ be a sequence of nonnegative IID RVs with $P\{X_1 > 0\} = 1$ and $E[X_1] = m$. Set $S_n = X_1 + \cdots + X_n$ and $T_x = \min\{n \geqslant 1: S_n > x\}$ or $T_x = \infty$ if no such n exists, where $x > 0$. Show that: (a) T_x is a stopping time WRT $\{X_n\}$, (b) $E[T_x] < \infty$ for every $x > 0$, (c) $E[T_x] \geqslant (x/m)$, (d) $E[T_x] \leqslant (x + m)/m$ and that $E[T_x] \simeq x/m$ as $x \to \infty$, if $P\{X_1 \leqslant M\} = 1$.

19. Let $\{X_n\}$ be a sequence of IID RVs and $T = \min\{n \geqslant 1: X_1 + \cdots + X_n > 0\}$ or $T = \infty$ if no such n exists. If $E[X_1] = 0$, show that $E[T] = \infty$.

9

Brownian Motion and Diffusion Stochastic Processes

9.1. Random Walk to Brownian Motion

In most of Chapter 2 we treated the simple RW (random walk) in which a particle takes, at the end of each unit of time, a unit step to either the right or left with equal probability. In this section we speed up the particle in such a way that it takes 2^k steps in one unit of time, with each step of length $2^{-k/2}$, and study the limiting behavior of the particle as $k \to \infty$. Let $\{X(t)\}$ be such a sped-up RW. Here $t \geq 0$ and varies over diadic rationals. As $k \to \infty$, we take $t \in R_+$. Similarly, the state space of the RW will become the real line R.

First let us observe the RW for one unit of time, that is, for $t \in [0, 1]$. Let there be 2^k steps in one unit of time and $X(0) = 0$. Then

$$X(1) = (J_1 + J_2 + \cdots + J_{2^k})2^{-k/2} \qquad (9.1.1)$$

where all J_i are independent RVs that are identically distributed according as $P\{J = 1\} = \frac{1}{2} = P\{J = -1\}$. Note that $E[J_i] = 0$ and $\text{var}(J_i) = 1$, for all i. Thus $X(1)$ is a centered and normalized sum $2^{-k/2}\Sigma_{i=1}^{2^k} J_i$ of IID RVs, where the centering constant is 0 and the normalizing constant is $1(2^{-k/2})$. Now appealing to the central limit theorem, we get

$$P\{a < X(1) \leq b\} = P\left\{a < 2^{-k/2} \sum_{i=1}^{2^k} J_i \leq b\right\}$$

$$\xrightarrow[k \to \infty]{} \int_a^b \left[e^{-x^2/2}/\sqrt{2\pi}\right] dx. \qquad (9.1.2)$$

Hence $X(1)$ is asymptotically normally distributed. Similarly, any one-dimensional distribution of $\{X'(t)\}$ is Gaussian, approximately.

Next let us take two time points s and t with $s < t$. Let l, m, p, and q be four positive integers such that $s = l2^{-p}$ and $t = m2^{-q}$. Then we claim that $(X(s), X(t))$ is a two-dimensional Gaussian random vector, approximately. To see this, it is easy to work with the characteristic functions. Now

$$X(s) = 2^{-k/2} \sum_{j=1}^{l2^{k-p}} J_j \quad \text{and} \quad X(t) = 2^{-k/2} \sum_{j=1}^{m2^{k-q}} J_j, \quad (9.1.3)$$

since there are 2^k steps in each unit of time. Also, the characteristic function of any step J is

$$E[e^{i\xi J}] = \tfrac{1}{2}e^{i\xi} + \tfrac{1}{2}e^{-i\xi} = \cos \xi. \quad (9.1.4)$$

Since $s < t$, $l2^{-p} < m2^{-q}$ and hence, from (9.1.3),

$$X(t) = X(s) + 2^{-k/2} \sum_{j=l2^{k-p}+1}^{m2^{k-q}} J_j. \quad (9.1.5)$$

Recalling that all J_n are IID RVs, we see from (9.1.3) and (9.1.5) that $X(s)$ and $X(t) - X(s)$ are independent. Therefore, it follows from (9.1.4) and (9.1.5) that

$$E[\exp\{i\xi X(s) + i\eta X(t)\}] = E[\exp\{i(\xi + \eta)X(s) + i\eta(X(t) - X(s))\}]$$

$$= E\left[\exp\left\{(\xi + \eta)2^{-k/2} \sum_{j=1}^{l2^{k-p}} J_j + i\eta 2^{-k/2} \sum_{j=l2^{k-p}+1}^{m2^{k-q}} J_j\right\}\right]$$

$$= \left[\{\cos 2^{-k/2}(\xi + \eta)\}^{2^k}\right]^s \left[(\cos 2^{-k/2}\eta)^{2^k}\right]^{t-s}. \quad (9.1.6)$$

Using the MacLaurin series for $\cos x$, we see that

$$\lim_{k \to \infty} [\cos 2^{-k/2} x]^{2^k} = \exp[-x^2/2]. \quad (9.1.7)$$

From (9.1.7) and (9.1.6) it follows that

$$\lim_{k \to \infty} E[\exp\{i\xi X(s) + i\eta X(t)\}] = \left(e^{-(\xi+\eta)^2/2}\right)^s \left(e^{-\eta^2/2}\right)^{t-s}$$

$$= \exp\{-2^{-1}(s\xi^2 + 2s\xi\eta + t\eta^2)\}. \quad (9.1.8)$$

Also note, for large k, that $X(s)$ and $X(t) - X(s)$ are approximately normal $N(0, s)$ and $N(0, t - s)$, respectively, where $N(\mu, \sigma^2)$ is the normal distribution with mean μ and variance σ^2. Therefore, (9.1.8) gives, as $k \to \infty$, the limiting characteristic function of $(X(s), X(t) - X(s))$. Also it follows from (9.1.8) that $(X(s), X(t)$ is approximately a two-dimensional Gaussian vector, since $X(s)$ and $X(t) - X(t)$ are independent and Gaussian.

In general, we have the following theorem.

Theorem 9.1.1. *Let* $\{X(t)\}$ *be a* RW *starting at 0, taking* 2^k *steps per unit time and having a jump length of* $2^{-k/2}$ *units. Then for* $0 < t_1 < \cdots < t_n$:

(i) $X(t_1), X(t_2) - X(t_1), \ldots, X(t_n) - X(t_{n-1})$ *are independent Gaussian* RV*s with* $(X(t_i) - X(t_{i-1})) \in N(0, t_i - t_{i-1})$, $i = 1, \ldots, n$.

(ii) *The n-vector* $(X(t_1), \ldots, X(t_n))$ *has, as* $k \to \infty$, *the limiting density function* $f(x_1, \ldots, x_n)$ *given by*

$$f(x_1, \ldots, x_n) = \prod_{i=1}^{n} [2\pi(t_i - t_{i-1})]^{-\frac{1}{2}} \exp\left\{-\frac{(x_i - x_{i-1})^2}{2(t_i - t_{i-1})}\right\}.$$

So, if we speed up the particle (a random walker) to take indefinitely large number of steps per unit time and jump an infinitesimally small distance each time, then in the limit we arrive at a process $X(t)$ described as follows: (1) $X(0) = 0$, (2) $X(t)$ is a process with independent increments, (3) each increment $X(t) - X(s)$ is $N(0, t - s)$. By Definition 6.4.10, the process $\{X(t)\}$ is a standard Brownian motion. Hence a sped-up simple symmetric RW converges to a standard Brownian motion.

9.2. Brownian Motion

The study of Brownian motion originated as an attempt to explain the physical phenomenon behind the rapid ceaseless irregular motion of a small particle suspended in a fluid. During 1827–1829, Robert Brown, a distinguished botanist, gave a step-by-step account of his experiments and discoveries on such a motion. Brown was not the first to observe this phenomenon, although the motion was named after him. The first explanation given to the Brownian phenomenon was that the particles were alive. Other earlier explanations were concerned with the attractions and repulsions among particles, their capillary action, their unstable equilibrium in the fluid in which they are immersed, and so on. But most of these explanations were refuted by Brown himself. After Brown's 1830 article hardly any literature appeared on this topic until 1857. Around 1860 Gouy put forward a kinetic theory explaining the Brownian motion. The salient point of his theory was that the motion is very active because of any combination of the following facts: (1) the particles are smaller and the fluid has lower viscosity and/or higher temperature, (2) the motion is ceaseless, the composition and density of the particle have no effects, and the ceaseless motion is irregular, composed of translations and rotations; and (3) the trajectory of the particle seems to have no tangent; that is, the velocity appears to be undefined at every point.

Despite the amount of research that went into understanding the Brownian phenomenon, Einstein was not aware of the work on Brownian motion, but he was the first to present a correct quantitative theory of Brownian motion. The kinetic theory behind the motion reasons that the chaotic motion of the particle is due to almost continuous bombardment of the particle by the molecules of the media.

Consider a small particle suspended in fluid. The position of the particle is described by a Cartesian coordinate system. The origin of the system is taken to be the initial position of the particle. The three coordinates of the location of the particle vary independently of each other. So, to fix an idea about the motion, let $X(t)$ denote the X-coordinate of the position at time t. Given a time interval (s, t), however small it may be, the particle experiences an enormous number of bombardments by the molecules of the fluid. Because the particle is heavier than the molecules, the effect of each collision is negligible. But there is an observable motion because of almost continuous bombardment of the particle. Also, each collision is independent of the others. So the displacement $X(t) - X(s)$ in the interval (s, t) is the sum of a very large number of independent infinitesimal displacements. Appealing to the central limit theorem, we can assume that the increment $X(t) - X(s)$ is Gaussian. Moreover, the increments $X(t) - X(s)$ and $X(u) - X(t)$ are independent for $s < t < u$. Now we are ready to define a Brownian motion (see Definition 6.4.10).

Definition 9.2.1. A *standard Brownian motion* $\{X(t)\}$, $t \geq 0$, is a stochastic process satisfying the following properties:

1. $X(0) = 0$ with probability one.
2. Every increment $X(t) - X(s)$ is normally distributed with mean zero and variance $\sigma^2 |t - s|$, for a fixed variance parameter σ.
3. For $0 = t_0 < t_1 < t_2 < \cdots < t_n < \infty$, the increments $(X(t_i) - X(t_{i-1}))$, $1 \leq i \leq n$, are independent and distributed as in (2).

Definition 9.2.2. A *Brownian motion with drift* is a stochastic process $\{X(t), t \geqslant 0\}$ with the following properties:

1. $X(0) = 0$ with probability one.
2. Every increment $(X(t) - X(s))$ is normally distributed with mean $\mu(t - s)$ and variance $\sigma^2 |t - s|$, for some fixed parameters μ and σ.
3. For $0 = t_0 < t_1 < \cdots < t_n < \infty$ the increments $X(t_1) - X(t_0), \ldots, X(t_n) - X(t_{n-1})$ are independent and distributed as described in (2).

The Brownian motion processes are also associated with the names Bachelier, Einstein, Levy, and Wiener. The statical treatment of Brownian motion was first given by Einstein and Smoluchowsky. Wiener and Levy presented the mathematical theory and extensively investigated the properties of a Brownian motion. We now summarize some of the elementary properties of a standard Brownian motion.

Property 9.2.3. A standard Brownian motion $X(t)$, $t \geqslant 0$, is a Gaussian process. This we have established in Example 6.4.11.

Property 9.2.4. A standard Brownian motion process $\{X(t), t \geqslant 0\}$ is a Markov process. This follows from the fact that $\{X(t)\}$ is a process with independent increments (see Theorem 4.2.1). The transition probability density of this Markov process is given by

$$f(t - s, y - x) = \frac{\partial}{\partial y} P\{X(t) \leqslant y | X(s) = x\}$$

$$= [2\pi(t - s)]^{-\frac{1}{2}} \exp\left\{ \frac{-(y - x)^2}{2(t - s)} \right\}, \qquad \text{(take } \sigma = 1\text{)}.$$

Notice the spatial homogeneity and temporal homogeneity of the Brownian motion since the transition density f depends on s, t, x, and y only through $t - s$ and $y - x$.

Property 9.2.5. If $\{X(t), t \geqslant 0\}$ is a standard Brownian motion, the following processes are also standard Brownian.

1. $X_1(t) = cX(t/c^2)$ for $t \geqslant 0$ and fixed $c > 0$;

2. $X_2(t) = X(t + h) - X(h)$ for fixed $h > 0$ and $t \geqslant 0$;

3. $X_3(t) = \begin{cases} tX(t^{-1}) & \text{if } t > 0 \\ 0 & \text{if } t = 0 \end{cases}$.

Clearly, each process $X_i(t)$, $i = 1, 2, 3$, is a process with independent increments and $X_i(0) = 0$. Also, each increment $X_i(t) - X_i(s)$ is a centered Gaussian RV. It remains to verify the condition on the variance of $X_i(t) - X_i(s)$. Now, for $s < t$,

$$E[\{X_1(t) - X_1(s)\}^2] = c^2 E\left[\left\{X\left(\frac{t}{c^2}\right) - X\left(\frac{s}{c^2}\right)\right\}^2\right]$$

$$= \frac{c^2 \sigma^2 (t - s)}{c^2} = \sigma^2 (t - s),$$

$$E[\{X_2(t) - X_2(s)\}^2] = E[\{X(t + h) - X(s + h)\}^2]$$

$$= \sigma^2(t + h - s - h) = \sigma^2(t - s),$$

$$E[\{X_3(t) - X_3(s)\}^2] = E[\{tX(t^{-1}) - sX(s^{-1})\}^2]$$

$$= s^2 E[\{X(s^{-1}) - X(t^{-1})\}^2] + (t - s)^2 E[(X(t^{-1}))^2]$$

$$= s^2 \sigma^2 (s^{-1} - t^{-1}) + (t - s)^2 \sigma^2 t^{-1}$$

$$= \sigma^2 (t - s).$$

Symmetrically treating the cases $s < t$ and $t < s$, we replace all $\sigma^2(t - s)$ above by $\sigma^2|t - s|$. This proves our claim (9.2.5).

Property 9.2.6. The covariance function $K(s, t)$ of a standard Brownian motion $X(t)$, $t \geq 0$, is given by $K(s, t) = \sigma^2 \min(s, t)$. The proof of this statement is identical to the corresponding formula for a Poisson process [see Example 6.1.1 (2)]. Note that $X(t)$ has stationary increments because of condition (2) in Definition 9.2.2.

Property 9.2.7. Let $\{X(t)\}$, $-\infty < t < \infty$, be a Brownian motion for $t \in R$ (Definition 9.2.1 remains the same). Then the Brownian increment process defined by

$$Y(t) = \epsilon^{-1}[X(t + \epsilon) - X(t)], \qquad t \in R,$$

is a stationary Gaussian process with covariance function

$$K_Y(t) = \begin{cases} \sigma^2 \epsilon^{-1}[1 - \epsilon^{-1}|t|], & |t| < \epsilon \\ 0, & |t| \geq \epsilon \end{cases}.$$

From the Gaussian nature of the increments of the Brownian motion $X(t)$ it easily follows that $\{Y(t)\}$ is a Gaussian process. Following the proof in Example 7.1.3 (7) for the Poisson increment process, one can prove the rest of the claim.

Let $\{Y(t), t \geqslant 0\}$ be an arbitrary stochastic process with $E[|Y(t)|] < \infty$ for all $t \geqslant 0$. Let $\mathcal{Q}_t^s$ be the σ-algebra generated by events of the form $\cap_{k=1}^n \{Y(t_k) \leqslant y_k\}$ for all $n \geqslant 1$ and $s \leqslant t_1 < t_2 < \cdots < t_n \leqslant t$. The process $Y(t), t \geqslant 0$, is called a *martingale* if for $s < t$,

$$E[Y(t)|\mathcal{Q}_s^0] = Y(s).$$

Property 9.2.8. Let $\{X(t), t \geqslant 0\}$ be a standard Brownian motion. Then $\{X(t), \mathcal{Q}_t, t \geqslant 0\}$ is a martingale, where $\mathcal{Q}_t = \sigma\{X(s): 0 \leqslant s \leqslant t\}$. For

$$E[X(t) - X(s)|\mathcal{Q}_s] = E[X(t) - X(s)] = 0$$

(since $X(t) - X(s)$ is independent of $\mathcal{Q}_s$, and $E[X(u)] = 0$).

Property 9.2.9. A standard Brownian motion $X(t), t \geqslant 0$, is continuous in the mean-square sense. For

$$E[|X(t) - X(s)|^2] = \sigma^2 |t - s| \to 0 \qquad \text{as} \quad t \to s.$$

Property 9.2.10. If $X(t), t \geqslant 0$, is a standard Brownian motion with $\sigma^2 = 1$, say, then the *reflected Brownian motion* defined by $Y(t) = |X(t)|$, $t \geqslant 0$, is a Markov process with

$$E[Y(t)] = (2t/\pi)^{\frac{1}{2}} \qquad \text{and} \quad \text{var}(X(t)) = t(1 - 2\pi^{-1}).$$

It follows from

$$P\{Y(t) \leqslant y|Y(t_k) = y_k, t_0 < t_1 < \cdots < t_n < t\}$$
$$= P\{-y \leqslant X(t) \leqslant +y|X(t_k) = \pm y_k, \qquad 0 \leqslant k \leqslant n\}$$
$$= P\{-y \leqslant X(t) \leqslant +y|X(t_k) = y_k, \qquad 0 \leqslant k \leqslant n\}, \quad \text{by symmetry,}$$
$$= P\{-y \leqslant X(t) \leqslant +y|X(t_n) = y_n\}, \qquad \text{by (9.2.4),}$$

that $Y(t)$ is a Markov process. Clearly, the transition probability of $Y(t)$ is given by

$$p(t; x, y) = Q(y - x, t) + Q(y + x, t),$$

where $Q(x, t) = (2\pi t)^{-\frac{1}{2}} \exp[-x^2/2t]$, since

$$P\{-y \leqslant X(t) \leqslant +y|X(t_n) = y_n\} = \int_{-y}^{y} Q(x - y_n, t - t_n) \, dx.$$

233

Next

$$E[Y(t)] = \int_{-\infty}^{\infty} |x| Q(x, t)\, dx$$

$$= 2 \int_0^{\infty} (2\pi t)^{-\frac{1}{2}} x e^{-x^2/2t}\, dx = \sqrt{\frac{2t}{\pi}},$$

$$\text{var}(Y(t)) = E[\{Y(t)\}^2] - \frac{2t}{\pi}$$

$$= E[|X(t)|^2] - \frac{2t}{\pi} = (1 - 2\pi^{-1})t.$$

Theorem 9.2.11. *Let* $X(t)$, $t \geqslant 0$, *be a standard Brownian motion* (take $\sigma = 1$). *Then, for any* $\lambda > 0$ *and time points* $0 = t_0 < t_1 < \cdots < t_n$:

(i) $P\left\{ \max_{0 \leqslant k \leqslant n} X(t_k) > \lambda \right\} \leqslant 2P\{X(t_n) > \lambda\},$

(ii) $P\left\{ \max_{0 \leqslant k \leqslant n} |X(t_k)| > \lambda \right\} \leqslant 2P\{|X(t_n)| > \lambda\}.$

PROOF. Define $\mathbf{t} = \min\{k: X(t_k) > \lambda\}$. The event $\{\mathbf{t} = k\}$ depends only on $X(t_0), \ldots, X(t_k)$. Since $X(t_n) - X(t_k)$, $0 \leqslant k < n$ is Gaussian, its distribution is symmetric about its mean 0 [i.e., $P\{X(t_n) - X(t_k) > 0\} = P\{X(t_n) - X(t_k) < 0\}$]. Then

$$P\left\{ \max_{0 \leqslant k \leqslant n} X(t_k) > \lambda, X(t_n) < \lambda \right\} = \sum_{m=1}^{n-1} P\{\mathbf{t} = m, X(t_n) < \lambda\}$$

$$\leqslant \sum_{m=1}^{n} P\{\mathbf{t} = m, X(t_n) - X(t_m) < 0\}$$

$$= \sum_{m=1}^{n} P\{\mathbf{t} = m\} P\{X(t_n) - X(t_m) < 0\}$$

$$= \sum_{m=1}^{n} P\{\mathbf{t} = m\} P\{X(t_n) - X(t_m) > 0\}$$

$$= \sum_{m=1}^{n} P\{\mathbf{t} = m, X(t_n) - X(t_m) > 0\}$$

$$\leqslant \sum_{m=1}^{n} P\{\mathbf{t} = m, X(t_n) > \lambda\}$$

$$= P\{X(t_n) > \lambda\}.$$

Also, $P\{\max_{0 \leqslant k \leqslant n} X(t_k) > \lambda, X(t_n) > \lambda\} = P\{X(t_n) > \lambda\}$. These imply (i). Part (ii) follows from (i) and the fact that $-X(t)$ is also a Brownian motion:

$$P\left\{\max_{0 \leqslant k \leqslant n} |X(t_k)| > \lambda\right\} = P\left\{\max_{0 \leqslant k \leqslant n} X(t_k) > \lambda\right\}$$

$$+ P\left\{\max_{0 \leqslant k \leqslant n} [-X(t_k)] > \lambda\right\}$$

$$\leqslant 2P\{X(t_n) > \lambda\} + 2P\{[-X(t_n)] > \lambda\}$$

$$= 2P\{|X(t_n)| > \lambda\}. \qquad \square$$

More generally, we have Theorem 9.2.12.

Theorem 9.2.12. *For each $T > 0$ and $\lambda > 0$ we have*

$$P\left\{\sup_{0 \leqslant t \leqslant T} X(t) > \lambda\right\} = 2P\{X(T) > \lambda\} = \sqrt{\frac{2}{\pi}} \int_\lambda^\infty e^{-x^2/2T} dx.$$

This theorem is a consequence of the so-called reflection principle. The proof of this principle depends on the fact that a Brownian motion is a strong Markov process; that is, it starts afresh at every stopping time (see Section 9.4). Let $X(t)$ be a (standard) Brownian motion and $\lambda > 0$ a fixed real number. Let $\mathbf{t}_\lambda$ be the first time t such that $X(t) \geqslant \lambda$. The hitting time $\mathbf{t}_\lambda$ is a stopping time. For all $t > \mathbf{t}_\lambda$, let us reflect the paths about the line $x = \lambda$, that is, set

$$X^*(t) = \begin{cases} X(t) & \text{if } t < \mathbf{t}_\lambda \\ 2\lambda - X(t) & \text{if } t > \mathbf{t}_\lambda \end{cases}.$$

Fix an interval $[0, T]$ and let $X(T) > \lambda$. Consider a sample path that hits λ for the first time at $\mathbf{t}_\lambda(\omega)$ and continues on such that $X(T) > \lambda$. To this path there corresponds another path (the path reflected for $t > \mathbf{t}_\lambda(\omega)$) such that $X^*(T) < \lambda$. Since a Brownian motion starts afresh at each stopping time, the behavior of these paths after $\mathbf{t}_\lambda$ is independent of what happened before $\mathbf{t}_\lambda$. Since $X(t)$ and $X^*(t)$ are reflections of each other for $t > \mathbf{t}_\lambda$ and the increments $X(t) - X(s)$, $\mathbf{t}_\lambda < s < t$, are normal $N(0, t - s)$ (and hence symmetric), the original path and its reflection beyond $\mathbf{t}_\lambda$ have same probability law. Now

$$\left\{\max_{0\leqslant t\leqslant T} X(t) > \lambda\right\} = \left\{\max_{0\leqslant t\leqslant T} X(t) > \lambda, X(T) > \lambda\right\}$$

$$\cup \left\{\max_{0\leqslant t\leqslant T} X(t) > \lambda, X(T) = \lambda\right\}$$

$$\cup \left\{\max_{0\leqslant tT} X(t) > \lambda, X(T) < \lambda\right\}.$$

Since $X(T)$ is Gaussian, a continuous distribution, the probability of the middle event on the RHS is zero. But each of the first and the last events is obtained from the other by reflection about the line $x = \lambda$ beyond the time t_λ and hence have equal probability. Noting that

$$\left\{\max_{0\leqslant t\leqslant T} X(t) > \lambda, X(T) > \lambda\right\} = \{X(T) > \lambda\},$$

we obtain Theorem 9.2.12. (This theorem is also needed in Section 9.4.)

9.3. Some Fundamental Path Properties of Brownian Motion

In Section 9.1 we saw that a sped-up RW converges to a Brownian motion. So we have to look at the sample paths of the RW to see what one can expect of the sample paths of Brownian motion. Speeding up an RW increases the number of steps per unit time, say, 2^k steps per second. But the step length is $2^{-k/2}$. Compared to the increase in the speed, the decrease in the step length is steady. Therefore, the peaks and the valleys in the sample path rise and fall sharply (see Figures 9.1 and 9.2). Figure 9.1 gives a typical sample path of an RW. By quadrupling the speed of the RW, the same sample path takes the form as that given in Figure 9.2.

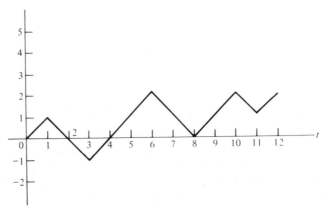

Figure 9.1

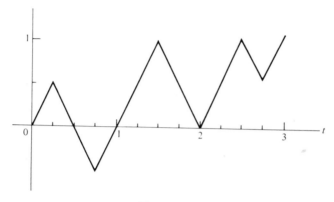

Figure 9.2

This suggests that the sample paths of a Brownian motion can be expected to be continuous but have no tangents at any point. In other words, the sample paths of a Brownian motion appear to be everywhere continuous and nowhere differentiable curves. This, in fact, is true. Let us continue to assume that our stochastic processes are separable. Also, let $X(t)$ be a standard Brownian motion with $\sigma = 1$.

Theorem 9.3.1. *A Brownian motion process* $X(t)$, $t \geq 0$, *is a continuous process; that is, almost all sample paths* $X_\omega(\cdot)$ *are continuous functions.*

PROOF. The result follows easily from Theorem 6.2.4. Indeed, take $0 < a < \frac{1}{2}$,

$$g(h) = |h|^a, \qquad q(h) = 2|h|^{\frac{1}{2}-a}\phi(|h|^{a-\frac{1}{2}}),$$

where $\phi(x)$ is the normal density $\phi(x) = (2\pi)^{-\frac{1}{2}}e^{-x^2/2}$. Then it is easy to verify that

$$\sum_{n \geq 1} g(2^{-n}) = \sum_{n \geq 1} 2^{-an} < \infty$$

and that the series

$$\sum_{n \geq 1} 2^n q(2^{-n}) = 2 \sum_{n \geq 1} 2^{n(1+2a)/2}\phi(2^{n(1-2a)/2})$$

is convergent. The increment $X(t + h) - X(t)$ is normally distributed $(N(0, |h|))$. Let Φ be this normal distribution with zero mean and variance $|h|$. Since

$$1 - \Phi(x) < x^{-1}\phi(x),$$

237

we have

$$P\{|X(t+h) - X(t)| \geqslant |h|^a\} = 2[1 - \Phi(|h|^{(2a-1)/2})]$$
$$\leqslant 2|h|^{(1-2a)/2}\phi(|h|^{(2a-1)/2}).$$

Thus all the conditions of Theorem 6.2.4 are satisfied. By our standing assumption, $X(t)$ is separable. Hence a (standard) Brownian motion is always a continuous process. $\qquad\square$

Theorem 9.3.2. *Almost all sample paths of a Brownian motion $X(t)$ are nowhere differentiable.*

PROOF. Let $X(t)$ be a Brownian motion and T a fixed time point. If a sample path $X_\omega(t)$ is differentiable from right at T, then the limit

$$\lim_{h\downarrow 0} h^{-1}[X(T+h) - X(T)]$$

exists. For such a sample path $X_\omega(\cdot)$ and the corresponding derivative X'_T, say, there is a δ such that

$$|h^{-1}[X(T+h) - X(T)] - X'_T| < 1$$

for $0 < h < \delta$, and consequently

$$|h^{-1}[X(T+h) - X(T)]| < X'_T + 1, \qquad (9.3.1)$$

if $0 < h < \delta$. Let A denote the collection of all $\omega \in \Omega$ such that $X_\omega(\cdot)$ is differentiable from right at T. We claim that $P(A) = 0$, and the conclusion of the theorem holds.

From (9.3.1) we see that we can find, for any $\omega \in A$, an $r > 0$ and an $m > 0$ such that $X'_T + 1 < r$ and $m^{-1} < \infty$. Therefore

$$A \subset \bigcup_{r \geqslant 1} \bigcup_{m \geqslant 1} \bigcap_{n \geqslant m} \{\omega: |n[X(T+n^{-1}) - X(T)]| < r\}.$$

Set $A_{nr} = \{\omega: |n[X(T+n^{-1}) - X(T)]| < r\}$. Then

$$P\{A_{nr}\} = P\{|X(T+n^{-1}) - X(T)| < n^{-1}r\}$$
$$= P\{-n^{-1}r < X(T+n^{-1}) - X(T) < n^{-1}r\}$$
$$= \int_{-r/n}^{r/n} (2\pi n^{-1})^{-\frac{1}{2}} \exp[-x^2/2n^{-1}]\,dx$$
$$= \int_{-rn^{-\frac{1}{2}}}^{rn^{-\frac{1}{2}}} (2\pi)^{-\frac{1}{2}} e^{-x^2/2}\,dy, \qquad \text{taking } y = n^{1/2}x.$$

But $\cap_{n \geqslant m} A_{nr} \subset A_{mr}$. Therefore

$$0 \leqslant P \left\{ \bigcap_{n \geqslant m} A_{nr} \right\} \leqslant P\{A_{mr}\} = \int_{-rm^{-\frac{1}{2}}}^{rm^{-\frac{1}{2}}} (2\pi)^{-\frac{1}{2}} e^{-y^2/2} \, dy$$

$$\xrightarrow[m \to \infty]{} 0,$$

since the integrals of continuous functions are continuous functions of the limits of integration. Thus $P\{\cap_{n \geqslant m} A_{nr}\} = 0$. Therefore, A as a subset of a countable union of sets of zero probability is itself of zero probability. This proves our theorem. $\qquad\square$

Heuristically, we followed the ideas of Dvoretski et al. (1961) to establish Theorem 9.3.2. Following the same idea, one can also prove Theorem 9.3.3.

Theorem 9.3.3. *For any $\alpha > \frac{1}{2}$, almost all sample paths of a Brownian motion $X(t)$ are nowhere Holder continuous with exponent α. Consequently, almost all sample paths of $X(t)$ are nowhere differentiable and hence have infinite variation on any finite interval.*

PROOF. Let T be an arbitrary positive integer. Choose a positive integer N such that $(2\alpha - 1) > (2/N)$. If a sample path $X_\omega(t)$ is Holder continuous with exponent α at some time point $t \in [0, T]$, then

$$|X(u, \omega) - X(t, \omega)| < H|u - t|^\alpha, \qquad \text{if } |u - t| \leqslant \frac{N}{n},$$

for some Holder constant $H > 0$ and an integer $n > 0$. Now fix a constant $H > 0$ and define a sequence A_n of events by

$$A_n = \{\omega: \text{ for some } t \in [0, T]$$

$$\text{we have } |X(u) - X(t)| < H|u - t|^\alpha$$

$$\text{whenever } |u - t| \leqslant N/n\}.$$

Note that $A_n \subset A_{n+1}, n \geqslant 1$. Let $A = \lim_{n \to \infty} A_n$. Since A_n is an increasing sequence, $P(A) = \lim_{n \to \infty} P(A_n)$. If we can show that each A_n is a subset of an event B_n of probability zero, then $P(A) = 0$, and this will prove our theorem.

Define, for $k = 0, 1, \ldots, nT$,

$$Y_k = \max_{1 \leqslant m \leqslant N} \left| X\left(\frac{k + m}{n}\right) - X\left(\frac{k + m - 1}{n}\right) \right|,$$

and

$$B_n = \left\{ \omega : Y_k(\omega) \leqslant 2H\left(\frac{N}{n}\right)^\alpha, \quad \text{for some} \quad k \right\}$$

$$= \bigcup_{k=0}^{nT} \left\{ \omega : Y_k(\omega) \leqslant 2H\left(\frac{N}{n}\right)^\alpha \right\}.$$

First note that $A_n \subset B_n$. For, if $\omega \in A_n$, we can find a largest integer $k > 0$ such that $(k/n) \leqslant t$ and

$$Y_k(\omega) \leqslant 2H\left(\frac{N}{n}\right)^\alpha.$$

Thus $A_n \subset B_n$.

Next we use the fact that $\{X(m/n) - X((m-1)/n)\}$, $1 \leqslant m \leqslant N$, are IID Gaussian RVs to estimate $P(B_n)$. Now

$$P(B_n) = P\left\{ \bigcup_{k=0}^{nT} \left\{ Y_k \leqslant 2H\left(\frac{N}{n}\right)^\alpha \right\} \right\} \leqslant \sum_{k=0}^{nT} P\left\{ Y_k \leqslant 2H\left(\frac{N}{n}\right)^\alpha \right\}$$

$$= (nT + 1)P\left\{ Y_0 \leqslant 2H\left(\frac{N}{n}\right)^\alpha \right\}$$

$$= (nT + 1)\left[P\left\{ |X(n^{-1}) - X(0)| \leqslant 2H\left(\frac{N}{n}\right)^\alpha \right\} \right]^N$$

$$= (nT + 1)\left[\int_{-2HN^\alpha/n^\alpha}^{2HN^\alpha/n^\alpha} \sqrt{\frac{n}{2\pi}} e^{-nx^2/2} \, dx \right]^N$$

$$= (nT + 1)\left[(2\pi)^{-\frac{1}{2}} n^{-(2\alpha-1)/2} \int_{-2HN^\alpha}^{2HN^\alpha} e^{-y^2/(2n^{2\alpha-1})} \, dy \right]^N$$

$$\to 0 \text{ as } n \to \infty, \text{ since } \frac{N(2\alpha - 1)}{2} > 1$$

Therefore

$$0 \leqslant P(A) = \lim_{n\to\infty} P(A_n) \leqslant \lim_{n\to\infty} P(B_n) = 0,$$

and hence almost all sample paths are nowhere Holder continuous for any exponent $\alpha > \frac{1}{2}$.

If a sample path is differentiable at some t, it is Holder continuous with $\alpha = 1$. Therefore, almost all sample paths are nowhere differentiable.

Since the bounded variation property of a path implies almost everywhere differentiability of that path, we see that almost all sample paths of a Brownian motion have infinite variation on any finite interval. This ends the proof.

$\square$

We have theoretically shown, by ignoring a set of probability zero, that the sample functions of a Brownian motion have infinite variation on any finite interval. We can actually compute this variation and show that it is infinite. This is a consequence of Theorem 9.3.4, which states that the squared variation of almost all paths on a finite interval $[a, b]$ is just $(b - a)$. This is an oddity in the sense that if f is a continuously differentiable function on $[a, b]$ and $\mathcal{P}_n$ is a sequence of successively finer partitions $\mathcal{P}_n: a = t_{n0} < t_{n1} < \cdots < t_{nN_n}$ with $|\mathcal{P}_n| \to 0$, as $n \to \infty$, then the squared variation

$$\lim_{n \to \infty} \sum_{k=1}^{N_n} [f(t_{nk}) - f(t_{n(k-1)})]^2 = 0;$$

(but we should also remember that Brownian paths are nowhere differentiable).

Theorem 9.3.4. *Let $\{X(t), t \geqslant 0\}$ be a Brownian motion and $\mathcal{P}_n$ the sequence of partitions of $[0, t]$*

$$\mathcal{P}_n: 0 < 2^{-n}t < 2^{-n+1}t < \cdots < t.$$

Then

$$\lim_{n \to \infty} \sum_{k=1}^{2^n} \left[X\left(\frac{kt}{2^n}\right) - X\left(\frac{(k - 1)t}{2^n}\right) \right]^2 = t, \qquad (9.3.2)$$

where the limit holds in the mean-square sense and also with probability one.

PROOF. First we show that (9.3.2) holds in the mean-square sense. To simplify writing, let us introduce some notation. Set

$$\Delta_{n,k} = X\left(\frac{kt}{2^n}\right) - X\left(\frac{(k - 1)t}{2^n}\right), \qquad 1 \leqslant k \leqslant 2^n,$$

$$D_{n,k} = \Delta_{n,k}^2 - 2^{-n}t, \qquad 1 \leqslant k \leqslant 2^n, \quad \text{and} \quad S_n = \sum_{k=1}^{2^n} D_{n,k}, \qquad n \geqslant 1.$$

So our claim is that $E[S_n^2] \to 0$ as $n \to \infty$. Since the increments $\Delta_{n,k}$ are independent and $N(0, 2^{-n}t)$, the variables $D_{n,k}$, $1 \leqslant k \leqslant 2^n$, are IID RVs with

$$E[D_{n,k}] = E[\Delta_{n,k}^2] - 2^{-n}t = 0 \quad \text{and} \quad E[D_{n,k}^2] = 2^{-2n}2t^2. \quad (9.3.3)$$

Therefore

$$E[S_n^2] = \sum_{k=1}^{2^n} E[D_{n,k}^2] = 2^{-2n}2^n 2t^2 = 2^{-n+1}t^2 \to 0 \quad (9.3.4)$$

as $n \to \infty$.

To see that (9.3.2) holds almost surely, we first appeal to Chebyshev's inequality and get, for any $\epsilon > 0$,

$$P\{|S_n| > \epsilon\} \leqslant \epsilon^{-2}\,\text{var}(S_n) = \epsilon^{-2}2t^2(2^{-n}), \quad \text{see (9.3.3)}.$$

But $\Sigma 2^{-n}$ is a convergent series, and hence by Borel–Cantelli lemma we obtain

$$P\{|S_n| > \epsilon \text{ infinitely often}\} = 0.$$

Therefore

$$P\left\{ \lim_{n\to\infty} S_n = 0 \right\} = 1.$$

This proves the theorem. $\qquad\qquad\qquad\qquad\qquad\qquad\qquad\qquad\qquad\qquad\square$

As remarked earlier, we can compute the variation

$$\lim_{n\to\infty} \sum_{k=1}^{2^n} \left| X\left(\frac{kt}{2^n}\right) - X\left(\frac{(k-1)t}{2^n}\right) \right|$$

using (9.3.2) and show that this limit is infinite. Indeed,

$$\sum_{k=1}^{2^n} |\Delta_{n,k}| \geqslant \left[\max_{1\leqslant m\leqslant 2^n} |\Delta_{n,m}| \right]^{-1} \sum_{k=1}^{2^n} [\Delta_{n,k}]^2. \quad (9.3.5)$$

The factor $\Sigma_{k=1}^{2^n}[\Delta_{n,k}]^2 \to t$. Since the sample paths of $X(t)$ are continuous, they are uniformly continuous on $[0, t]$. Hence

$$\max_{1\leqslant m\leqslant 2^n} |\Delta_{n,m}| \to 0 \quad \text{as} \quad n \to \infty.$$

This proves that $\lim_{n \to \infty} \Sigma_{k=1}^{2^n} |\Delta_{n,k}| = \infty$.

Next we state, without proof, the celebrated laws of iterated logarithm.

Theorem 9.3.5. (Laws of Iterated Logarithm). *For a Brownian motion we have*

$$P\left\{ \overline{\lim_{t \downarrow 0}} (2t \log \log t^{-1})^{-\frac{1}{2}} X(t) = 1 \right\} = 1$$

$$P\left\{ \overline{\lim_{t \uparrow \infty}} (2t \log \log t)^{-\frac{1}{2}} X(t) = 1 \right\} = 1$$

$$P\left\{ \underline{\lim_{t \downarrow 0}} (2t \log \log t^{-1})^{-\frac{1}{2}} X(t) = -1 \right\} = 1$$

$$P\left\{ \underline{\lim_{t \uparrow \infty}} (2t \log \log t)^{-\frac{1}{2}} X(t) = -1 \right\} = 1.$$

9.4. Examples and Further Properties

Let $\{X(t)\}$ be a Brownian motion and $\mathcal{Q}_t$ the σ-algebra generated by $\{X(s), 0 \leqslant s \leqslant t\}$, that is, by the events of the form

$$\{\omega: X(s_1) \in B_1, \ldots, X(s_n) \in B_n\}$$

for all $n \geqslant 1$, time points $0 \leqslant s_1 < s_2 < \cdots < s_n \leqslant t$, and Borel sets $B_1, \ldots, B_n \in \mathcal{B}(R)$. Then $\{\mathcal{Q}_t, t \geqslant 0\}$ is an increasing family of σ-algebras. A mapping $\mathbf{t}: \Omega \to \bar{R}_+$ is called a *stopping time* WRT $\mathcal{Q}_t, t \geqslant 0$, if $\{\omega: \mathbf{t}(\omega) \leqslant t\} \in \mathcal{Q}_t$ for all $t \geqslant 0$. As in the discrete-time case we define $\mathcal{Q}_{\mathbf{t}}$ by

$$\mathcal{Q}_{\mathbf{t}} = \{A \in \mathcal{Q}: A \cap \{\mathbf{t} \leqslant t\} \in \mathcal{Q}_t, t \geqslant 0\}.$$

Theorem 9.4.1. (Strong Markov Property). *Let* $\mathbf{t}$ *be a stopping time* WRT *a Brownian motion* $\{X(t)\}$, *that is,* WRT $\{\mathcal{Q}_t\}$. *Define*

$$Y(t) = X(t + \mathbf{t}) - X(\mathbf{t}), \qquad t \geqslant 0.$$

Then $\{Y(t), t \geqslant 0\}$ *is a Brownian motion. That is, a Brownian motion starts afresh at any stopping time.*

PROOF. We present the proof only in the case where $\mathbf{t}$ is at most countably valued.

If $\mathbf{t} = t$, a constant, then the theorem is obvious. Now let $\mathbf{t}$ be countably valued, that is, $\mathbf{t}: \Omega \to \{t_1, t_2, \ldots\}$. If $A \in \mathcal{Q}_{\mathbf{t}}$, then $A \cap \{\mathbf{t} = t_k\} \in \mathcal{Q}_{t_k}$.

243

Also, if $\mathcal{F}$ is the σ-algebra generated by $Y_k = X(t + t_k) - X(t_k)$, $t \geqslant 0$, then the events in $\mathcal{F}$ are independent of events in $\mathcal{C}_{t_k}$ because the Brownian motion is a process with independent increments. Therefore, for any time points $0 \leqslant s_1 < s_2 < \cdots < s_n$, real numbers $x_1, \ldots, x_n$, and $A \in \mathcal{C}_t$, we have

$$P\{Y(s_1) \leqslant x_1, \ldots, Y(s_n) \leqslant x_n, A\}$$

$$= \sum_{i \geqslant 1} P\{Y(s_1) \leqslant x_1, \ldots, Y(s_n) \leqslant x_n, \mathbf{t} = t_i, A\}$$

$$= \sum_{i \geqslant 1} P\{(X(s_k + t_i) - X(t_i)) \leqslant x_k, 1 \leqslant k \leqslant n, \mathbf{t} = t_i, A\}$$

$$= \sum_{i \geqslant 1} P\{[X(s_k + t_i) - X(t_i)] \leqslant x_k, 1 \leqslant k \leqslant n\} P\{\mathbf{t} = t_i, A\}$$

$$= P\{X(s_k) \leqslant x_k, 1 \leqslant k \leqslant n\} \sum_{i \geqslant 1} P\{\mathbf{t} = t_i, A\}$$

$$= P\{X(s_k) \leqslant x_k, 1 \leqslant k \leqslant n\} P(A).$$

By taking $A = \Omega$ we see that

$$P\{Y(s_k) \leqslant x_k, 1 \leqslant k \leqslant n\} = P\{X(s_k) \leqslant x_k, 1 \leqslant k \leqslant n\}, \qquad n \geqslant 1.$$

Thus the processes $\{Y(t)\}$ and $\{X(t)\}$ have the same joint distributions. Since $X(t)$ is a continuous process (by being a Brownian motion), we see that $Y(t)$ is also a continuous process. Hence $\{Y(t)\}$ is a Brownian motion. $\qquad \square$

We heuristically discussed the reflection principle in Section 9.2. A rigorous proof uses Theorem 9.4.1 and is beyond the scope of this textbook. But the reflection principle has several interesting applications, and we discuss one or two examples of it.

Examples 9.4.2

EXAMPLE 1. Find $P_0\{X(s) < \lambda$ for all s with $0 \leqslant s \leqslant t\}$, where P_a is the probability measure corresponding to the Brownian paths starting from a, and $\lambda > 0$.

First note that $P\{X(t) = \lambda\} = 0$, since $X(t)$ is Gaussian. Therefore

$$1 = P_0\{X(t) > \lambda\} + P_0\{X(s) < \lambda, 0 \leqslant s \leqslant t\}$$

$$+ P_0\{X(t) < \lambda \text{ and } X(s) = \lambda \text{ for some } s < t\}$$

$$= 2P_0\{X(t) > \lambda\} + P_0\{X(s) < \lambda, 0 \leqslant s \leqslant t\} \qquad \text{(why?)},$$

and hence

$$P\{X(s) < \lambda, 0 \leqslant s \leqslant t\} = 1 - 2P\{X(t) > \lambda\}$$

$$= 1 - 2\int_\lambda^\infty (2\pi t)^{-\frac{1}{2}} e^{-x^2/2t} dx$$

$$= 1 - \sqrt{\frac{2}{\pi}} \int_{\lambda t^{-\frac{1}{2}}}^\infty e^{-y^2/2} dy, \qquad (9.4.1)$$

by taking $y = xt^{-\frac{1}{2}}$.

EXAMPLE 2. Show that

$$\lim_{t \to \infty} P_0\{X(s) < \lambda \text{ for all } s \text{ with } 0 \leqslant s \leqslant t\} = 0.$$

Letting $t \to \infty$ in (9.4.1), we get

$$\lim_{t \to \infty} P_0\{X(s)\lambda, 0 \leqslant s \leqslant t\} = 1 - \sqrt{\frac{2}{\pi}} \int_0^\infty e^{-y^2/2} dy$$

$$= 1 - 2\int_0^\infty (2\pi)^{-1/2} e^{-y^2/2} dy$$

$$= 1 - 2 \cdot \tfrac{1}{2}$$

$$= 0.$$

In other words, we cannot keep Brownian paths below any level λ for an indefinite period of time.

EXAMPLE 3. Show that

$$P_0\{X(s) \leqslant s\lambda \text{ for all } s \geqslant t^{-1}\} = 1 - \sqrt{\frac{2}{\pi t}} \int_\lambda^\infty e^{-x^2/2t} dx \qquad (9.4.2)$$

Recall that the process $X_3(t)$ defined by

$$X_3(t) = \begin{cases} tX(1/t), & t > 0 \\ 0, & t = 0 \end{cases}$$

is also a standard Brownian motion (Property 9.2.5). Set $\sigma = s^{-1}$, $s > 0$, and note that $s \geqslant t^{-1}$ if and only if $0 < \sigma \leqslant t$. Also, $X(s) = sX_3(s^{-1})$. Therefore

$$P_0\{X(s) \leqslant s\lambda \text{ for all } s \geqslant t^{-1}\} = P_0\{sX_3(s^{-1}) \leqslant s\lambda \text{ for all } s \geqslant t^{-1}\}$$

$$= P_0\{X_3(\sigma) \leqslant \lambda \text{ for } 0 \leqslant \sigma \leqslant t\}$$

$$= 1 - \sqrt{\frac{2}{\pi t}} \int_\lambda^\infty e^{-x^2/2t} dx,$$

by applying Example 1 to the Brownian motion $X_3(t)$.

EXAMPLE 4. Let t_λ be the hitting or the first passage time of the level $\lambda > 0$ by the (standard) Brownian motion $X(t)$: $t_\lambda = \min\{t \geqslant 0: X(t) \geqslant \lambda\}$. Find the density function $f_\lambda(t)$ of t_λ.

If $t_\lambda > t$, then $X(s) < \lambda$ for all s with $0 \leqslant s \leqslant t$, and conversely. Hence, for $t > 0$,

$$P_0\{t_\lambda > t\} = P_0\{X(s) < \lambda, 0 \leqslant s \leqslant t\}$$

and consequently

$$P_0\{t_\lambda \leqslant t\} = \sqrt{\frac{2}{\pi}} \int_{\lambda/\sqrt{t}}^\infty e^{-x^2/2} dx,$$

from (9.4.1). Therefore

$$f_\lambda(t) = \frac{d}{dt} \sqrt{\frac{2}{\pi}} \int_{\lambda/\sqrt{t}}^\infty e^{-x^2/2} dx$$

$$= (2\pi)^{-1/2} \lambda t^{-3/2} e^{-\lambda^2/2t}, \qquad t > 0. \qquad (9.4.3)$$

Theorem 9.4.3. (Zero-Crossing Theorem). *Let $\{X(t)\}$, $t \geqslant 0$, be a standard Brownian motion with $X(0) = 0$. Then the probability p that $\{X(t)\}$ has at least one zero in an interval (a, b) is given by*

$$p = \frac{2}{\pi} \arccos \sqrt{\frac{a}{b}} . \qquad (9.4.4)$$

Consequently, if t is the largest zero of $X(t)$ with $t \leqslant T$, then

$$P_0\{t < a\} = \frac{2}{\pi} \arcsin \sqrt{\frac{a}{T}} . \qquad (9.4.5)$$

PROOF. The Brownian particle $X(t)$ starts from 0, and $|X(a)| = \lambda$ for some nonnegative λ. We want to compute the probability that the particle will then cross the zero level at some point in the time interval (a, b). Let p_λ denote the probability that the particle located at λ at time a [i.e., $X(a) = \lambda$] will cross the zero level in the time interval (a, b). Then

$$p = \int_0^\infty p_\lambda P_0\{|X(a)| = \lambda\} d\lambda \qquad (9.4.6)$$

So we first compute p_λ. Now

$$
\begin{aligned}
p_\lambda &= P\left\{\min_{a \le t \le b} X(t) \le 0 \,|\, X(a) = \lambda\right\} \\
&= P\left\{\min_{a \le t \le b} X(t) \le -\lambda \,|\, X(a) = 0\right\}, \qquad \text{by spatial homogeneity,} \\
&= P\left\{\max_{a \le t \le b} X(t) \ge \lambda \,|\, X(a) = 0\right\}, \qquad \text{by symmetry,} \\
&= P_0\left\{\max_{0 \le t \le b-a} X(t) \ge \lambda\right\}, \qquad \text{by temporal homogeneity,} \\
&= P_0\{\mathbf{t}_\lambda \le b - a\}, \qquad \text{where } \mathbf{t}_\lambda \text{ is the hitting time of } \lambda, \\
&= (2\pi)^{-1/2} \lambda \int_0^{b-a} t^{-3/2} e^{-\lambda^2/2t} \, dt, \qquad \text{by Example 9.4.2(4).} \tag{9.4.7}
\end{aligned}
$$

Using this in (9.4.6), we have

$$
\begin{aligned}
p &= \int_0^\infty p_\lambda \, P_0\{|X(a)| = \lambda\} \, d\lambda = 2 \int_0^\infty (2\pi a)^{-\frac{1}{2}} e^{-\lambda^2/2a} p_\lambda \, d\lambda \\
&= \sqrt{\frac{2}{\pi a}} \int_0^\infty (2\pi)^{-1/2} \lambda \int_0^{b-a} t^{-3/2} e^{-\lambda^2/2t} e^{-\lambda^2/2a} \, dt \, d\lambda \\
&= \pi^{-1} a^{-1/2} \int_0^{b-a} t^{-3/2} \left\{ \int_0^\infty \lambda \exp[-\lambda^2(t^{-1} + a^{-1})/2] \, d\lambda \right\} dt \\
&= \pi^{-1} a^{-1/2} \int_0^{b-a} t^{-3/2} \{at/(t + a)\} \, dt,
\end{aligned}
$$

by setting $u = \lambda^2(t + a)/2at$ in the inner integral,

$$
\begin{aligned}
&= \pi^{-1} a^{1/2} \int_0^{b-a} t^{-1/2}(t + a)^{-1} \, dt \\
&= 2\pi^{-1} \int_0^{[(b-a)/a]^{1/2}} \frac{ds}{1 + s^2}, \qquad \text{by setting } t = as^2, \\
&= 2\pi^{-1} \arctan[(b - a)/a]^{1/2}. \tag{9.4.8}
\end{aligned}
$$

Set $\alpha = \arctan[(b - a)/a]^{1/2}$. Then $\tan \alpha = [(b - a)/a]^{1/2}$ and $\cos \alpha = [a/b]^{1/2}$. Hence

$$
p = \frac{2}{\pi} \arctan\left[\frac{b - a}{a}\right]^{1/2} = \frac{2}{\pi} \arccos\left[\frac{a}{b}\right]^{1/2}.
$$

This proves (9.4.4).

To see (9.4.5), note that if $\mathbf{t} < a$, there is no zero in (a, T), and conversely. Therefore

$$P\{\mathbf{t} < a\} = 1 - p = 1 - \frac{2}{\pi} \arccos \left[\frac{a}{b} \right]^{1/2} = \frac{2}{\pi} \arcsin \left[\frac{a}{b} \right]^{1/2},$$

by (9.4.4) and the relation $(\pi/2) = \cos^{-1}\theta + \sin^{-1}\theta$. This completes the proof. □

Examples 9.4.4

EXAMPLE 1. *Ornstein–Uhlenbeck Velocity.* In Section 9.3 we established that the Brownian paths are nowhere differentiable; that is, the particle has infinite velocity at all times. Recall that the mean square velocity is also infinite at all times. To handle this situation, Ornstein and Uhlenbeck introduced a process as a model for the velocities of particles performing a Brownian motion. Let $X(t)$ be a standard Brownian motion. The process $V(t)$ defined by

$$V(t) = e^{-t} X(e^{2t}), \qquad t \in R, \tag{9.4.9}$$

is called a *Ornstein–Uhlenbeck process.* Show that $V(t)$ is a Gaussian stationary process.

Let $a_1, \ldots, a_n$ be some real constants and $t_1 < t_2 < \cdots < t_n$ be arbitrarily chosen time points. Then

$$\sum_{k=1}^{n} a_k V(t_k) = \sum_{k=1}^{n} a_k e^{-t_k} X(e^{2t_k}),$$

is a Gaussian variable since $X(t)$ is a Gaussian process. Hence $V(t)$ is a Gaussian process. Let us next compute the mean and covariance functions of $V(t)$.

$$m_V(t) = E[V(t)] = E[e^{-t}X(e^{2t})] = e^{-t} \cdot 0 = 0, \tag{9.4.10}$$

$$K_V(s, t) = e^{-s}e^{-t}E[X(e^{2s})X(e^{2t})], \qquad \text{where we let } s < t,$$

$$= e^{-s}e^{-t}\min(e^{2s}, e^{2t})$$

$$= e^{-t+s} = e^{-|t-s|}. \tag{9.4.11}$$

By symmetry in s and t, $K_V(s, t) = e^{-|t-s|}$ for all s, t. Because $K_V(s, t)$ is a function of $|t - s|$, $V(t)$ is wide sense stationary. Combining this with the Gaussian nature of $\{V(t)\}$, we see that $V(t)$ is also strictly stationary. The

Ornstein–Uhlenbeck process $V(t)$ is just the Brownian motion $X(t)$ transformed appropriately into a stationary process. The transformation is done by rescaling the time (such that zero is taken into $-\infty$) and normalizing by e^{-t}. Since $X(t)$ is a Markov process and the transformation is one-to-one, it is easy to see that the Ornstein–Uhlenbeck process $V(t)$ is a Markov process.

Let us now obtain the transition probability density function $f(s, x; t, y)$ of $V(t)$. Since the Brownian motion $X(t)$ is a process with independent increments, it is clear that the RVs $V(s)$ and $V(t) - e^{-(t-s)} V(s)$ are independent for $s < t$. Therefore

$$
\begin{aligned}
E[V(t)|V(s)] &= E[V(t) - e^{-(t-s)} V(s) + e^{-(t-s)} V(s)|V(s)] \\
&= E[V(t) - e^{-(t-s)} V(s)] + e^{-(t-s)} V(s) \\
&= e^{-(t-s)} V(s), \quad \text{since } m_V(t) \equiv 0 \quad \text{by (9.4.10), (9.4.12)}
\end{aligned}
$$

and also

$$
\begin{aligned}
E[\{V(t) - e^{-(t-s)} V(s)\}^2 | V(s)] &= E[\{V(t) - e^{-(t-s)} V(s)\}^2] \\
&= 1 - e^{-2(t-s)}, \quad s < t, \quad (9.4.13)
\end{aligned}
$$

from (9.4.11). Since $V(t)$ is a Gaussian process, the conditional distribution of $V(t)$ given $V(s)$ is also Gaussian with conditional mean and variance given by (9.4.12) and (9.4.13). Hence

$$
\begin{aligned}
f(s, x; t, y) &= \frac{\partial}{\partial y} P\{V(t) \leqslant y | V(s) = x\} \\
&= [2\pi(1 - e^{-2(t-s)})]^{-\frac{1}{2}} \exp\left[-\frac{(y - e^{-(t-s)} x)^2}{2(1 - e^{-2(t-s)})} \right]. \quad (9.4.14)
\end{aligned}
$$

[Further discussion is continued in Example 9.6.7 (3).]

EXAMPLE 2. Let $f(s)$ be a continuous function on $[0, t]$ and set $Y(t) = \int_0^t f(s) X(s) \, ds$, where $X(s)$ is a Brownian motion. Compute the mean and variance of $Y(t)$,

$$
m_Y(t) = E[Y(t)] = E \int_0^t f(s) X(s) \, ds = \int_0^t f(s) E[X(s)] \, ds = 0,
$$

where the interchange of E and $\int$ operations can be justified. We suggest (but not as the justification), that the reader think of these operations as finite sums. Since $m_Y(t) = 0$,

$$\text{var}(Y(t)) = E[(Y(t))^2] = E\left[\left\{\int_0^t f(s)X(s)\,ds\right\}^2\right]$$

$$= E\left[\int_0^t \int_0^t f(r)f(s)X(r)X(s)\,dr\,ds\right]$$

$$= \int_0^t \int_0^t f(r)f(s)E[X(r)X(s)]\,dr\,ds$$

$$= \int_0^t \int_0^t f(r)f(s)\min(r,s)\,dr\,ds$$

$$= \int_0^t f(s)\int_0^s f(r)r\,dr\,ds + \int_0^t sf(s)\int_s^t f(r)\,dr\,ds.$$

EXAMPLE 3. Show that the process

$$Y(t) = \int_t^{t+1}(X(s) - X(t))\,ds, \qquad t \in R,$$

where $X(t)$ is a standard Brownian motion, is a wide-sense-stationary process.

$$m_Y(t) = E\int_t^{t+1}[X(s) - X(t)]\,ds = \int_t^{t+1} E[X(s) - X(t)]\,ds = 0. \quad (9.4.15)$$

Therefore

$$K_Y(u,v) = E[Y(u)Y(v)]$$

$$= E\left[\int_u^{u+1}[X(s) - X(u)]\,ds \int_v^{v+1}[X(t) - X(v)]\,dt\right].$$

If $u < u + 1 \leqslant v < v + 1$, that is, if $|v - u| > 1$, then $K_Y(u,v) = 0$ since $X(t)$ is a process with independent increments. Therefore, let $u \leqslant v < u + 1 \leqslant v + 1$. Then

$$K_Y(u,v) = E\left[\left(\int_u^v + \int_v^{u+1}\right)\left(\int_v^{u+1} + \int_{u+1}^{v+1}\right)\right]$$

$$= E\left[\left\{\int_v^{u+1}[X(s) - X(v)]\,ds\right\}^2\right]$$

$$= E\left[\int_v^{u+1}\int_v^{u+1}[X(s) - X(v)][X(t) - X(v)]\,ds\,dt\right]$$

$$= \int_v^{u+1}\int_v^{u+1}\min(s - v, t - v)\,ds\,dt$$

$$= \int_v^{u+1}\left\{\int_v^t (s - v)\,ds + \int_t^{u+1}(t - v)\,ds\right\}dt$$

$$= \int_v^{u+1} \{(t-v)(2u-v-t+2)/2\} \, dt$$

$$= 2^{-1}\{u[(u+1)^2 - v^2] - \tfrac{v}{2}[(u+1)^2 - v^2] + \tfrac{1}{3}[(u+1)^3 - v^3]$$

$$+[(u+1)^2 - v^2] - 2uv(u+1-v) + v^2(u+1-v)$$

$$+\tfrac{v}{2}[(u+1)^2 - v^2] - 2v(u+1-v)\}$$

$$= 3^{-1}[1 - (v-u)]^3 = 3^{-1}[1 - |v-u|]^3.$$

The same expression holds for $v < u$. Therefore,

$$K_Y(u,v) = \tfrac{1}{3}(1 - |v-u|)^3 = K_Y(|v-u|).$$

Hence $Y(t)$ is a wide-sense-stationary process.

9.5. White Noise and Stochastic Integrals

Consider a mechanical or electronic system. Incessant fluctuations in such devices limit their sensitivity and are called *noise*. For example, the fluctuating current called *shot noise* in vacuum tubes is due to the random emissions of electrons from the heated cathode. Similarly, the thermal agitations of conduction electrons in various resistors in an electric network cause fluctuations in the voltage across the ends of the resistor. The fluctuating voltage is called the *thermal noise*. These types of noise are examples of white-noise processes. The white-noise processes are defined in several ways. Let us motivate one of them.

We need first the notion of δ-function(al). A (generalized) function $\delta(x)$ defined by the evaluation relation

$$\int_R f(x)\delta(x - x_0) \, dx = f(x_0), \qquad (9.5.1)$$

where f is a continuous function on R, is called the δ-*functional*. One advantage in introducing this function is that we can define the derivatives of a function $g(x)$ at its discontinuity points. For example, let $g(x)$ be the unit step function with unit jump at the origin. Then

$$\delta(x) = \frac{dg(x)}{dx}.$$

We have seen in Chapter 4 that the electron emissions can be modeled by Poisson processes. Consider the emissions occurring at random times

9. Brownian Motion and Diffusion Stochastic Processes

$W_1, W_2, \ldots$. The total current flowing through a vacuum tube is due to the superpositions of current pulses. So one defines a shot noise by means of the process

$$X(t) = \sum_k h(t - W_k), \qquad (9.5.2)$$

where $h(t)$ is a real-valued function called *impulse response*. If $N(t), t \in R$, is a Poisson process, then the *Poisson impulses process* is defined by

$$Y(t) = \sum_k \delta(t - W_k) = \frac{dN(t)}{dt}. \qquad (9.5.3)$$

If $h(t)$ is the impulse-response function corresponding to the input Poisson impulses, then the shot noise is the resulting output. Now set

$$h(t) = \begin{cases} 2eT^{-2}t & \text{if } 0 < t < T \\ 0 & \text{otherwise} \end{cases}, \qquad (9.5.4)$$

where $-e$ is the charge of each electron and T is the transit time that each electron takes to move from the cathode to anode. The Campbell's theorem states that

$$E[X(t)] = \lambda \int h(t)\,dt = \lambda e$$

$$\text{var}(X(t)) = \lambda \int h^2(t)\,dt = \frac{4\lambda e^2}{3T}$$

$$K_X(t, t + u) = \lambda \int h(s)h(s + u)\,ds$$

$$= \begin{cases} \dfrac{4\lambda e^2}{3T}\left[1 - \dfrac{3|u|}{2T} + \dfrac{|u|^3}{2T^3}\right] & \text{for } |u| \leqslant T \\ 0 & \text{otherwise} \end{cases},$$

where λ is the Poisson intensity parameter. If the transit time T is very negligible, say, of order 10^{-N}, for suitable N, one can show that the spectral density function of the shot noise process is

$$f(\xi) = \text{constant},$$

a flat spectrum.

Similarly, a thermal noise in a resistor is defined as a normal process with zero mean function and the δ-functional as the covariance function. The δ-functional has a flat spectrum that in the thermal noise case is given by $f(\xi) = \alpha kT$, where α is a constant, k is the Boltzmann constant, and T is the absolute temperature.

252

There are several stationary processes with a flat spectrum. In particular, the process representing the energy distribution in white light from an incandescent body has a flat spectrum. With this analogy, a *white noise* can be defined as a stationary process with constant spectral density.

Definition 9.5.1. A *white noise* $W(t)$, $t \in R$, is a stationary process, either: (1) with constant spectral density, (2) whose covariance is the Dirac δ-functional, or (3) with covariance and spectral density given respectively by

$$K(t) = Ne^{-N|t|} \quad \text{and} \quad f(x) = \frac{1}{\pi} \frac{1}{1 + (x/N)^2},$$

where N is infinitely large.

9.5.2. A Formal Representation of White Noise

Let $X(t)$, $-\infty < t < \infty$, be a standard Brownian motion. We proved in Chapter 6 that $X(t)$ is not mean-square differentiable. But let us formally consider the mean-square derivative $\dot{X}(t)$ of $X(t)$. Then

$$K_{\dot{X}}(s, t) = \frac{\partial^2 K_X(s, t)}{\partial s\, \partial t}$$

$$= \frac{\partial^2 \min(s, t)}{\partial s\, \partial t}$$

$$= \frac{\partial H(s - t)}{\partial s}$$

$$= \delta(s - t),$$

where H is the Heavyside function

$$H(s - t) = \begin{cases} 1 & \text{for } s > t \\ 0 & \text{for } s < t \end{cases}.$$

That is, the covariance function of $\dot{X}(t)$ is the Dirac δ-functional. So, by Definition 9.5.1, we can treat $\dot{X}(t)$ as a white noise. Thus the white-noise process $W(t)$ can be formally represented as

$$W(t) = \frac{dX(t)}{dt}.$$

This representation suggests another definition of white-noise processes. Actually, many probabilists prefer to work with the following definition.

Definition 9.5.3. A white-noise process $W(t)$ is the formal derivative of a Brownian motion $X(t)$ where $W(t) = \dot{X}(t)$ is to be treated as a functional that acts on continuously differentiable functions as follows:

$$\int_a^b f(t)W(t)\,dt = \int_a^b f(t)\,dX(t) = f(t)X(t)\big|_a^b - \int_a^b X(t)\,df(t). \quad (9.5.5)$$

Since almost no sample path of $X(t)$ is differentiable, it is not possible to define the integral $\int_a^b f(t)\,dX(t)$ by the usual recipe. Thus (9.5.5) gives also a definition of the integration WRT to a Brownian motion. There is extensive work available on such integrals. It is possible to define a stochastic integral $\int_a^b f(t,\omega)\,dX(t,\omega)$ in a suitable sense for a sufficiently large class of processes $f(t,\omega)$. But the theory of the so-called Itô integral is beyond the level of this textbook. Bernstein and Wiener did the initial work on stochastic integrals and equations. We simply call the integral $\int_a^b f(t)\,dX(t)$ a *stochastic integral*. Theorem 9.5.4 lists some of the first and basic properties of the stochastic integrals.

Theorem 9.5.4. *Let $C^1[a,b]$ denote the class of all (once) continuously differentiable functions, and let $X(t)$ be a standard Brownian motion, where $[a,b] \subset R_+$. Then:*

(i)   *The stochastic integral $\int_a^b f(t)\,dX(t)$ is a Gaussian RV with zero mean.*

(ii) $E\left[\int_a^b f(t)\,dX(t) \int_a^b g(t)\,dX(t) \right] = \int_a^b f(t)g(t)\,dt$

$$\hspace{4cm} (9.5.6)$$

$$\textit{for } f, g \in C^1,$$

in particular,

$$\mathrm{var}\left(\int_a^b f(t)\,dX(t) \right) = \int_a^b f^2(t)\,dt, \qquad (9.5.7)$$

(iii) $E\left[\int_a^b f(t)\,dX(t) \int_c^d g(t)\,dX(t) \right] = 0 \qquad \textit{if } a < b \leqslant c < d,$

$$\hspace{10cm} (9.5.8)$$

(iv) $E\left[\int_a^b f(t)\,dX(t) \int_a^c g(t)\,dX(t) \right] = \int_a^{\min(b,c)} f(t)g(t)\,dt. \qquad (9.5.9)$

PROOF. (i) Since the Brownian motion $X(t)$ is a Gaussian process, it is now clear from (9.5.5) that $\int_a^b f(t)\,dX(t)$ is a Gaussian variable with zero mean function.

(ii)
$$E\left[\int_a^b f(t)\,dX(t)\int_a^b g(t)\,dX(t)\right]$$

$$= E\left[\left\{f(b)X(b) - f(a)X(a) - \int_a^b X(t)f'(t)\,dt\right\}\right.$$
$$\left.\times\left\{g(b)X(b) - g(a)X(a) - \int_a^b X(t)g'(t)\,dt\right\}\right]$$

$$= E\left[\left\{f(b)(X(b) - X(a)) - \int_a^b f'(t)(X(t) - X(a))\,dt\right\}\right.$$
$$\left.\times\left\{g(b)(X(b) - X(a)) - \int_a^b g'(t)(X(t) - X(a))\,dt\right\}\right]$$

$$= E[f(b)g(b)(X(b) - X(a))^2]$$
$$+ E\left[\int_a^b f'(s)(X(s) - X(a))\,ds\int_a^b g'(t)(X(t) - X(a))\,dt\right]$$
$$- E\left[f(b)(X(b) - X(a))\int_a^b g'(t)(X(t) - X(a))\,dt\right]$$
$$- E\left[g(b)(X(b) - X(a))\int_a^b f'(t)(X(t) - X(a))\,dt\right]$$

$$= E_1 + E_2 - E_3 - E_4, \qquad \text{call them so .} \qquad (9.5.10)$$

Now

$$E_1 = f(b)g(b)E[(X(b) - X(a))^2] = f(b)g(b)(b - a) = \int_a^b f(b)g(b)\,dt,$$
$$(9.5.11)$$

$$E_2 = \int_a^b \int_a^b f'(s)g'(t)E[(X(s) - X(a))(X(t) - X(a))]\,ds\,dt$$
$$= \int_a^b f'(s)\int_a^b g'(t)\min(s - a, t - a)\,dt\,ds$$
$$= \int_a^b f'(s)\left\{\int_a^s (t - a)g'(t)\,dt + \int_s^b (s - a)g'(t)\,dt\right\}\,ds$$
$$= \int_a^b f'(s)\left\{(t - a)g(t)|_a^s - \int_a^s g(t)\,dt + (s - a)(g(b) - g(s))\right\}\,ds$$

$$= \int_a^b f'(s) \left\{ (s-a)g(b) - \int_a^s g(t)\,dt \right\} ds$$

$$= \int_a^b f'(s) \int_a^s (g(b) - g(t))\,dt\,ds$$

$$= \int_a^b (g(b) - g(t)) \int_t^b f'(s)\,ds\,dt$$

$$= \int_a^b (g(b) - g(t))(f(b) - f(t))\,dt, \qquad (9.5.12)$$

$$E_3 = \int_a^b f(b)g'(t) E[(X(b) - X(a))(X(t) - X(a))]\,dt$$

$$= \int_a^b f(b)g'(t) \min(b-a, t-a)\,dt$$

$$= \int_a^b f(b)g'(t)(t-a)\,dt$$

$$= f(b) \int_a^b (t-a)\,dg(t)$$

$$= f(b) \left[(t-a)g(t)\big|_a^b - \int_a^b g(t)\,dt \right]$$

$$= f(b) \left[(b-a)g(b) - \int_a^b g(t)\,dt \right]$$

$$= f(b) \int_a^b (g(b) - g(t))\,dt, \qquad (9.5.13)$$

and similarly

$$E_4 = g(b) \int_a^b (f(b) - f(t))\,dt. \qquad (9.5.14)$$

Using (9.5.11)–(9.5.14) in (9.5.10), we get

$$E\left[\int_a^b f(t)\,dX(t) \int_a^b g(t)\,dX(t) \right]$$

$$= \int_a^b [f(b)g(b) + (g(b) - g(t))(f(b) - f(t)) - f(b)(g(b) - g(t))$$

$$- g(b)(f(b) - f(t))]\,dt$$

$$= \int_a^b f(t)g(t)\,dt.$$

(iii) This is a simple consequence of (9.5.5).

(iv) Let $b \leqslant c$. The case where $c \leqslant b$ follows similarly.

$$E\left[\int_a^b f(t)\,dX(t) \int_a^c g(t)\,dX(t) \right]$$

$$= E\left[\int_a^b f(t)\,dx(t)\left\{ \int_a^b g(t)\,dX(t) + \int_b^c g(t)\,dX(t) \right\} \right]$$

$$= \int_a^b f(t)g(t)\,dt, \qquad \text{by (ii) and (iii)},$$

$$= \int_a^{\min(b,c)} f(t)g(t)\,dt, \qquad \text{since } b \leqslant c \text{ by assumption}.$$

This completes the proof. □

To mention a point about defining a stochastic integral as the limit of a sum $\sum_{k=0}^{n-1} f(t_k,\omega)[X(t_{k+1},\omega) - X(t_k,\omega)]$, consider a mean-square continuous process $f(t,\omega)$, $t \in [a,b]$ and a sequence $\{\mathcal{P}_n\}$, $n \geqslant 1$, of successively finer partitions of $[a,b]$ such that the mesh $|\mathcal{P}_n| \to 0$ as $n \to \infty$. Define

$$J_n(\omega) = \sum_{k=0}^{n-1} f(t_k^n,\omega)[X(t_{k+1}^n,\omega) - X(t_k^n,\omega)]. \qquad (9.5.15)$$

Let us also assume now that $f(t,\omega)$ is nonanticipatory of the Brownian motion $X(t,\omega)$; that is, $f(t)$ is independent of the increments $X(v) - X(u)$ for all $a \leqslant t \leqslant u < v \leqslant b$. Then

$$E[f(t_k^n)[X(t_{k+1}^n) - X(t_k^n)]]^2 = E[f(t_k^n)]^2 E[X(t_{k+1}^n) - X(t_k^n)]^2 < \infty.$$

Consequently, each term in the sum (9.5.15) is a second-order RV, and thus $\{J_n\}$, $n \geqslant 1$, forms a sequence of second-order RVs.

Definition 9.5.5. If the sequence J_n converges in the mean-square sense to an RV J as $n \to \infty$, then the limit is called the *Itô integral* of $f(t)$ WRT $X(t)$ and is denoted by

$$J(\omega) = \int_a^b f(t,\omega)\,dX(t,\omega). \qquad (9.5.16)$$

The Itô integral is not an ordinary mean-square integral as that defined in Chapter 6. Not all of the usual integration formulas extend to the Itô integral. Moreover, the limit J may depend on the choice of partition points $\{t_k^n\}$ of the interval $[a, b]$. As pointed out earlier, a detailed discussion of Itô integrals is beyond our present scope. Since several standard calculus formulas fail for the Itô integral case, many applied scientists hesitate to use Itô calculus. To remedy the situation, Stratonovich introduced an integral that bears his name. Although it enjoys the standard integration properties, the Stratonovich integral fails to possess some important probabilistic properties that the Itô integral satisfies. Which calculus to use depends on what kind of properties one investigates. A detailed account of the differences between Itô and Stratonovich calculi can be found in Mortensen (1968). We simply point out the two anomalies mentioned at the beginning of this paragraph. Example 9.5.6 is due to Doob. [See also McShane (1974).]

Example 9.5.6. Show that, for an Itô integral,

$$\int_a^b X(t)\,dX(t) = \tfrac{1}{2}[X^2(b) - X^2(a)] - \tfrac{1}{2}(b - a). \qquad (9.5.17)$$

The extra term $(b - a)/2$ is the anomalous term that we do not have in the standard calculus.

We have seen in Property 9.2.9 that $X(t)$ is mean-square continuous. Since a Brownian motion is a process with independent increments, $X(t)$ is independent of $X(v) - X(u)$ for all $a \leqslant t \leqslant u < v \leqslant b$. Hence the integral $\int_a^b X(t)\,dX(t)$ exists (uniquely). Let $\mathscr{P}: a = t_0 < t_1 < \cdots < t_n = b$ be a partition of $[a, b]$ such that $\lim_{n\to\infty} |\mathscr{P}| = 0$. Note that

$$\int_a^b X(t)\,dX(t) = \lim_{n\to\infty} E\left[\sum_{k=0}^{n-1} X(t_k)\{X(t_{k+1}) - X(t_k)\}\right]^2$$

$$= \lim_{n\to\infty} E\left[\tfrac{1}{2}\sum_{k=0}^{n-1} \{X^2(t_{k+1}) - X^2(t_k)\} - \{X(t_{k+1}) - X(t_k)\}^2\right]^2$$

Now,

$$E\left[\tfrac{1}{2}\left\{\sum_{k=0}^{n-1} (X^2(t_{k+1}) - X^2(t_k)) - (X(t_{k+1}) - X(t_k))^2\right.\right.$$

$$\left.\left. - [(X^2(b) - X^2(a)) - (b - a)]\right\}^2\right]$$

$$= E\left[\tfrac{1}{2}\left\{\sum_{k=0}^{n-1} \{X(t_{k+1}) - X(t_k)\}^2 - (b - a)\right\}^2\right]$$

$$\to 0, \qquad \text{as } n \to \infty,$$

by Theorem 9.3.4 on the squared variation. This proves (9.5.17).

Now define $I_n = \Sigma X(t_{k+1})[X(t_{k+1}) - X(t_k)]$ and $J_n = \Sigma X(t_k)[X(t_{k+1}) - X(t_k)]$ corresponding to the partition $\mathscr{P}$ as in Example 9.5.6. But, as seen above,

$$E[(I_n - J_n)^2] = E[\{\Sigma [(X(t_{k+1}) - X(t_k)]^2\}^2]$$

$$\to (b - a), \quad \text{by Theorem 9.3.4.}$$

Hence the sequences I_n and J_n do not converge to the same limit.

9.6. Diffusion Process and Kolmogorov Equations

Definition 9.6.1. A stochastic process $X(t)$, $t \geqslant 0$, on a (complete) probability space $(\Omega, \mathcal{C}, P)$ is called a *Markov process* with state space R, the real line, if for all time points $0 \leqslant t_0 < t_1 < \cdots < t_n < \infty$ and any $x \in R$ we have

$$P\{X(t_n) \leqslant x | X(t_0), \ldots, X(t_{n-1})\} = P\{X(t_n) \leqslant x | X(t_{n-1})\}. \quad (9.6.1)$$

Set, for $s < t$,

$$F(s, x; t, y) = P\{X(t) \leqslant y | X(s) = x\}. \quad (9.6.2)$$

The function $F(\cdot, \cdot; \cdot, \cdot)$ is called the *transition distribution function* of the Markov process $X(t)$.

Since F is a distribution function, the following properties are satisfied for any $x, y \in R$ and $s, t \in R_+$ with $s < t$:

$$F(s, x; t, y) \geqslant 0, \lim_{y \to -\infty} F(s, x; t, y) = 0, \lim_{y \to \infty} F(s, x; t, y) = 1, \quad (9.6.3)$$

and $F(s, x; t, y)$ is continuous from right as a function of y.

Let us assume that $F(s, x; t, y)$ is continuous WRT s, t, and x. Now let $s < t < u$. As in the case of (5.1.3), we obtain the Chapman–Kolmogorov equation

$$F(s, x; u, y) = \int F(s, x; t, dz) F(t, z; u, y). \quad (9.6.4)$$

We also set

$$\lim_{t \downarrow s} F(s, x; t, y) = \lim_{s \uparrow t} F(s, x; t, y) = \begin{cases} 1 & \text{if } y \geqslant x \\ 0 & \text{if } y < x \end{cases}. \quad (9.6.5)$$

If F is partially differentiable WRT y, the partial derivative $f(s, x; t, y) = (\partial/\partial y)F(s, x; t, y)$ is called the *transition density function* of $X(t)$. The Chapman–Kolmogorov equation now takes the form

$$f(s, x; u, y) = \int f(s, x; t, z) f(t, z; u, y)\, dz. \tag{9.6.6}$$

Consider now a continuous-time purely discontinuous or jump Markov process. In Chapter 5 we saw that such a process does not change its state in a negligible amount of time with high probability. But when it does move, it jumps by a noticeable distance. However, consider a Brownian motion. Since the Brownian particle is heavier than the interacting molecules of the media, any appreciable change during any small time interval occurs with negligible probability. Such processes can be called *purely continuous processes* or *Khintchin continuous processes*.

Definition 9.6.2. A Markov process $X(t)$, $t \geqslant 0$, with state space R is said to be *Khintchin continuous* if for any fixed $\delta > 0$ the following condition holds:

$$\lim_{\Delta t \to 0} \frac{1}{\Delta t} \int_{|y-x|>\delta} F(t, x; t + \Delta t, dy) = 0. \tag{9.6.7}$$

Definition 9.6.3. A Markov process $X(t)$, $t \geqslant 0$, with state space R is called a *diffusion process* if there exist real-valued functions $M(t, x)$ and $S(t, x)$ on $R_+ \times R$ such that for any fixed $\delta > 0$ the following conditions are satisfied:

$$\lim_{\Delta t \to 0} \frac{1}{\Delta t} \int_{|y-x|>\delta} F(t, x; t + \Delta t, dy) = 0, \tag{9.6.8}$$

$$\lim_{\Delta t \to 0} \frac{1}{\Delta t} \int_{|y-x|\leqslant\delta} (y - x) F(t, x; t + \Delta t, dy) = M(t, x), \tag{9.6.9}$$

$$\lim_{\Delta t \to 0} \frac{1}{\Delta t} \int_{|y-x|\leqslant\delta} (y - x)^2 F(t, x; t + \Delta t, dy) = S(t, x). \tag{9.6.10}$$

9.6.4. Physical Meaning of Conditions (9.6.8) – (9.6.10)

To explain the physical meaning of the coefficients $M(t, x)$ and $S(t, x)$, let us now modify Khintchin's continuity condition as follows:

$$\lim_{\Delta t \to 0} \frac{1}{\Delta t} \int_{|y-x|\geqslant\delta} (y - x)^2 F(t, x; t + \Delta t, dy) = 0. \tag{9.6.11}$$

It is clear that (9.6.11) implies (9.6.8). Now, in the light of this, we can replace conditions (9.6.9) and (9.6.10) by

$$\lim_{\Delta t \to 0} \frac{1}{\Delta t} \int (y - x) F(t, x; t + \Delta t, dy) = M(t, x) \qquad (9.6.12)$$

and

$$\lim_{\Delta t \to 0} \frac{1}{\Delta t} \int (y - x)^2 F(t, x't + \Delta t, dy) = S(t, x), \qquad (9.6.13)$$

respectively. Then

$$\int (y - x) F(t, x; t + \Delta t, dy) = E[X(t + \Delta t) - X(t)], \qquad (9.6.14)$$

the expectation of variation of $X(t)$ during a time period Δt, and

$$\int (y - x)^2 F(t, x; t + \Delta t, dy) = E[(X(t + \Delta t) - X(t))^2], \qquad (9.6.15)$$

the expectation of squared variation of $X(t)$ during Δt. From (9.6.12) and (9.6.14) it is clear that $M(t, x)$ is the mean rate of change of $X(t)$ and is called the *drift coefficient*. Noting that the squared variation is proportional to the kinetic energy we see that from (9.6.13) and (9.6.15) that $S(t, x)$ is proportional to the mean kinetic energy of the system and is called the *diffusion coefficient*.

Theorem 9.6.5. (Kolmogorov's Backward Equation.) *Let $X(t)$, $t \geqslant 0$, be a diffusion process with drift $M(s, x)$, diffusion coefficient $S(s, x)$, and transition distribution function $F(s, x; t, y)$ such that the partial derivatives*

$$\frac{\partial}{\partial x} F(s, x; t, y) \quad \text{and} \quad \frac{\partial^2}{\partial x^2} F(s, x; t, y)$$

exist and are continuous for all s, x, y and $t > s$. Then $F(s, x; t, y)$ satisfies Kolmogorov's backward equation:

$$-\frac{\partial}{\partial s} F(s, x; t, y) = M(s, x) \frac{\partial}{\partial x} F(s, x; t, y) + \frac{S(s, x)}{2} \frac{\partial^2}{\partial x^2} F(s, x; t, y).$$
$$(9.6.16)$$

PROOF. Since F is a distribution function, we have

$$F(s, x; t, y) = \int F(s, x; t, y) F(s - \Delta s, x; s, dz). \qquad (9.6.17)$$

9. Brownian Motion and Diffusion Stochastic Processes

By the Chapman–Kolmogorov equation,

$$F(s - \Delta s, x; t, y) = \int F(s - \Delta s, x; s, dz) F(s, z; t, y). \qquad (9.6.18)$$

From (9.6.17) and (9.6.18),

$$\frac{1}{\Delta s}[F(s - \Delta s, x; t, y) - F(s, x; t, y)]$$

$$= \frac{1}{\Delta s} \int [F(s, z; t, y) - F(s, x; t, y)] F(s - \Delta s, x; s, dz)$$

$$= \frac{1}{\Delta s} \int_{|z-x|>\delta} [F(s, z; t, y) - F(s, x; t, y)] F(s - \Delta s, x; s, dz)$$

$$+ \frac{1}{\Delta s} \int_{|z-x|\leqslant\delta} [F(s, z; t, y) - F(s, x; t, y)] F(s - \Delta s, x; s, dz)$$

$$= \frac{1}{\Delta s} \int_{|z-x|>\delta} [F(s, z; t, y) - F(s, x; t, y)] F(s - \Delta s, x; s, dz)$$

$$+ \frac{\partial}{\partial x} F(s, x; t, y) \frac{1}{\Delta s} \int_{|z-x|\leqslant\delta} (z - x) F(s - \Delta s, x; s, dz)$$

$$+ \frac{1}{2} \frac{\partial^2}{\partial x^2} F(s, x; t, y) \frac{1}{\Delta s} \int_{|z-x|\leqslant\delta} [(z - x)^2 + o(z - x)^2] F(x - \Delta s, x; s, dz),$$
$$(9.6.19)$$

by the hypothesis and Taylor's formula.

Now let $\Delta s \to 0$ in (9.6.19). The first term converges to 0 by Khintchin's continuity condition (9.6.7). By condition (9.6.9), the second term approaches $M(s, x)(\partial F/\partial x)$ as $\Delta s \to 0$. Letting $\Delta s \to 0$ and $\delta \to 0$ in the third term, we have the limit $\frac{1}{2}S(s, x)(\partial^2 F/\partial x^2)$. So the RHS of (9.6.19) has the limit

$$M(s, x) \frac{\partial}{\partial x} F(s, x; t, y) + S(s, x) \frac{\partial^2}{\partial x^2} F(s, x; t, y).$$

Therefore, the LHS of (9.6.19) exists as $\Delta s \to 0$ and equals

$$-\frac{\partial}{\partial s} F(s, x; t, y),$$

and hence we obtain (9.6.16). $\qquad\qquad\qquad\qquad\qquad\qquad\qquad\qquad$ □

Next we derive Kolmogorov's forward equation, which is also known among physicists as the *Fokker–Planck equation*.

Theorem 9.6.6. (Kolmogorov's Forward Equation.) *Let* $X(t)$, $t \geqslant 0$, *be a diffusion process with drift* $M(t, y)$, *diffusion* $S(t, y)$, *and transition distribution function* $F(s, x; t, y)$ *such that the following partial derivatives exist and are continuous:*

$$\frac{\partial}{\partial x} F(s, x; t, y), \quad \frac{\partial^2}{\partial x^2} F(s, x; t, y), \quad \frac{\partial}{\partial t} f(s, x; t, y),$$

$$\frac{\partial}{\partial y}[M(t, y)f(s, x; t, y)], \quad \text{and} \quad \frac{\partial^2}{\partial y^2}[S(t, y)f(s, x; t, y)],$$

where $f(s, x; t, y)$ *is the transition density function* $f = (\partial F / \partial y)$, *which is also assumed to exist. Then* $f(s, x; t, y)$ *satisfies the forward equation*

$$\frac{\partial}{\partial t} f(s, x; t, y) = -\frac{\partial}{\partial y}[M(t, y)f(s, x; t, y)] + \tfrac{1}{2}\frac{\partial^2}{\partial y^2}[S(t, y)f(s, x; t, y)].$$

$$(9.6.20)$$

PROOF. Let $g(y)$ be a nonnegative and twice continuously differentiable function vanishing outside a compact interval $[a, b]$. Then, by the continuity of g and its derivatives, we have

$$0 = g(a) = g(b) = g'(a) = g'(b) = g''(a) = g''(b). \quad (9.6.21)$$

By the differentiability assumption and Taylor's formula, we have

$$g(z) - g(y) = (z - y)g'(y) + \tfrac{1}{2}(z - y)^2 g''(y) + o((z - y)^2). \quad (9.6.22)$$

From Khintchin continuity and boundedness of $g(\cdot)$, we obtain

$$\int_{|y-z|>\delta} f(t, y; t + \Delta t, z)g(z) \, dz = o(\Delta t). \quad (9.6.23)$$

Now

$$\int_a^b g(y)\frac{\partial}{\partial t} f(s, x; t, y) \, dy$$

$$= \frac{\partial}{\partial t} \int_a^b f(s, x; t, y)g(y) \, dy$$

$$= \lim_{\Delta t \to 0} \int_a^b (\Delta t)^{-1}[f(s, x; t + \Delta t, y) - f(s, x; t, y)]g(y) \, dy$$

$$= \lim_{\Delta t \to 0} (\Delta t)^{-1}\left\{ \int_a^b \int_R f(s, x; t, z)f(t, z; t + \Delta t, y)g(y) \, dz \, dy \right.$$

263

$$- \int_a^b f(s,x;t,y)g(y)\,dy \Big\}, \qquad \text{by (9.6.6)},$$

$$= \lim_{\Delta t \to 0} (\Delta t)^{-1} \int_R f(s,x;t,y) \left[\int_a^b f(t,y;t+\Delta t,z)g(z)\,dz - g(y) \right] dy,$$

by changing the order of integration,

$$= \lim_{\Delta t \to 0} (\Delta t)^{-1} \int_R f(s,x;t,y) \left\{ \int_{|y-z|\leqslant\delta} f(t,y;t+\Delta t,z)g(z)\,dz - g(y) \right\} dy,$$

by (9.6.23),

$$= \lim_{\Delta t \to 0} (\Delta t)^{-1} \int_R f(s,x;t,y) \Bigg\{ g'(y) \int_{|y-z|\leqslant\delta} (z-y)f(t,y;t+\Delta t,z)\,dz$$

$$+ 2^{-1}g''(y) \int_{|y-z|\leqslant\delta} [(z-y)^2 + o((z-y)^2)]$$

$$\times f(t,y;t+\Delta t,z)\,dz + o(\Delta t) \Bigg\} dy,$$

$$= \int_R f(s,x;t,y)[M(t,y)g'(y) + 2^{-1}S(t,y)g''(y)]\,dy,$$

by using the uniform limits (9.6.9) and (9.6.10),

$$= \int_a^b f(s,x;t,y)[M(t,y)g'(y) + 2^{-1}S(t,y)g''(y)]\,dy,$$

since $g'(y)$ and $g''(y)$ vanish outside (a,b),

$$= \int_a^b g(y) \left\{ -\frac{\partial}{\partial y}[M(t,y)f(s,x;t,y)] + 2^{-1}\frac{\partial^2}{\partial y^2}[S(t,y)f(s,x;t,y)] \right\} dy,$$

using integration by parts .

Therefore

$$\int_a^b g(y) \left\{ \frac{\partial}{\partial t}f(s,x;t,y) + \frac{\partial}{\partial y}[M(t,y)f(s,x;t,y)] \right.$$

$$\left. - 2^{-1}\frac{\partial^2}{\partial y^2}[S(t,y)f(s,x;t,y)] \right\} dy = 0. \qquad (9.6.24)$$

Since $g(\cdot)$ is arbitrarily chosen, we claim that relation (9.6.24) yields the forward equation (9.6.20). Suppose that this is not true. We now show that this leads to a contradiction. If (9.6.24) does not imply (9.6.20), then we can find a $(s, x; t, y)$ such that the expression in the double bracket $\{\cdots\}$ of (9.6.24) is nonzero at $(s, x; t, y)$. By the hypotheses the expression in $\{\cdots\}$ is a continuous function, and hence we can find an interval (c, d) on which the function in $\{\cdots\}$ retains its sign. If $(c, d) \subset [a, b]$, let us choose $g(\cdot)$ such that it satisfies the properties stipulated earlier except that it now vanishes outside $[c, d]$ and is positive in (c, d). Then, the integral $\int_c^d \cdots$ in (9.6.24) is nonzero. This is the contradiction we are looking for. Hence (9.6.24) gives us (9.6.20), and this completes the proof. $\qquad\square$

If the transition density function $f(s, x; t, y)$ exists, then we can rewrite the backward equation (9.6.16) as

$$\frac{\partial}{\partial s} f(s, x; t, y) + M(s, x)\frac{\partial}{\partial x} f(s, x: t, y) + \tfrac{1}{2}S(s, x)\frac{\partial^2}{\partial x^2} f(s, x; t, y) = 0.$$

$$(9.6.25)$$

Examples 9.6.7

EXAMPLE 1. *Random Walk to Diffusion.* By speeding up the simple symmetric random walk, we obtained a standard Brownian motion as the limit. If the RW is nonsymmetric, let p and q be the probabilities of jumps to the right and left, respectively. In this case we obtain a Brownian motion with drift, and the corresponding transition density $f(x; t, y)$ is

$$f(x; t, y) = \frac{1}{(2\pi t)^{1/2}} \exp\left\{-\frac{(y - x - t)^2}{2t}\right\} = f(t, y - x).$$

Let $\{X_n\}$ be an unrestricted simple RW on the integral lattice. Then the n-step transition $p(n; x, y)$ is given by

$$p(n; x, y) = P\{J_1 + \cdots + J_n = y - x\}$$

$$= \binom{n}{(n + y - x)/2} p^{(n+y-x)/2} q^{(n-y+x)/2} = p(n, y - x),$$

a binomial distribution. Let us speed up the RW with 2^k jumps per unit time and each step length of $2^{-k/2}$ units. Let p and q be such that

$$p = \tfrac{1}{2} + 2^{-k/2} \quad \text{and} \quad q = \tfrac{1}{2} - 2^{-k/2}.$$

In the limiting case we obtain the transition density given above.

Of course, the n-step transition probability $p(n, x)$ satisfies the recurrence relation

$$p(n + 1, x) = pp(n, x - 1) + qp(n, x + 1),$$

265

which in the case of sped-up RW becomes

$$f(t + \Delta t, x) = pf(t, x - \Delta x) + qf(t, x + \Delta x),$$

where $\Delta x = 2^{-k/2}$, $\Delta t = 2^{-k}$. Using the choice of p and q shown above, we obtain

$$f(t, x) + \Delta t \frac{\partial f}{\partial t} + o(\Delta t)$$

$$= \left(\frac{1}{2} + \Delta x\right)\left[f(t, x) - \Delta x \frac{\partial f}{\partial x} + \frac{1}{2}(\Delta x)^2 \frac{\partial^2 f}{\partial x^2}\right]$$

$$+ \left(\frac{1}{2} - \Delta x\right)\left[f(t, x) + \Delta x \frac{\partial f}{\partial x} + \frac{1}{2}(\Delta x)^2 \frac{\partial^2 f}{\partial x^2}\right] + o((\Delta x)^2),$$

where again $\Delta x = 2^{-k/2}$ and $\Delta t = 2^{-k}$. Therefore

$$\frac{\partial f}{\partial t} = -\frac{(\Delta x)^2}{\Delta t} \frac{\partial f}{\partial x} + \frac{1}{2} \frac{(\Delta x)^2}{\Delta t} \frac{\partial^2 f}{\partial x^2} + o(1).$$

Proceeding to the limit, we obtain

$$\frac{\partial f}{\partial t} = -\frac{\partial f}{\partial x} + \frac{1}{2} \frac{\partial^2 f}{\partial x^2}. \tag{9.6.26}$$

Equation (9.6.26) is the diffusion equation corresponding to a Brownian motion. This limiting argument can be found in Kac (1947).

EXAMPLE 2. *Spatially Homogeneous Diffusion.* Let the transition density function $f(s, x; t, y)$ be dependent on s, t, and on x and y through $y - x$; that is, the physical process under consideration is spatially homogeneous. Then the drift and diffusion coefficients $M(s, x)$ and $S(s, x)$ are independent of the position x at time s and hence $M(s, x) = M(s)$ and $S(s, x) = S(s)$. Now equations (9.6.20) and (9.6.25) take the forms

$$\frac{\partial f}{\partial t} = -M(t)\frac{\partial f}{\partial y} + \frac{1}{2}S(t)\frac{\partial^2 f}{\partial y^2}$$

$$\frac{\partial f}{\partial s} = -M(s)\frac{\partial f}{\partial x} - \frac{1}{2}S(s)\frac{\partial^2 f}{\partial x^2}, \tag{9.6.27}$$

respectively. Let us now proceed to solve system (9.6.27).

In Property 9.2.4 we saw that the Brownian transition density is both temporally and spatially homogeneous. In such cases M and S are constants. So let us consider now a constant drift M and diffusion S. In particular, let $M = 0$ and $S = 1$. Then system (9.6.27) becomes

$$\frac{\partial f}{\partial t} = +\frac{1}{2}\frac{\partial^2 f}{\partial y^2} \quad \text{and} \quad \frac{\partial f}{\partial s} = -\frac{1}{2}\frac{\partial^2 f}{\partial x^2}. \tag{9.6.28}$$

The first equation in this system is the heat equation, and the second is its adjoint equation [see (9.6.26) also]. We first solve this system and then by proper transformation obtain the solution of system (9.6.27).

By spatial and temporal homogeneity, we can write $f(0, x; t, y)$ as $f(x, t, y)$. Let us solve (9.6.28) under the boundary conditions

$$f(x, t, y) \to 0 \quad \text{and} \quad \frac{\partial}{\partial y} f(x, t, y) \to 0, \quad \text{as} \quad |y| \to \infty. \tag{9.6.29}$$

Set

$$\phi(\theta; t, x) = \int_R e^{i\theta y} f(x, t, y)\, dy. \tag{9.6.30}$$

Then

$$\int_R e^{i\theta y} \frac{\partial f}{\partial y} dy = [e^{i\theta y} f]_{-\infty}^{\infty} - i\theta \int_R e^{i\theta y} f(x, t, y)\, dy$$

$$= -i\theta\phi(\theta; t, x), \quad \text{by 9.6.29,}$$

and

$$\int_R e^{i\theta y} \frac{\partial^2 f}{\partial y^2} dy = \left[e^{i\theta y} \frac{\partial f}{\partial y} \right]_{-\infty}^{\infty} - i\theta \int_R e^{i\theta y} \frac{\partial f}{\partial y} dy$$

$$= -i\theta(-i\theta\phi(\theta; t, x)) = -\theta^2 \phi(\theta; t, x). \tag{9.6.31}$$

Therefore, taking the Fourier transform on both sides of the heat equation $(\partial f/\partial t) = \frac{1}{2}(\partial^2 f/\partial y^2)$ and using (9.6.31), we obtain

$$\frac{\partial \phi}{\partial t} = -\frac{1}{2}\theta^2 \phi \tag{9.6.32}$$

whose solution (by separating the variables ϕ and t), is

$$\phi(\theta; t, x) = a \exp\left[-\frac{\theta^2 t}{2} \right]. \tag{9.6.33}$$

Since f is a density function, $f(s, x; s, y) = \delta(y - x)$, the Dirac δ-function, and consequently we have

$$\phi(\theta; 0, x) = \int_R e^{i\theta y} \delta(y - x)\, dy = e^{i\theta x}, \quad \text{by (9.5.1).} \tag{9.6.34}$$

9. Brownian Motion and Diffusion Stochastic Processes

From (9.6.33) and (9.6.34) we obtain $a = e^{i\theta x}$ and

$$\phi(\theta; t, x) = \exp \frac{i\theta x - t\theta^2}{2}, \tag{9.6.35}$$

which is merely the characteristic function of a Gaussian density with mean x and variance t. Hence

$$f(x; t, y) = (2\pi t)^{-1/2} \exp\left\{ \frac{-(y - x)^2}{2t} \right\}, \quad y \in R, \tag{9.6.36}$$

which is the transition density of the Brownian motion (see Property 9.2.4).
Let us now set

$$\xi = x - \int_0^s M(u)\, du, \qquad \eta = y - \int_0^t S(u)\, du$$

$$\sigma = \int_0^s M(u)\, du \quad \text{and} \quad \tau = \int_0^t S(u)\, du.$$

This transformation reduces system (9.6.27) into (9.6.28). Therefore, the solution $f(s, x; t, y)$ corresponding to system (9.6.27) is given by

$$f(s, x; t, y) = (2\pi D^2)^{-1/2} \exp\left\{ \frac{-(y - x - m)^2}{2D^2} \right\}, \tag{9.6.37}$$

where

$$m = \int_s^t M(u)\, du \quad \text{and} \quad D^2 = \int_s^t S(u)\, du.$$

EXAMPLE 3. (*Ornstein–Uhlenbeck Velocity*). In Example 9.4.4 (1) we defined the Ornstein–Uhlenbeck process $V(t)$ as a transformation of a Brownian motion and computed its transition density function $f(s, x; t, y)$. It is shown there that

$$f(s, x; t, y) = [2\pi(1 - e^{-2(t-s)})]^{-\frac{1}{2}} \exp\left\{ -\frac{(y - e^{-(t-s)}x)^2}{2(1 - e^{-2(t-s)})} \right\}. \tag{9.6.38}$$

Here we derive the Kolmogorov equations corresponding to this f. From (9.6.38) we first have

$$E[V(t + h) - V(t)|V(t)] = e^{-h}V(t) - V(t) = -hV(t) + o(h^2), \tag{9.6.39}$$

268

and then

$$
\begin{aligned}
&E[\{V(t+h) - V(t)\}^2 | V(t)] \\
&\quad = E[\{V(t+h) - e^{-h}V(t) + V(t)(e^{-h} - 1)\}^2 | V(t)] \\
&\quad = 1 - e^{-2h} + V^2(t)(e^{-h} - 1)^2 \\
&\quad = 2h + o(h^2).
\end{aligned}
\tag{9.6.40}
$$

Therefore, the transition density function of the Ornstein–Uhlenbeck process satisfies the following Kolmogorov equations:

$$
\frac{\partial f}{\partial t} = \frac{\partial}{\partial y}(yf) + \frac{\partial^2 f}{\partial y^2},
\tag{9.6.41}
$$

$$
-\frac{\partial f}{\partial s} = -x\frac{\partial f}{\partial x} + \frac{\partial^2 f}{\partial x^2},
\tag{9.6.42}
$$

These equations follow from (9.6.39)–(9.6.40) and the physical meaning discussed in Section 9.6.4. The converse is also true: one can start with (9.6.41), say, employ the Fourier transform or characteristic function method used in Example 2, and show that the transition function $f(s, x; t, y)$ is given by relation (9.6.38). We leave this as an exercise.

EXAMPLE 4. *Ornstein–Uhlenbeck's Position-Velocity Two-Dimensional Process.* Let $V(t)$ be the Ornstein–Uhlenbeck velocity process. Then the position of the particle is given by

$$
U(t) = \int_0^t V(s)\,ds.
\tag{9.6.43}
$$

Compute the mean and covariance functions of $U(t)$ and derive the Kolmogorov equations of the two-dimensional process $(U(t), V(t))$,

$$
m_U(t) = \int_0^t E[V(s)]\,ds = 0
$$

and thus

$$
\begin{aligned}
K_U(s, t) &= E[U(s)U(t)] \\
&= \int_0^t \int_0^s E[V(p)V(q)]\,dp\,dq \\
&= \int_0^t \int_0^s e^{-|p-q|}\,dp\,dq, \qquad \text{by (9.4.11.)} \\
&= \int_{s\wedge t}^{s\vee t} \int_0^{s\wedge t} e^{-(p-q)}\,dq\,dp + 2\int_0^{s\wedge t} \int_0^s e^{-(p-q)}\,dq\,dp \\
&= 2(s \wedge t) + e^{-(s\wedge t)} + e^{-(s\vee t)} - e^{-|t-s|} - 1,
\end{aligned}
\tag{9.6.44}
$$

where $s \wedge t = \min(s, t)$ and $s \vee t = \max(s, t)$.

By itself $U(t)$ is not a Markov process, but the two-dimensional process $(U(t), V(t))$ is. From (9.6.39) and (9.6.40) we obtain

$$E[V(t + h) - V(t)|U(t), V(t)] = -hV(t) + o(h^2) \qquad (9.6.45)$$

$$E[\{V(t + h) - V(t)\}^2|U(t), V(t)] = 2h + o(h^2). \qquad (9.6.46)$$

Next

$$
\begin{aligned}
E[U(t + h) - U(t)|U(t), V(t)] &= E\left[\int_t^{t+h} V(s)\,ds\middle|V(t)\right] \\
&= E\left[\int_t^{t+h} \{V(s) - e^{-(s-t)}V(t) + e^{-(s-t)}V(t)\}\,ds\middle|V(t)\right] \\
&= E\left[\int_t^{t+h} \{V(s) - e^{-(s-t)}V(t)\}\,ds\middle|V(t)\right] + \int_t^{t+h} e^{-(s-t)}V(t)\,ds \\
&= hV(t) + o(h^2),
\end{aligned}
\qquad (9.6.47)
$$

$$E[\{U(t + h) - U(t)\}^2|U(t), V(t)] = E\left[\left\{\int_t^{t+h} V(s)\,ds\right\}^2\middle|V(t)\right]$$
$$= o(h^2) \qquad (9.6.48)$$

and hence it follows from (9.6.46) that

$$E[(U(t + h) - U(t))(V(t + h) - V(t))|U(t), V(t)] = o(h). \qquad (9.6.49)$$

Therefore, the forward equation for the transition density $f(s, \xi, \eta; t, x, y)$ of $(U(t), V(t))$ is given by

$$\frac{\partial f}{\partial t} = -\frac{\partial}{\partial x}(yf) + \frac{\partial}{\partial y}(yf) + \frac{\partial^2 f}{\partial y^2},$$

and the corresponding backward equation is

$$\frac{\partial f}{\partial s} = -y\frac{\partial f}{\partial x} + y\frac{\partial f}{\partial y} - \frac{\partial^2 f}{\partial y^2}.$$

EXAMPLE 5. *Bernstein Equation.* Let $\{X_n\}$ be a simple symmetric random walk with jumps $\{J_n\}$. Then X_n satisfies the stochastic difference equation

$$\Delta X_n = X_{n+1} - X_n = J_{n+1}, \qquad X_0 = 0, \quad n \geqslant 1. \qquad (9.6.50)$$

To pass to a continuous-time analog, we replace J_n by a process $J(t)$ such that: (1) $J(t)$ has the same distribution for all $t \geq 0$, and (2) for any finite sequence t_k, $1 \leq k \leq n$, of time points, the RVs $\{J(t_k)\}$ are independent. Now we can extend (9.6.50) by defining a process $X(t)$ as follows:

$$\Delta X(t) = X(t + \Delta t) - X(t) = J(t)\sqrt{\Delta t}.$$

To accommodate a process with nonzero drift and nonunit diffusion or variance, we can define

$$\Delta X(t) = M(t, X(t))\Delta t + \sqrt{S(t, X(t))} J(t)\sqrt{\Delta t}. \tag{9.6.51}$$

Equations of this type are due to Bernstein. Let $X(t)$ be a Markov process satisfying Bernstein's equation (9.6.51). Let the process $J(t)$ satisfy the conditions

$$E[J(t)] = 0, \qquad E[J^2(t)] = 1, \qquad E[|J(t)|^3] = o((\Delta t)^{-\frac{1}{2}}), \tag{9.6.52}$$

as $\Delta t \downarrow 0$. Assume that the transition density function $f(s, x; t, y)$ of the Markov process $X(t)$ exists and possesses continuous partial derivatives

$$\frac{\partial f}{\partial t}, \qquad \frac{\partial}{\partial y}(Mf), \qquad \text{and} \qquad \frac{\partial^2}{\partial y^2}(Sf).$$

Then show that

$$\frac{\partial f}{\partial t} = -\frac{\partial}{\partial y}(Mf) + \frac{1}{2}\frac{\partial^2}{\partial y^2}(Sf). \tag{9.6.53}$$

The idea of proof is the same as that in the derivation of the forward equation. Let $g(x)$ be a thrice continuously differentiable function that vanishes outside a compact interval $[a, b]$. Since (9.6.53) is a forward equation, we simply write $f(t, y)$ in place of $f(s, x; t, y)$. Define

$$h(t) = E[g(X(t))] = \int g(y)f(t, y)\, dy.$$

Then

$$h'(t) = \frac{d}{dt}\int g(y)f(t, y)\, dy = \int g(y)\frac{\partial}{\partial t}f(t, y)\, dy. \tag{9.6.54}$$

Also

$$h'(t) = \lim_{\Delta t \to 0} E[(\Delta t)^{-1}\{g(X(t + \Delta t)) - g(X(t))\}]$$

$$= \lim_{\Delta t \to 0} E[(\Delta t)^{-1}\{g'(X(t))\Delta X(t) + \tfrac{1}{2}g''(X(t))(\Delta X(t))^2$$

$$+ \frac{1}{3!}g'''(X(t))(\Delta X(t))^3 + o(\Delta(X(t))^3)\}]$$

$$= \lim_{\Delta t \to 0} E[(\Delta t)^{-1}\{g'(X(t))[M\Delta t + \sqrt{S}J\sqrt{\Delta t}\,]$$

$$+ \tfrac{1}{2}g''(X(t))[M\Delta t + \sqrt{S}J\sqrt{\Delta t}\,]^2$$

$$+ \tfrac{1}{2}g'''(X(t))[M\Delta t + \sqrt{S}J\sqrt{\Delta t}\,]^3$$

$$+ o[M\Delta t + \sqrt{S}J\sqrt{\Delta t}\,]^3\}], \qquad \text{using (9.6.51)},$$

$$= E\left[Mg'(X(t)) + \frac{S}{2}g''(X(t))\right], \qquad \text{by conditions (9.6.52)},$$

$$= \int [M(t,y)g'(y) + \tfrac{1}{2}S(t,y)g''(y)]f(t,y)\,dy$$

$$= \int g(y)\left[-\frac{\partial}{\partial y}(Mf) + \frac{1}{2}\frac{\partial^2}{\partial y^2}(Sf)\right]dy. \tag{9.6.55}$$

Combining (9.6.54) and (9.6.55), we obtain

$$0 = \int g(y)\left\{-\frac{\partial f}{\partial t} - \frac{\partial}{\partial y}(Mf) + \frac{1}{2}\frac{\partial^2}{\partial y^2}(Sf)\right\}dy,$$

which yields (9.6.53).

EXAMPLE 6. Let $X(t)$ be a diffusion process whose transition density $f(s,x;t,y)$ satisfies the following forward equation:

$$\frac{\partial f}{\partial t} = \frac{\partial}{\partial y}\left\{\left(by - \frac{a^2}{2y}\right)f\right\} + \frac{a^2}{2}\frac{\partial^2 f}{\partial y^2}, \qquad y > 0. \tag{9.6.56}$$

Using the method of separation of variables, solve this equation for $f(s,x;t,y)$.

Since we are working with the forward equation, we write $f(t,y)$ instead of $f(s,x;t,y)$. Set $f(t,y) = g(t)h(y)$. Using this in (9.6.56), we get

$$\frac{1}{g}\frac{dg}{dt} = \frac{1}{h}\left\{\frac{d}{dy}\left\{\left(by - \frac{a^2}{2y}\right)h\right\} + \frac{a^2}{2}\frac{d^2h}{dy^2}\right\}.$$

Note that the LHS is independent of y and the RHS is independent of t. So both sides must be equal to a constant $-\lambda$, say. Then

$$\frac{dg}{dt} + \lambda g = 0$$

$$\frac{a^2}{2}\frac{d^2h}{dy^2} + \frac{d}{dy}\left\{\left(by - \frac{a^2}{2y}\right)h\right\} + \lambda h = 0.$$

Clearly

$$g(t) = e^{-\lambda(t-s)}. \tag{9.6.57}$$

The second equation has a solution vanishing at infinity corresponding only to the discrete values of $\lambda = 2nb$, $n = 0, 1, \ldots$. In such case the solution is given by

$$h(y) = \frac{1}{n!}\frac{2b^2 y}{a^2}\exp\left(-\frac{b^2 y^2}{a^2}\right)L_n\left(\frac{b^2 y^2}{a^2}\right), \tag{9.6.58}$$

where

$$L_n(x) = e^x\frac{d^n}{dx^n}e^{-x}x^n$$

are the orthogonal Laguerre polynomials. Thus from (9.6.57) and (9.6.58),

$$f(s, x; t, y) = \sum_{n \geqslant 0} c_n e^{-2nb(t-s)}\frac{1}{n!}\frac{2b^2 y}{a^2}e^{-b^2 y^2/a^2}L_n\left(\frac{b^2 y^2}{a^2}\right),$$

where the constants c_n are to be evaluated from the condition $f(s, x; s, y) = \delta(y - x)$ [as in Example 9.6.7 (2)]. Therefore

$$\delta(y - x) = \sum_{n \geqslant 0} c_n\frac{1}{n!}\frac{2b^2 y}{a^2}e^{-b^2 y^2/a^2}L_n\left(\frac{b^2 y^2}{a^2}\right).$$

Since the Dirac δ-function is an evaluation-type functional [see (9.5.1)], multiply both sides by

$$y\exp\left\{\frac{-b^2 y^2}{a^2}\right\}L_n\left(\frac{b^2 y^2}{a^2}\right)$$

and integrate. Since the Laguerre polynomials are orthogonal, we obtain

$$c_n = \frac{1}{n!}\frac{2b^2 x}{a^2}\exp\left\{-\frac{b^2 x^2}{a^2}\right\}L_n\left(\frac{b^2 x^2}{a^2}\right).$$

273

Hence

$$f(s, x; t, y) = \frac{4b^4 xy}{a^4} \exp\left[-\frac{b^2(x^2 + y^2)}{a^2}\right]$$

$$\times \sum_{n \geqslant 0} (n!)^{-2} L_n\left(\frac{b^2 x^2}{a^2}\right) L_n\left(\frac{b^2 y^2}{a^2}\right) e^{-2nb(t-s)}.$$

EXAMPLE 7. *Ehrenfests Diffusion.* In Chapter 3 we saw the importance of the Ehrenfests chain in explaining the recurrence in the dynamical system and thermodynamical irreversibility. Here we derive the diffusion limit to this chain. For this and other examples we refer again to Kac (1947). Let $\{X_n\}$ be the Ehrenfests chain on the state space $-N \leqslant k \leqslant N$ and with transition probabilities given by

$$p(k, k - 1) = \tfrac{1}{2}\left(1 + \frac{k}{N}\right), \qquad p(k, k + 1) = \frac{1}{2}\left(1 - \frac{k}{N}\right) \tag{9.6.59}$$

$$p(N, N - 1) = 1 = p(-N, -N + 1).$$

Clearly

$$p(n + 1; k, m) = p(n; k, m - 1)p(m - 1, m) + p(n; k, m + 1)p(m + 1, m)$$

$$= \frac{N - m + 1}{2N} p(n; k, m - 1) + \frac{N + m + 1}{2N} p(n; k, m + 1). \tag{9.6.60}$$

Let us speed up this nonhomogeneous RW. Let $N \to \infty$ such that

$$\frac{(\Delta x)^2}{\Delta t} = \sigma^2 \qquad \text{and} \qquad N\Delta t \to \nu > 0. \tag{9.6.61}$$

Let us pass from the n-step transition probability to the transition density $f(t, y) = f(x; t, y)$ by taking $n\Delta t = t$, $m\Delta x = x$:

$$f(t + \Delta t, y) = \frac{N - m + 1}{2N} f(t, y - \Delta y) + \frac{N + m + 1}{2N} f(t, y + \Delta y).$$

Then

$$f(t, y) + \Delta t \frac{\partial f}{\partial t} + o(\Delta t)$$

$$= \frac{N - m + 1}{2N}\left\{f(t, y) - \Delta y \frac{\partial f}{\partial y} + \tfrac{1}{2}(\Delta y)^2 \frac{\partial^2 f}{\partial y^2}\right\}$$

$$+ \frac{N + m + 1}{2N}\left\{f(t, y) + \Delta y \frac{\partial f}{\partial y} + \tfrac{1}{2}(\Delta y)^2 \frac{\partial^2 f}{\partial y^2}\right\} + o(\Delta y)^2,$$

from which we obtain

$$\frac{\partial f}{\partial t} = \nu \frac{\partial}{\partial y}(yf) + \frac{\sigma^2}{2}\frac{\partial^2 f}{\partial y^2}.$$

9.7. First Passage Time

Diffusion processes arise very naturally in many physical, chemical, and engineering problems. They are also very successfully utilized to model phenomena in population dynamics, genetics, epidemiology, ecology, and neural network. One of the important problems in diffusion modeling is the first-passage-time problems, say, the problem of finding the distribution of the random time of absorption or crossing a given level. For example, in population genetics one is interested in the probability of "fixation," that is, the gene frequency $X(t)$ hitting either the zero or one level. In stochastic neuron modeling, the first-passage-time distribution provides the theoretical distribution of the distances between successive spikes released by the neuron. In statistics, first-passage problems arise in Wald's sequential sampling and test of goodness of fit. On such a useful topic we present first some of the simple results of Darling and Siegert (1953) and Siegert (1951).

The reverse problem arises in neural network. Suppose that a function is chosen for representing the experimental neuron's interspike distances. The problem now is to investigate whether this function is the first-passage-time distribution for some diffusion process. In this respect we quote some recent results of Capocelli and Ricciardi (1972).

Let $X(t)$ be a temporally homogeneous diffusion process with transition distribution $F(x; t, y)$ and transition density $f(x; t, y)$ (if it exists). Let $X(0) = x$, and let λ and ρ be two extended reals that are fixed to represent some left or lower and right levels such that $-\infty \leqslant \lambda < x < \rho \leqslant \infty$. Define $\mathbf{t}(\omega) = \mathbf{t}_{\lambda\rho}(\omega)$ by

$$\mathbf{t}(\omega) = \mathbf{t}^x(\omega) = \inf\{t: X(t, \omega) \leqslant \lambda \quad \text{or} \quad X(t, \omega) \geqslant \rho\}. \tag{9.7.1}$$

Then $\mathbf{t}$ is called the *first passage time* or *exit time* of $X(t)$ from the interval (λ, ρ). Let

$$\begin{aligned}
\mathbf{t}^x_\lambda &= \inf\{t: X(t) \leqslant \lambda\} && \text{if } \rho = \infty, \\
\mathbf{t}^x_\rho &= \inf\{t: X(t) \geqslant \rho\} && \text{if } \lambda = -\infty.
\end{aligned} \tag{9.7.2}$$

Since we are interested in the exit time $\mathbf{t}$ from the interval (λ, ρ), we consider the process $X(t)$ up to that time and then *kill* it by absorption, that is, one works with the *absorbed process*

$$Y(t) = \begin{cases} X(t) & \text{for } t < \mathbf{t} \\ X(\mathbf{t}) & \text{for } t \geqslant \mathbf{t} \end{cases}. \tag{9.7.3}$$

Next set

$$G(t; x) = P_x\{\mathbf{t} \leqslant t\}, \qquad t \geqslant 0, \tag{9.7.4}$$

$$H(x; t, y) = P_x\{Y(t) \leqslant y\} = P\{Y(t) \leqslant y | Y(0) = x\}. \tag{9.7.5}$$

Theorem 9.7.1. (Siegert). *Let* $X(t)$, $t \geqslant 0$, *be a temporally homogeneous diffusion process with symmetric transition distribution* $F(x; t, y)$, *that is,*

$$F(x; t, y) = 1 - F(-x; t, -y), \tag{9.7.6}$$

and $\mathbf{t}_0^x$, *with* $x > 0$, *be the exit time through the lower level* $\lambda = 0$, *($\rho = \infty$). Then*

$$G(t; x) = P_x\{\mathbf{t} \leqslant t\} = 2F(x; t, 0). \tag{9.7.7}$$

PROOF. Since the transition distribution is symmetric,

$$F(0; t, 0) = 1 - F(0; t, 0), \qquad \text{by (9.7.6).} \tag{9.7.8}$$

But, from the well-known renewal argument

$$F(x; t, 0) = \int_0^t G(ds; x) F(0; t - s, 0)$$

$$= \tfrac{1}{2} G(t; x), \qquad \text{from (9.7.8),}$$

which is nothing but (9.7.7). $\qquad\qquad\square$

Let us now assume that the densities f and g of F and G, respectively, exist. Denote their Laplace transforms by ϕ and γ, that is,

$$\phi(\theta; x, y) = \int_0^\infty e^{-\theta t} f(x; t, y)\, dy, \qquad \gamma(\theta; x) = \int_0^\infty e^{-\theta t} g(t; x)\, dt.$$

Let $\lambda = -\infty$, so that $\mathbf{t} = \mathbf{t}_\rho^x$, and $x < \rho < y$. Again by the renewal principle,

$$f(x; t, y) = \int_0^t g_\rho(s; x) f(\rho; t - s, y)\, ds, \tag{9.7.9}$$

where we have used the notation $g_\rho(s; x)$ to denote the density function of $\mathbf{t}_\rho^x$. Taking the Laplace transform on both sides of (9.7.9), we obtain

$$\int_0^\infty e^{-\theta t} f(x; t, y)\, dt = \int_0^\infty g_\rho(s; x) \int_s^\infty e^{-\theta t} f(\rho; t - s, y)\, dt\, ds$$

$$= \int_0^\infty e^{-\theta s} g_\rho(s; x)\, ds \int_0^\infty e^{-\theta t} f(\rho; t, y)\, dt$$

Thus

$$\phi(\theta; x, y) = \gamma_\rho(\theta; x)\phi(\theta; \rho, y), \qquad x < \rho < y. \qquad (9.7.10)$$

Relation (9.7.10) is a formal solution to the first-passage-time problem, since $g_\rho(t; x)$ can be obtained from (9.7.10) by inverting the transform γ_ρ, and here ϕ is, at least theoretically, known.

Theorem 9.7.2. (Darling–Siegert). *Let* $X(t)$, $t \geqslant 0$, *be a temporally homogeneous diffusion process satisfying the Kolmogorov equation*

$$\frac{\partial f}{\partial t} = M(x)\frac{\partial f}{\partial x} + \tfrac{1}{2}S(x)\frac{\partial^2 f}{\partial x^2} \qquad (9.7.11)$$

under the conditions

$$f(x; 0, y) = \delta(y - x), \quad f(x; t, y) \to 0 \qquad \text{as } |x| \to \infty. \quad (9.7.12)$$

Then γ_ρ *can be expressed as*

$$\gamma_\rho(\theta, x) = \frac{\xi(x)}{\xi(\rho)}, \qquad x < \rho, \qquad (9.7.13)$$

where $\xi(x)$ *is a solution of the equation*

$$\theta\psi = M(x)\frac{\partial\psi}{\partial x} + \tfrac{1}{2}S(x)\frac{\partial^2\psi}{\partial x^2}, \qquad \psi(x) \to 0 \text{ as } x \to -\infty. \quad (9.7.14)$$

PROOF. Taking the Laplace transform on both sides of equation (9.7.11) and using condition (9.7.12), we see that $\psi(x; \theta) = \phi(\theta; x, y)$ satisfies the equation

$$\theta\psi = M(x)\frac{\partial\psi}{\partial x} + \tfrac{1}{2}S(x)\frac{\partial^2\psi}{\partial x^2},$$

where $-\psi$ is the so-called Green's solution. Let $\xi(x)$ and $\eta(x)$ be two linearly independent solutions of this equation with $\xi(-\infty) = 0 = \eta(\infty)$. Then

$$\psi(x; \theta) = \phi(\theta; x, y) = \begin{cases} \xi(x)\eta(y) & \text{if } x < y \\ \xi(y)\eta(x) & \text{if } x > y \end{cases}, \qquad (9.7.15)$$

except possibly for a constant factor. Now using (9.7.15) in (9.7.10) we get

$$\gamma_\rho(\theta; x) = \frac{\xi(x)\eta(y)}{\xi(\rho)\eta(y)} = \frac{\xi(x)}{\xi(\rho)},$$

which is what we wanted to establish. $\qquad\qquad\qquad\square$

Remark 9.7.3. In Theorem 9.7.2 we worked with the passage time of exit through one barrier ρ. Now let both λ and ρ be finite, and $\mathbf{t}$ the time of exit from (λ, ρ). If $\gamma(\theta; x)$ denotes the Laplace transform of the density g of $\mathbf{t}$, it can be shown, as in Theorem 9.7.2, that γ satisfies the equation

$$\theta\gamma = M(x)\frac{\partial\gamma}{\partial x} + \tfrac{1}{2}S(x)\frac{\partial^2\gamma}{\partial x^2}, \qquad x \in (\lambda, \rho), \qquad (9.7.16)$$

with the obvious boundary conditions $\gamma(\lambda) = 1 = \gamma(\rho)$. Let g^λ and g^ρ be the density functions of exit time through the boundaries λ and ρ, respectively, before hitting the other. Then

$$g(t; x) = g^\lambda(t; x) + g^\rho(t; x).$$

Solving (9.7.16) with the boundary conditions $\gamma(\lambda) = 1$ and $\gamma(\rho) = 0$, we obtain $\gamma^\lambda(\theta; x)$. Solving (9.7.16) with $\gamma(\rho) = 1$ and $\gamma(\lambda) = 0$, we obtain γ^ρ.

Applications of first-passage-time distributions to Brownian motion, the Ornstein–Uhlenbeck process, Wald's sequential sampling, and the "goodness of fit" test can be found in Darling and Siegert (1953).

9.7.4. Converse Problem

It was pointed out earlier that the reverse of the first-passage problem arises in certain diffusion modeling of neuron spike activity. This converse problem is as follows. Given a complex function $\gamma(\theta; x, y)$ such that

$$0 < \gamma(0; x, y) \leqslant 1 \qquad \text{and} \qquad \lim_{x \to y} \gamma(\theta; x, y) = 1, \qquad (9.7.17)$$

under what conditions γ is the Laplace transform of a first passage time density function $g(t; x, y)$ of a continuous temporally homogeneous diffusion process satisfying the Kolmogorov equation (9.7.11)? Corresponding to this converse problem, we simply state, without proof, the following theorems due to Capocelli and Ricciardi.

Theorem 9.7.5. *For $\gamma(\theta; x, y)$ to be the Laplace transform of the first-passage time of a temporally homogeneous diffusion process, it is necessary that the drift and diffusion coefficients be given by*

$$M = \frac{\psi_0}{\eta_0} \text{ and } S = 2\left(\phi_0 - \psi_0\frac{\xi_0}{\eta_0}\right), \qquad (9.7.18)$$

where

$$\phi_0 = \text{Re}\{\theta\gamma/\gamma''\}, \qquad \psi_0 = \text{Im}\{\theta\gamma/\gamma''\},$$

$$\xi_0 = \text{Re}\{\gamma'/\gamma''\}, \qquad \eta_0 = \text{Im}\{\gamma'/\gamma''\},$$

with $\gamma' = (\partial\gamma/\partial x)$, $\gamma'' = (\partial^2\gamma/\partial x^2)$, $\text{Re}(\cdot) = $ *real part of* $(\cdot)$, *and* $\text{Im}(\cdot)$ *= imaginary part of* $(\cdot)$.

The boundary points λ and ρ have been classified by Feller as follows. Let $S(x) > 0$ for $x \in (\lambda, \rho)$. Define

$$h(x) = \exp\left\{-\int_a^x [2S(u)/M(u)]\,du\right\}, \qquad (9.7.19)$$

the so-called *Hille function*, where $a \in (\lambda, \rho)$. Set

$$H(x) = \frac{2}{S(x)h(x)}$$

$$\sigma_1 = \iint_{\lambda < y < x < a} H(x)h(y)\,dx\,dy, \qquad \sigma_2 = \iint_{a < x < y < \rho} H(x)h(y)\,dx\,dy$$

$$\mu_1 = \iint_{\lambda < y < x < a} h(x)H(y)\,dx\,dy, \qquad \mu_2 = \iint_{a < x < y < \rho} h(x)H(y)\,dx\,dy.$$

$$(9.7.20)$$

To simplify writing, set $b_1 = \lambda$ and $b_2 = \rho$. Then the boundary b_i, $i = 1, 2$, is called:

regular	if $\sigma_i < \infty$, $\mu_i < \infty$,
an exit boundary	if $\sigma_i < \infty$, $\mu_i = \infty$,
an entrance boundary	if $\sigma_i = \infty$, $\mu_i < \infty$,
natural	if $\sigma_i = \infty$, $\mu_i = \infty$. $\qquad$ (9.7.21)

The regular or exit boundaries are called *accessible* and the others, *inaccessible*. The probabilistic meaning of the classification is that the probability of reaching an accessible boundary in finite time is positive, whereas the same probability is zero for an inaccessible boundary.

We assume the following conditions to hold in the next two theorems: (1) $S(x) > 0$, (2) M and dS/dx are continuous in (λ, ρ) (with M and S as in Theorem 9.7.5), and (3) M and S depend only on x.

Theorem 9.7.6. *If λ is an inaccessible boundary, then $\gamma(\theta; x, \rho)$ is the Laplace transform of the time of first passage through the boundary ρ for the diffusion process with drift and diffusion as given in Theorem 9.7.5.*

Theorem 9.7.7. *Let* λ *be an accessible boundary. If* $\gamma(\theta; \lambda, \rho) = 0$, *then* $\gamma(\theta; x, \rho)$ *is the Laplace transform of the time of first passage through* ρ *without hitting* λ, *and the drift and diffusion coefficients of the corresponding process are given as in Theorem 9.7.5.*

Remark 9.7.8. The conditions (1)–(3) stipulated above are essential. It is important that the quantities M and S in (9.7.18) be functions of x alone. For example, take

$$\gamma(\theta; x, \rho) = \exp\{(x - \rho)\theta^2\}, \qquad x \leqslant \rho.$$

Then γ satisfies the conditions in (9.7.17) but M and S in (9.7.18) depend on θ, and there does not exist any diffusion possessing this γ as the Laplace transform of a first passage time.

Examples 9.7.9

EXAMPLE 1. *Brownian Motion.* Consider the function

$$\gamma(\theta; x, \rho) = \exp\left\{ \frac{-(\rho - x)(-\mu + \sqrt{\mu^2 + 2\theta\sigma^2})}{\sigma^2} \right\}, \qquad (9.7.22)$$

for $\mu > 0$ and $-\infty < x \leqslant \rho < \infty$. This γ clearly satisfies (9.7.17). Let us compute M and S using (9.7.18). Set $\theta = (\alpha + i\beta)$ and $(\mu^2 + 2\theta\sigma^2)^{\frac{1}{2}} = \xi + i\eta$. From (9.7.22),

$$\frac{d\gamma}{dx} = \frac{[-\mu + (\mu^2 + 2\theta\sigma^2)^{\frac{1}{2}}]\gamma}{\sigma^2}$$

$$\frac{d^2\gamma}{dx^2} = \frac{[-\mu + (\mu^2 + 2\theta\sigma^2)^{\frac{1}{2}}]^2\gamma}{\sigma^4}.$$

and therefore

$$\frac{\theta\gamma}{\gamma''} = \frac{\theta\sigma^4}{-\mu + (\mu^2 + 2\theta\sigma^2)^{\frac{1}{2}}}$$

$$= \frac{\mu^2 + \theta\sigma^2 + \mu(\mu^2 + 2\theta\sigma^2)^{\frac{1}{2}}}{2\theta}$$

$$= [2(\alpha^2 + \beta^2)]^{-1}\mu\left\{ \left[\alpha\mu + \alpha\xi + \beta\eta + \left(\frac{\sigma^2}{\mu}\right)(\alpha^2 + \beta^2) \right] \right.$$

$$\left. + i(-\mu\beta + \alpha\eta - \xi\beta) \right\}$$

$$\frac{\gamma'}{\gamma''} = (2\theta)^{-1}[\mu + (\mu^2 + 2\theta\sigma^2)^{\frac{1}{2}}]$$

$$= [2(\alpha^2 + \beta^2)]^{-1}[\alpha\mu + \alpha\xi + \beta\eta + i(-\beta\mu + \alpha\eta - \xi\beta)],$$

from which [and also (9.7.18)] we get

$$M = \mu \quad \text{and} \quad S = \sigma^2.$$

Since $\sigma_1 = \infty$ in (9.7.20), $\lambda = -\infty$ is inaccessible, and hence by Theorem 9.7.6 we see that the γ in (9.7.22) is the Laplace transform of the first passage time of a diffusion with drift μ and diffusion σ^2. Inverting (9.7.22), we obtain

$$g(t; x, \rho) = \frac{\rho - x}{\sigma(2\pi t^3)^{\frac{1}{2}}} \exp\left\{\frac{-(\rho - x - \mu t)^2}{2\sigma^2 t}\right\}$$

EXAMPLE 2. *Diffusion with Lognormal Transition.* Consider the function

$$\gamma(\theta; x, \rho) = \begin{cases} (x/\rho)^{\sqrt{\theta}} & \text{if } 0 < x \leqslant \rho \text{ or } \rho \leqslant x < 0 \\ (\rho/x)^{\sqrt{\theta}} & \text{if } 0 < \rho \leqslant x \text{ or } x \leqslant \rho < 0 \end{cases}. \quad (9.7.23)$$

This function satisfies conditions (9.7.17). Also

$$\frac{\theta\gamma}{\gamma''} = \frac{\pm\sqrt{\theta}x^2}{(\pm\sqrt{\theta} + 1)}, \quad \frac{\gamma'}{\gamma''} = \frac{-x}{(\pm\sqrt{\theta} + 1)}. \quad (9.7.24)$$

Setting $\pm\sqrt{\theta} = \alpha + i\beta$ in (9.7.24), we get

$$\frac{\theta\gamma}{\gamma''} = \frac{\alpha^2 + \beta^2 + \alpha}{(\alpha + 1)^2 + \beta^2}x^2 + i\frac{\beta}{(\alpha + 1)^2 + \beta^2}x^2 \quad (9.7.25)$$

$$\frac{\gamma'}{\gamma''} = \frac{-(\alpha + 1)}{(\alpha + 1)^2 + \beta^2}x + i\frac{\beta}{(\alpha + 1)^2 + \beta^2}x. \quad (9.7.26)$$

From (9.7.25) and (9.7.26) one computes

$$M(x) = x \quad \text{and} \quad S(x) = 2x^2. \quad (9.7.27)$$

We leave it as an exercise to show that the origin 0 is a natural and hence inaccessible boundary for M and S given by (9.7.27). Inverting (9.7.23), we get

$$g(t; x, \rho) = \frac{|\log(\rho/x)|}{2\sqrt{\pi t^3}} \exp\left\{-\frac{\log^2(\rho/x)}{4t}\right\}, \quad x\rho > 0. \quad (9.7.28)$$

Let us compute the transition density $f(x; t, y)$ corresponding to this diffusion process with drift and diffusion given by (9.7.27). We begin with the forward equation

$$\frac{\partial f}{\partial t} = -\frac{\partial}{\partial y}(yf) + \frac{\partial^2}{\partial y^2}(y^2 f), \qquad (9.7.29)$$

with the obvious condition $\lim_{t \downarrow 0} f(x; t, y) = \delta(y - x)$. This is a singular diffusion equation in any interval containing the natural boundary 0. We solve (9.7.29) separately in the intervals $(-\infty, 0)$ and $(0, \infty)$. We proceed by separating the variables $f(t, y) = \xi(t)\eta(y)$, where we dropped the variable x from $f(x; t, y)$ for the convenience of writing and also since we are considering the forward equation. Then

$$\frac{d\xi}{dt} + \lambda^2 \xi = 0, \qquad (9.7.30)$$

$$y^2 \frac{d^2\eta}{dy^2} + 3y\frac{d\eta}{dy} + (1 + \lambda^2)\eta = 0, \qquad (9.7.31)$$

where λ is an arbitrary constant. From (9.7.30)

$$\xi(t) = ae^{-\lambda^2 t}.$$

Equation (9.7.31) is the well-known totally Fuchsian equation. Its general solution is

$$\zeta(y) = by^{r_1} + cy^{r_2},$$

where b and c are constants and r_i values are the solutions of the indicial equation $r(r - 1) + 3r + 1 + \lambda^2 = 0$. Note that

$$r_k = -1 \pm i\lambda, \qquad k = 1, 2.$$

Now let $y \in (0, \infty)$. Then a particular solution of (9.7.29) is $f_0(t, y) = y^{-1}\exp[-(i\lambda \log y + \lambda^2 t)]$, $y > 0$, and hence the general solution is given by

$$f(t, y) = \frac{1}{y(2\pi)^{\frac{1}{2}}} \int \alpha(\lambda)\exp[-(i\lambda \log y + \lambda^2 t)]\, d\lambda, \qquad y > 0, \qquad (9.7.32)$$

where $\alpha(\lambda)$ is an arbitrary function to be determined by the initial condition. From $f(x; 0, y) = \delta(y - x)$

$$e^z \delta(e^z - x) = (2\pi)^{-\frac{1}{2}} \int \alpha(\lambda)e^{-i\lambda z}\, d\lambda,$$

from which

$$\alpha(\lambda) = (2\pi)^{-\frac{1}{2}} \int_0^\infty e^{i\lambda \log y} \delta(y - x) \, dy$$

$$= (2\pi)^{-\frac{1}{2}} \exp[i\lambda \log x], \qquad x > 0, \qquad (9.7.33)$$

and $\alpha(\lambda) = 0$ for $x < 0$. Therefore, from (9.7.32) and (9.7.33),

$$f(x; t, y) = \frac{1}{2y(\pi t)^{\frac{1}{2}}} \exp\left[-\frac{\log^2(y/x)}{4t}\right], \qquad x > 0, y > 0. \quad (9.7.34)$$

Similarly, on $(-\infty, 0)$,

$$f(x; t, y) = \frac{1}{2(-y)(\pi t)^{\frac{1}{2}}} \exp\left[-\frac{\log^2(y/x)}{4t}\right], \qquad x < 0, y < 0. \quad (9.7.35)$$

From (9.7.34) and (9.7.35),

$$f(x; t, y) = \frac{1}{2|y|(\pi t)^{\frac{1}{2}}} \exp\left[-\frac{\log^2(y/x)}{4t}\right], \qquad xy > 0, \quad (9.7.36)$$

which is the well-known lognormal distribution.

Exercises

1. Draw the graph of the first 48 steps of a sample path of a simple symmetric RW. On the same graph paper draw the graphs of the same path, but now with: (a) two steps per second and with step length $1/\sqrt{2}$, (b) four steps per second and with step length $1/2$, (c) 16 steps per second and with step length $1/4$.

2. Let $\{X_n\}$, $n \geqslant 1$, be a sequence of IID RVs distributed according to $P\{X = 1\} = \frac{1}{2} = P\{X = -1\}$. Let $t > 0$ take diadic rational values and $n = 2^k$, $k = 1$, $2, \ldots$. Show that the limiting distribution of

$$\mu + \sigma n^{-1/2}(X_1 + \cdots + X_{nt})$$

is $N(\mu, \sigma^2 t)$.

3. Let $\{X_n\}$ be a sequence of IID RVs with $E[X_n] = 0$ and $\mathrm{var}(X_n) = 1$. Let n and t be as above with nt an integer. Find the limiting distribution of $n^{-\frac{1}{2}}(X_1 + \cdots + X_{nt})$.

4. Let $\{X(t), t \geqslant 0\}$ be a standard Brownian motion with $\sigma = 1$. If $T(\omega)$ is the amount of time in $[0, 1]$ for which $X(t, \omega) \geqslant 0$, show that, for $t \in [0, 1]$,

$$P\{T \leqslant t\} = \pi^{-1} \int_0^t [s(1 - s)]^{-1/2} \, ds = 2\pi^{-1} \arcsin \sqrt{s}.$$

5. Find the distribution of $X(1) + \cdots + X(n)$ for a positive integer n, where $X(t)$ is a standard Brownian motion with $\sigma = 1$. [Answer: $N(0, n(n + 1)(2n + 1)/6)$.]

6. Let $X(t)$ be a Brownian motion with drift μ and diffusion coefficient σ^2. The geometric Brownian motion associated with $X(t)$ is defined by $Y(t) = \exp[X(t)]$, $t \geqslant 0$. Show that:

(a) $E[Y(t)|Y(0) = y] = y \exp\left[t\left(\mu + \dfrac{\sigma^2}{2}\right)\right]$

(b) $\mathrm{var}[Y(t)|Y(0) = y] = y^2(e^{t\sigma^2} - 1)\exp\left[2t\left(\mu + \dfrac{\sigma^2}{2}\right)\right].$

7. Find the correlation between $X(t)$, $0 \leqslant t \leqslant 1$, and $\int_0^1 X(s)\,ds$, where $X(t)$ is a standard Brownian motion with $\sigma = 1$.

8. Let $X(t)$ be a standard Brownian motion. Show that the processes $Y(t) = [X^2(t) - t]$ and $Z(t) = \exp[aX(t) - a^2t/2]$ are martingales.

9. Let $X(t)$ be a standard Brownian motion with variance parameter σ^2 and $X(0) = x$. Let $A < x < B$ and T be the hitting time of A or B. Then show that

$$P_x\{X(T) = B\} = \frac{x - A}{B - A}, \qquad E_x[T] = \frac{(B - x)(x - A)}{\sigma^2}.$$

10. Let $X(t)$ be a Brownian motion with drift $\mu \neq 0$ and diffusion parameter σ^2. Show, for any real a, that $\xi(t) = \exp\{aX(t) - (a\mu + a^2\sigma^2/2)t\}$ is a martingale. Suppose that $X(0) = 0$ and $A < 0 < B$. Then show that the probability that the particle hits B before hitting A is given by $(1 - e^{CA})/[e^{CB} - e^{CA}]$, where $C = (-2\mu/\sigma^2)$.

11. For a Brownian motion with drift μ and $\sigma = 1$, define $Y(t) = \int_0^t [X(s)]^2\,ds$. Show that $E[Y_t] = (t^2 + 2\mu t)/2$.

12. Let $X(t)$ be a standard Brownian motion with $\sigma = 1$ and $X(0) = x$. If $a > 0$ and $b > 0$, show that

$$P_x\{X(t) \leqslant at + b \text{ for all } t \geqslant 0\} = 1 - e^{-2a(b-x)} \qquad \text{for } x \leqslant b.$$

13. Establish the identity

$$E\left[\exp\left\{a\int_0^t sX(s)\,ds\right\}\right] = e^{a^2t^5/15}, \qquad a \in R.$$

14. Establish the identity

$$E\left[\exp\left\{a\int_0^t g(s)X(s)\,ds\right\}\right] = \exp\left[a^2\int_0^t g(u)\int_0^u vg(v)\,dv\,du\right],$$

where $g(\cdot)$ is a continuous function on $[0, \infty)$.

15. Let $0 < s \leqslant t \leqslant u$. Given that the Brownian motion $X(\cdot) \neq 0$ in (s, t), show that the (conditional) probability that $X(\cdot)$ is not zero in (s, u) is given by $(\arcsin \sqrt{s/u})/(\arcsin \sqrt{s/t})$.

16. For a standard Brownian motion, establish that

$$P\left\{ \sup_{0 \leqslant s \leqslant t} |X(s)| > \epsilon \right\} \leqslant \frac{t}{\epsilon^2}, \qquad \epsilon > 0.$$

Using this inequality, show that $\lim_{t \to \infty} t^{-1} X(t) = 0$ with probability one.

17. Exponential martingales play a basic role in the theory of Brownian motion and diffusion processes (see Exercises 8–10). Let $g(a, x, t) = \exp[ax - a^2 t/2]$. Take $a > 0$, $a = \sqrt{2b}$. Let $T = \min\{t: |X(t)| = A\}$. Using the martingale $g(a, X(t), t) + g(-a, X(t), t)$, establish that

$$E[e^{-bT}] = [\cosh \sqrt{2b} A]^{-1}.$$

18. Show that $h(x, t) = \int_R (2\pi)^{-\frac{1}{2}} g(a, x, t)\, da = t^{-\frac{1}{2}} e^{-x^2/2t}$ $t > 0$, where $g(a, x, t) = \exp[ax - a^2 t/2]$. Show that $h(X(t), t + A)$ is a martingale for $A > 0$. Now show that

$$P\{|X(t)| \geqslant [2(A + t)\log(A + t)^{\frac{1}{2}}]^{\frac{1}{2}} \text{ for some } t \geqslant 0\} \leqslant A^{-\frac{1}{2}}.$$

19. Show that the transition density function

$$p(t, x, y) = (2\pi t)^{-\frac{1}{2}} \exp\left[\frac{-(x - y)^2}{2t}\right]$$

solves the heat equation $(\partial p/\partial t) = \frac{1}{2}(\partial^2 p/\partial x^2)$.

20. Compute the drift and diffusion coefficients for the geometric Brownian motion defined in Exercise 6.

21. Prove that the Ornstein–Uhlenbeck velocity process $V(t)$ does not possess independent increments.

22. A population contains N individuals. Each individual is of one or the other of two genotypes. At the end of each unit of time a randomly chosen individual dies and is replaced by a new individual of randomly chosen genotype. Let X_n denote the MC representing the number of persons of first genotype having the transition probabilities

$$p(x, x - 1) = \frac{x(N - x)}{N^2} = p(x, x + 1), \qquad p(x, x) = \frac{N^2 - 2x(N - x)}{N^2},$$

and $p(x, y) = 0$ if $|x - y| > 1$. To obtain a diffusion approximation $X(t)$, let us take N sufficiently large and speed up the removal and replacement of dead individuals. Now let $X(t)$ denote the frequency of individuals of first genotype and $f(t, x)$ the density function of $X(t)$. Show that

$$\frac{\partial}{\partial t}f(t, x) = \frac{\partial^2}{\partial x^2}[x(1 - x)f(t, x)], \qquad x \in (0, 1)$$

(Watterson 1961).

23. Consider the equation

$$\frac{\partial f}{\partial t} + \tfrac{1}{2}S(t, x)\frac{\partial^2 f}{\partial x^2} + M(t, x)\frac{\partial f}{\partial x} = 0,$$

where the diffusion coefficient $S(t, x)$ is differentiable WRT x and $0 < S(t, x) < \infty$. Under the change of variable $y = \int_0^x [S(t, u)]^{-1/2}\, du$, show that the preceding equation reduces to the form

$$\frac{\partial g}{\partial t} + \tfrac{1}{2}\frac{\partial^2 g}{\partial y^2} + A(t, y)\frac{\partial g}{\partial y} = 0.$$

24. Let A be an absorbing barrier for a homogeneous diffusion process with constant M and S. Then show that the transition density is given by

$$f(0, 0; t, x) = [2\pi S]^{-\frac{1}{2}}\left[\exp\left\{\frac{-(x - Mt)^2}{2St}\right\}\right.$$
$$\left. - \exp\left\{\frac{2MA}{S} - \frac{(x - 2A - Mt)^2}{2St}\right\}\right].$$

25. Consider a diffusion process with drift $M(t, x) = -m(t)$ and $S(t, x) = S(t)$. By solving the forward equation, show that the transition density $f(s, x; t, y)$ is given by

$$[2\pi g(s, t)]^{-1/2}\exp\left[\frac{-(y - h(s, t)x)^2}{2g(s, t)}\right],$$

where g and h solve

$$\frac{\partial g}{\partial t} = S(t) + 2M(t)g \qquad \text{and} \qquad \frac{\partial h}{\partial t} = M(t)h,$$

respectively.

Bibliography

Bailey, N. T. J. *The Mathematical Theory of Epidemics*, Hafner, New York, 1957.

Bailey, N. T. J. A simple stochastic epidemic: a complete solution in terms of known functions, *Biometrika* 50: 235–240 (1963).

Barber, M. N., and Ninham, B. W. *Random and Restricted Walks*, Gordon and Breach, New York, 1970.

Bartholomay, A. F. Stochastic models for chemical reactions: I. Theory of the unimolecular reaction process, *Bull. Math. Biophys.* 20: 175–190 (1958).

Bartlett, M. S.: *Stochastic Population Models in Ecology and Epidemiology*, John Wiley and Sons, New York, 1960.

Bellman, R., and Harris, T. E. Recurrence time for the Ehrenfests model, *Pacific J. Math.* 1: 179–193 (1951).

Bharucha-Reid, A. T., and Landau, H. G. A suggested chain process for radiation damage, *Bull. Math. Biophys.* 13: 153–163 (1951).

Bharucha-Reid, A. T. *An Introduction to the Stochastic Theory of Epidemics and Some Related Statistical Problems*, USAF School of Aviation Medicine, Randolph Field, Texas, 1957.

Bharucha-Reid, A. T. *Elements of the Theory of Markov Processes and Their Applications*, McGraw-Hill Book Company, New York, 1960.

Capocelli, R. M., and Ricciardi, L. M.: On the inverse of the first passage time probability problem, *J. Appl. Prob.* 9: 270–287 (1972).

Chung, K. L. *Markov Chains with Stationary Transition Probabilities*, 2nd ed., Springer-Verlag, New York, 1967.

Cox, D. R., and Miller, H. D. *The Theory of Stochastic Processes*, Methuen, London, 1970.

Cramer, H.: On stochastic processes whose trajectories have no discontinuities of the second kind, *Ann. Math.* 71: 85–92 (1966).

Bibliography

Cramer, H., and Leadbetter, M. R. *Stationary and Related Stochastic Processes*, John Wiley and Sons, New York, 1967.

Darling, D. A., and Siegert, A. J. F. First passage problem for a continuous Markov process, *Ann. Math. Stat.* 24: 624–639 (1953).

Doob, J. L. *Stochastic Processes*, John Wiley and Sons, New York, 1953.

Dubin, N. *A Stochastic Model for Immunological Feedback in Carcinogenesis*, Lecture notes in *Biomath*, vol. 9, Springer-Verlag, New York, 1976.

Dvoretski, A., Erdos, P., and Kakutani, S. Nonincrease everywhere of the Brownian motion, *Proceedings of IVth Symposium on Mathematical Statistics and Problems, Berkeley, Calif.*, vol. II, 1961, pp. 103–116.

Dynkin, E. B. *Theory of Markov Processes*, vols. I and II, Springer-Verlag, New York, 1965.

Evans, R. D. *The Atomic Nucleus*, McGraw-Hill Book Company, New York, 1965.

Feller, W. Zur theorie der stochastischeu prozesse (Existenz und Eindeutigkeitssatze), *Math. Ann.* 113: 113–160 (1936).

Feller, W. On the integrodifferential equations of completely discontinuous Markov processes, *Transact. Am. Math. Soc.* 48: 488–515 (1940).

Feller, W. The parabolic differential equations and the associated semigroups of transformations, *Ann. Math.* 55: 468–519 (1952).

Feller, W. *An Introduction to Probability Theory and Its Applications*, vol. I, 2nd ed., 1957, vol. II, 2nd ed., 1970, John Wiley and Sons, New York.

Foster, F. G. On stochastic matrices associated with certain queueing processes, *Ann. Math. Stat.* 24: 355–360 (1953).

Gerstein, G. L., and Mandelbrot, D. Random Walk models for the spike activity of a single neuron, *Biophys. J.* 4: 41–68 (1964).

Gihman, I. I., and Skorohod, A. V. *Theory of Stochastic Processes*, vol. I (1974); vol. II (1975), Springer-Verlag, New York.

Harris, T. E. *The Theory of Branching Processes*, Springer-Verlag, New York, 1963.

Hoel, P. G., Port, S. C., and Stone, C. J. *Introduction to Stochastic Processes*, Houghton Mifflin, New York, 1972.

Itô, K., and McKean, H. P. *Diffusion Processes and Their Sample Paths*, Springer-Verlag, Berlin, 1965.

Jacques, J. A. Tracer Kinetics, *Principles of Nuclear Medicine*, H. N. Wagner, Ed., Saunders, Philadelphia, 1968.

Kac, M. Random walk and the theory of Brownian motion, *Am. Math. Monthly* 54: 369–417 (1947).

Karlin, S., and McGregor, J. Linear growth, birth and death processes, *J. Math. Mech.* 7: 643–662 (1958).

Karlin, S., and Taylor, H. M. *A First Course in Stochastic Processes*, 2nd ed., Academic Press, New York, 1975.

Karp, S., O'Neill, E. L., and Gagliardi, R. M. Communication theory for the free space optical channel, *Proc. IEEE* 58: 1611–1626 (1970).

Kendall, D. G. Some problems in the theory of comets, I, II; *Proceedings of IVth Symposium on Mathematical Statistics and Problems, Berkeley, Calif.*, vol. III, 1961, pp. 99–120, 121–147.

Kolmogorov, A. N. Über die analytischen methoden in der Wahrscheinlichkeitsrechnung, *Math. Ann.* 104: 415–458 (1931).

Levy, P. *Processus Stochastiques et Mouvement Brownian*, Gauthier-Villars, Paris, 1948.

Loeve, M. *Probability Theory*, 3rd ed., Van Nostrand, New York, 1963.

Ludwig, D., (1) Final size distributions for epidemics, (2) Qualitative behavior of stochastic epidemics, *Math. Biosci.* 23: 33–73 (1975).

MacArthur, R. H., and Wilson, E. O., *Theory of Island Biogeography*, Princeton Univ. Press, Princeton, N.J., 1967.

McShane, E. J., *Stochastic Calculus and Stochastic Models*, Academic Press, New York, 1974.

Meyer, P. A. *Probability and Potentials*, Blaisdell Publ. Co. (Division of Ginn and Co.), Waltham, Massachusetts, 1966.

Montroll, E. W. Random Walks on lattices. III. Calculation of first passage times with application to exciton trapping on photosynthetic units, *J. Math. Phys.* 10: 753–765 (1969).

Mortensen, R. E. *Mathematical Problems of Modeling Stochastic Non-linear Dynamic Systems*, NASA CR-1168. NASA, Washington, D.C., 1968.

Neveu, J. *Mathematical Foundations of the Calculus of Probability*, Holden-Day, San Francisco, 1965.

Papoulis, A. *Probability, Random Variables, and Stochastic Processes*, McGraw-Hill Book Company, New York, 1965.

Parzen, E. *Stochastic Processes*, Holden-Day, San Francisco, 1962.

Prabhu, N. U. *Stochastic Processes, Basic Theory and its Applications*, Macmillan, New York, 1966.

Rosenblatt, M. *Random Processes*, 2nd ed., Springer-Verlag, New York, 1974.

Rutherford, E., Chadwick, J., and Ellis, C. D. *Radiations from Radioactive Substances*, Cambridge University Press, New York, 1930.

Selivanov, B. I. Some explicit formulas in the theory of branching random processes with discrete time and one type of particles, *Theor. Probl. Appl.* 14: 336–342 (1969).

Sheppard, C. W. *Basic Principles of the Tracer Method*, John Wiley and Sons, New York, 1962.

Siebert, W. M. Frequency discrimination in the auditory system: Place or periodicity mechanisms?, *Proc. IEEE* 58: 723–730 (1970).

Siegert, A. J. F. On the first passage time probability problem, *Phys. Rev.* 81: 617–623 (1951).

Snyder, D. L. *Random Point Processes*, John Wiley and Sons, New York, 1975.

Spitzer, F. *Principles of Random Walk*, Van Nostrand, New York, 1964.

Wagner, H. N., Ed., *Principles of Nuclear Medicine*, W. B. Saunders, Philadelphia, 1968.

Watterson, G. A. Markov chains with absorbing states, a genetic example, *Ann. Math. Stat.* 32: 716–729 (1961).

Wax, N., Ed. *Selected Papers on Noise and Stochastic Processes*, Dover Publications, New York, 1954.

Wonham, W. M., and Fuller, A. T. Probability densities of the smoothed random telegraph signal, *J. Electron. Control* 4: 567–576 (1958).

Yaglom, A. M. *An Introduction to the Theory of Stationary Random Functions*, Prentice-Hall, Englewood Cliffs, N.J., 1962.

Index

Index